Mathematics at Berkeley

Mathematics at Berkeley

A History

Calvin C. Moore

CRC Press is an imprint of the
Taylor & Francis Group, an **informa** business

AN A K PETERS BOOK

CRC Press
Taylor & Francis Group
6000 Broken Sound Parkway NW, Suite 300
Boca Raton, FL 33487-2742

ISBN-13: 9781568813028 (hbk)

Visit the Taylor & Francis Web site at
http://www.taylorandfrancis.com

and the CRC Press Web site at
http://www.crcpress.com

Library of Congress Cataloging-in-Publication Data

Moore, C. C. (Calvin C.), 1936-
   Mathematics at Berkeley : a history / Calvin C. Moore.
      p. cm.
   Includes bibliographical references and index.
   ISBN-13: 978-1-56881-302-8 (alk. paper)
   ISBN-10: 1-56881-302-3 (alk. paper)
   1. University of California, Berkeley. Dept. of Mathematics--History. 2. Mathematical Sciences Research Institute (Berkeley, Calif.)--History. 3. Mathematics--Study and teach-ing (Higher)--California. I. Title.
   QA13.5.C23C356 2006
   510.71'179467--dc22   2006012021

For Doris

Table of Contents

Preface

The origins of this book lie in a conference that the mathematics department at Berkeley sponsored in August 2000 on computational number theory in honor of the Lehmer family. The honorees, Derrick Norman Lehmer (1869–1938), Derrick Henry Lehmer (1904–1991), and Emma Trotskaya Lehmer (1906–) have contributed enormously to number theory and to Berkeley, and it was these contributions that the conference recognized. Since the conference fell on the exact centenary of the arrival of Derrick Norman in Berkeley as a fresh PhD from the University of Chicago to begin his faculty career at Berkeley, it was my idea, as department chair, to present, as a kind of preamble to the conference, a description of the Berkeley mathematics department in 1900. Such a project involved historical research on the university and the development of the mathematics program and the faculty, beginning with its origins in the 1850s. What was presented to the conference was in essence a summary of the first three chapters of this book.

Friends and colleagues encouraged me to extend the narrative forward in time, but then the question became where to stop. The year 1933–1934 was one possibility, since this was when Griffith Evans was recruited to Berkeley with a charge to make over the department and bring in a strong research component that had been lacking. Robin Ryder had already told this story very well [Ryder], but it seemed useful to embed these events in the larger context of what preceded it and what followed it. What I have written here overlaps with Ryder's narrative, since we both used the same primary sources in the University Archives in Bancroft Library on campus. The next possible cutoff was 1949, when Evans stepped down, but that would exclude the events of the 1950s when Berkeley first emerged as one of the very top mathematics departments in the country. Then 1960 seemed a good cutoff, particularly since I arrived in Berkeley as a new faculty member in 1961, and my own curiosity about how the department had evolved up until that time would be satisfied. In addition, as I often commented jocularly to colleagues, too many people were still alive to be able safely to venture beyond this point. But with trepidation, I did so for two reasons.

First, the primary audience that I was writing for were the many colleagues in the department who shared my curiosity about the evolution of mathematics at Berkeley before they arrived, as well as colleagues in the Center for Studies in Higher Education at Berkeley who are professionals working on the history of the university. Because most current departmental colleagues had arrived years after I had, their historical interests would not be served by stopping in 1960.

Second, my original idea about the mode of publication was simply posting it as a file on the department's web page. It was in a conversation with my old friend Klaus Peters that he suggested and argued that commercial publication of the manuscript (with his company, of course!) would make an attractive and appealing book that would be of interest to a much wider group of mathematicians and historians of science and higher education. He argued also for an extension of the scope of book beyond 1960. Both of these considerations led to my decision to continue the narrative to what I found to be a natural stopping place of about 1985 so as to include the establishment and early years of the Mathematical Sciences Research Institute (MSRI), in which I was personally involved. This was also the time when I took extended leave from the department to serve in an administrative position in the university. So the narrative goes through 1985, and there is a very brief final chapter covering some events since 1985.

In writing this history, I feel a bit like someone practicing a profession without a license, and I hope that professional historian colleagues and friends will forgive whatever lapses from established standards and expectations occur. A number of articles about the history of selected mathematics departments in the United States have appeared, especially in the American Mathematical Society's (AMS) centennial volumes in 1988, but to my knowledge this is the first book-length such study to appear. In addition, there exist a number of book-length manuscripts devoted to the history of various Berkeley departments, including physics, chemistry, mechanical engineering, and classics, and some of these have been privately published.

In writing about the history of the mathematics department and its faculty, I have relied on a number of primary sources. These have included the University Archives, housed in the Bancroft Library, and I would like to thank both Bill Roberts, university archivist, now emeritus, and his successor, David Farrell, for their invaluable help in identifying and locating documents and for their encouragement in this project. I also want to acknowledge the help and kindness of the entire staff of Bancroft Library. Another source of primary documents has been the minutes of the University of California Board of Regents, and I would like to thank Anne Shaw, associate secretary of the regents, for her help in identifying and locating documents and in arranging permission to quote from them. The other main source was departmental records, but these records become pretty thin as one searches back beyond about 50 years. The personal papers of mathematics faculty on deposit in the Bancroft Library were another source of information. University catalogues and the memorial biographical essays on deceased UC faculty (now online) were most helpful, as were discussions with colleagues. In many of the later chapters, especially Chapters 14 through 20, I have also relied on my own personal knowledge of events. My own involvement in events during this period is narrated in the third person, rather than the first person, since this is a history of the department and not a personal memoir.

The history of the department has to be embedded in the history of the campus and the university of which it is a part, as well as the politics of the State of California, and finally, the history of the department must be seen in the context of the history of mathematics in the United States over the last 150 years. It was never my intention to delve into primary sources for these narratives, but rather I have relied on secondary sources and the work of others for these narratives. Important secondary sources that I have used include Samuel Willey's *The History of the College of California*, William Warren Ferrier's *Origin and Development of the University of California*, Verne Stadtman's *Centennial History*, and *The Centennial Record*, John Douglas's *The California Idea*, David Gardner's *The California Oath Controversy*, Clark Kerr's memoirs *The Gold and the Blue* (two volumes), and David Gardner's memoir *Earning My Degree*. Three of Constance Reid's biographies of mathematicians—*Courant in Göttingen and New York*; *Neyman from Life*, and *Julia: A Life in Mathematics*—were of great help. Likewise, the Feffermans' recently published biography of Alfred Tarski and Steve Batterson's biography of Steve Smale were of great help. The essays in the AMS centennial volumes on various aspects of the history of American mathematics, as well as the Parshall and Rowe book *The Emergence of the American Mathematical Community*, were essential. See the list of references at the end of the book for the complete citations of these works. The Mathematics Genealogy Project has been invaluable as a source for listings of doctoral students, and even though the tabulation of doctoral students is a moving target for many faculty mentioned, I have used its most up-to-date numbers from mid-August 2006.

I have profited immensely from discussions with Erich Lehmann, who is serving as historian of the Berkeley statistics department. His personal knowledge dates back to 1940, when he enrolled as a student in the mathematics department. He has read and helpfully commented on drafts of the manuscript, as have Constance Reid, Helen Crosby Lewy, Albert Bowker, Beresford Parlett, Richard Karp, George Bergman, and Steve Finacom, as well as many other Berkeley colleagues. Conversations with Lensey Namioka, who is currently writing a biography of S. S. Chern, were also helpful. I am indebted to David Kramer whose judicious and thoughtful editing of the manuscript greatly improved its readability. In the end, any errors or oversights are, of course, my responsibility. I want to thank George Bergman for permission to reproduce a number of his photographs for inclusion in the book, and I thank the University Archives of the Bancroft Library for permission to reproduce photographs from the archives. Faye Yeager's work in digitizing the photographs was of great help to me.

In the publication agreement with A K Peters, I am pleased that we were able to work out a way in which commercial publication can occur along with web publication. Finally, all royalties from the sale of this book will go directly to the UC Berkeley Foundation and are dedicated to support for graduate fellowships in the mathematics department.

Introduction

This is the story of mathematics at Berkeley, a story that concerns almost entirely the Department of Mathematics at the University of California at Berkeley, but which also includes some discussion in Chapter 1 of mathematics instruction in California prior to the founding of the University of California, and also discussion in Chapter 19 of the Mathematical Sciences Research Institute (MSRI). This story is embedded in the history of the university, and of the social, political, and financial history of the state of California, as well as the history of American mathematics. These contextual factors are important because events and developments external to the department in these sectors influenced and shaped some key events in the history of the department.

A major theme in the story will be how a department in a state university that was for much of its early history devoted primarily to teaching developed into a major research center that is ranked among a small group of the very best departments in the country and the world. In this connection, it was said of Benjamin Peirce, the leading American mathematician of the mid-nineteenth century, who spent his career at Harvard, that before his time it never occurred to anyone that "mathematical research was one of the things for which a mathematics department existed." Realization of this function at Berkeley took place over time, but the story is marked by three periods in which substantial change took place rapidly and abruptly, which might even be analogized to phase changes in matter; these three critical times in the history of the department occurred in 1881–1882, in 1933–1934, and then finally in 1957–1958.

The University of California, which was chartered in 1868, resulted from the fusion of two institutional precursors, or "parents". One of them was the College of California, which was conceptualized by its founders, many of whom were Congregational ministers and Yale graduates, as a premier undergraduate college (or even perhaps more explicitly the "Yale of the West"). It was to be a Christian nonsectarian, college, unlike the sectarian colleges founded contemporaneously by the Methodists, namely the College of the Pacific, and the College founded by the Jesuits in Santa Clara. The College of California was chartered in 1855, shortly after California entered the Union, but because there were virtually no students prepared for collegiate studies, it started out as an academy—the Contra Costa Academy—located in downtown Oakland. By 1860, enough students had received the necessary preparation so that collegiate instruction could begin. The college curriculum was rigid, with no electives and a decidedly classical cast, with Latin, Greek, and Mathematics in the first two years. Collegiate instruction began in 1860. The mathematics instructor listed was the Rev. Francis Hodgson, MA, whose credentials are unknown other than what is contained in his title. The College of California graduated its first class of four students in 1864. In the 1850s, the college had acquired a magnificent parcel of land a few miles north of Oakland in open country as its future permanent home. In 1866, the college land and it environs were designated by the college as Berkeley, in honor of the Irish philosopher and bishop George Berkeley.

Meanwhile, the state was trying to figure out how to establish the Agricultural and Mechanical College, funds for which were provided in the Morrill Act. The legislature had chartered this college in 1866, but it existed only on paper. The plan that emerged in 1868 was to fold the College of California, shorn of its religious connections and the Agricultural and Mechanical College together into a single university, the University of California, that would comprehend all areas of knowledge. This was a wise and forward-looking decision, but it engendered warfare in the early years of the university between the competing parental traditions (polytechnic versus liberal arts) for the heart and soul of the university. The Board of Regents of the university spent the year 1868–1869 hiring the initial faculty, a task they took upon themselves, and the university opened in 1869 in the College of California building in Oakland with 12 new students plus continuing students from the College of California. The university began construction of buildings on its Berkeley site, which it inherited from the College of California, and the university moved from Oakland to Berkeley in 1873.

When the University of California (UC) opened for business in 1869, the first chair of mathematics hired by the regents was a West Point graduate who had never taught mathematics at the collegiate level, William Welcker. He imported the West Point mathematics curriculum, which had been widely adopted throughout the country for many decades, but which was seen by then as somewhat dated and as resembling more a drill room. Welcker was fired in 1881 by the regents, who wanted a more "scholarly" department with a more up-to-date curriculum. This action, which was widely reported in the newspapers, was one of the final shots in the battle over the mission of the university, and the first turning point in the direction that the university was to take as a comprehensive, not a polytechnic, institution. It should be remarked that the mathematics department at the time consisted of one professor and a few assistants. These early years of mathematics at Berkeley are described in Chapter 2.

Irving Stringham, who succeeded Welcker in 1882, was selected by a committee of regents, headed by Horatio Stebbins, minister of the First Unitarian Church in San Francisco and a man with many Harvard connections and friendships. Stringham had sterling credentials: he had studied under Benjamin Peirce at Harvard as an undergraduate, J. J. Sylvester at Johns Hopkins for his PhD, and Felix Klein at Leipzig for postdoctoral studies. He was joined in 1890 by Mellen Haskell, who after receiving his bachelor s and master s degrees from Harvard went to Germany and received his doctorate under Klein at G ttingen. Stringham and Haskell modernized the curriculum and added breadth and depth to the course offerings. Their research interests were in the Klein tradition of function theory, elliptic functions, and classical algebraic geometry. Derrick Norman Lehmer, who joined the department in 1900, with a PhD from Chicago under Eliakim Hastings Moore, was perhaps the most visible member of the department from the outside. The first PhD degree granted by the department was in 1901. There was a gap until 1909, after which there were about two a year on average until the 1930s.

Stringham, and then Haskell after Stringham s death in 1909, together ran the department for over 50 years, from 1882 to 1933, as described in Chapters 3 and 4. The tradition at that time was chair for life. The emphasis was on teaching, with some attention to research but not very much, and the department became ingrown: 10 of the 12 appointments to faculty positions from 1913 to 1933 were Berkeley PhDs, including all 9 after 1918. And one of the two non-Berkeley doctorates appointed in the period was Florian Cajori, who was a scholar of the history of mathematics education. Haskell was scheduled to retire in 1933. But a year earlier, in 1932, some chairs of other science departments, notably chemistry, physics, and astronomy, together with President Sproul, his provost, and the campus Budget Committee, were all persuaded that the department needed to be seriously reconstituted and reorganized. It had fallen far behind these other departments in its research standing and intellectual distinction. A campus committee to review the department proposed the dismissal of a number of junior faculty not oriented toward research and that a distinguished leader from outside be brought in to remake the department. This proposal was unusual, coming in the depths of the Great Depression during a virtual hiring freeze. The decision was imposed from above by campus leadership and without consultation with the department, an action that produced some unhappiness. This was the second turning point, or phase change, in our narrative and is the major topic of Chapter 5.

Joel Hildebrand (chemistry), a high-profile figure on campus, was deputized to undertake national talent scouting tour in February 1933, and he identified Griffith Evans, then at Rice University, as a potential candidate to lead the department's renewal. He also identified several potential junior faculty members who might be hired (including Charles Morrey). The campus approved the Evans appointment, but it took Evans a year to disengage from Rice and make the move to Berkeley. He arrived in 1934. Evans held a Harvard AB and PhD (under Bôcher) and had been a postdoctoral student with Vito Volterra in Rome. Integral equations was one of his specialties. Harvard had tried to hire him away from Rice in 1925, but the call failed, much to Harvard's dismay. It is interesting to note that then Harvard hired Marston Morse as their second choice.

Evans served as chair at Berkeley for 15 years, until 1949, and he remade the department as described in Chapters 6 and 7. Among the notable early hires in the 1930s and early 1940s were Jerzy Neyman (statistics), Alfred Tarski (logic), Charles Morrey and Hans Lewy (analysis and partial differential equations), and Derrick Henry Lehmer (number theory). Working with Neyman, Evans made a number of stellar appointments in statistics, including that of Erich Lehmann. Evans made 21 appointments in all and significantly transformed the department's stature. Neyman was a builder and entrepreneur, and from the day he arrived in 1938, he wanted a separate statistics department. Evans vigorously resisted him, and it took Neyman 17 years, until 1955, to realize his ambition, as described in Chapter 11, but indeed only after Evans's retirement. This narrative does not trace the history of statistics at Berkeley beyond 1955. In the late 1940s and early 1950s, the department pursued development in actuarial science, applied mathematics, and computation. These initiatives, which are described in Chapter 8, all came to naught; in particular, in 1948, Berkeley lost the competition to host the National Bureau of Standards Computational Center (the Institute for Numerical Analysis) to UCLA, and the effort in 1949–1950 to bring Richard Courant and his entire group at New York University to Berkeley ended in failure. Efforts in actuarial science never got off the ground. The California loyalty oath controversy, which is described in Chapter 9, extended from 1949 to 1953 and had a serious impact on the department, resulting in faculty losses, interference with some of the initiatives mentioned above, and academic growth being placed in suspended animation for several years.

In spite of the obvious successes in hiring and the rise in reputation of the department under Evans, its image on campus in the 1950s appeared to some still frozen in time from 1930 as a teaching and service department. Located in Wheeler Hall and then Dwinelle Hall, among the humanities departments, it was remote from the other science departments, which were prospering. Its size was small, and the campus had been reluctant to invest in it in spite of vigorous protestations by the department. Within the American mathematics community, the department was informally ranked somewhere in the top ten mathematics departments in the country, but it is not clear exactly where. Harvard, Princeton, and Chicago had been the traditional top-three departments for many decades, and Berkeley was by no means their equal. The department was unbalanced, with considerable strength in analysis and logic (and of course statistics, which had separated), but lacked appropriate numbers of faculty in geometry and topology, algebra, and applied mathematics, and had difficulty recruiting in these areas.

This all changed in 1957 when under John Kelley's leadership, the department proposed a plan of action and convinced the campus leadership to invest very heavily in mathematics, including a number of senior appointment as well as substantial overall growth. Clark Kerr, who was the Berkeley chancellor at the time, wrote in his memoirs:

> Mathematics had quality of faculty under the guidance of Griffith Evans, but not quantity. The sciences and engineering had treated it mostly as a "service" department for training their students. I was convinced, although I was a non mathematical economist, that mathematics should and would be as central a department in a great research university of the future as philosophy had been in the past. Philosophy, once the "mother" of so many other departments, itself had become one of its more specialized children. Quantitative methods were rapidly getting more emphasis across the board (as they had gradually and intermittently since Pythagoras). Thus, I concluded, if a campus were to have one preeminent department in modern times, it should be mathematics. Also statistics was a new department at Berkeley under the excellent leadership of Jerzy Neyman, and deserving of expansion for related reasons. [Kerr 2001, p. 85]

The department grew from 19 in 1955 to 75 in 1967, a quadrupling in size in 12 years, including a series of strategic hires in 1958–1960 of established mathematicians under John Kelley, notably Chern, Spanier, Smale, Hochschild, Rosenlicht, and Kato. These appointments provided strength in geometry and topology, algebra, and applied mathematics for the first time. With the imbalance among mathematical subfields finally ameliorated, the department then broke through to the very top in rankings by the early 1960s, coming in a close second to Harvard and ahead of Princeton in the 1964 ACE survey. Thus 1957–1958 was the third turning point for the department, and these events are chronicled in Chapter 13. It is interesting to note that the change at the first turning point, in 1881–1882, was imposed on the department from the Board of Regents; the second change, in 1933–1934, was imposed on the department by colleagues in other science departments and the campus academic leadership. By contrast, the momentum for the third change came from within the department itself, and the transformation was the result of its ability to persuade the campus leadership to act.

The department had moved into Campbell Hall in 1959 along with the Departments of Astronomy and Statistics, a move that finally brought the department into proximity with the other science departments; this move can be seen as signaling recognition of mathematics as a scientific discipline and that interaction with other science departments was a valued academic goal. Campbell Hall was much too small, and early planning had already been underway in 1956 for a dedicated mathematical sciences building. This project was subject to many delays, and the building was not completed until 1971. The result was Evans Hall (named in honor of Griffith Evans), which one sees looming over the central glade of the campus. The story of the planning for Evans Hall and campus academic planning in the 1960s is related in Chapter 15.

The story of computing and computer science on the Berkeley campus and its relationship to mathematics is a convoluted one that is related in Chapter 14. There are interesting parallels and contrasts with how statistics developed. After nearly 15 years of wrangling, the organizational issues were finally settled in 1973 by then-chancellor Bowker.

In a real sense, the mathematics department had by 1973 achieved a steady-state configuration that has not changed subsequently in any substantive way. The rapid growth that began in the late 1950s and extended through 1967–1968 was absorbed and digested, so to speak, by 1973. Initial gains in geometry, topology, algebra, and applied mathematics were solidified and built on. The 1960s were a time of social and political turmoil, and this affected the department, but by 1973, the unrest of the "sixties" had abated. These events through 1973 are described in Chapters 16 and 17. Since 1973, the department has oscillated somewhat in faculty size, but this has been less the result of conscious academic planning than of fluctuations in the economy of the state and resulting oscillations in the university's budget. Chapter 18 carries the story of the department through about 1985. There was one signal event during this final period: the founding of the Mathematical Sciences Research Institute (MSRI) in 1982, and Chapter 19 is devoted to the beginning years of MSRI. A brief end note (Chapter 20) contains an outline of events since 1985. The appendices reproduce in their entirety some key documents mentioned in the narrative.

Chapter 1

Creating a University

At the time of its creation in 1868, the University of California (UC) was the offspring of two quite different parent institutions. The legacies of these two parents pulled the institution in different directions during its early years, leading to conflict about the mission of the new university from both within and outside the university. This conflict played a significant role in the early years of the development of mathematics at Berkeley.

One parent was the College of California, a private college founded in Oakland in 1855 by immigrants to California from New England. The early leaders of the college included a number of Congregational and New School Presbyterian ministers who came to California with the goal of creating such a college. They had in mind specifically creating a college in the pacific that would ultimately be the equal of Harvard or Yale. Indeed, Yale College was overrepresented among the early founders and faculty, which included the Rev. Henry Durant (Yale '27), the Rev. Martin Kellogg (Yale '50 and class valedictorian), and the Rev. Samuel Hopkins Willey (Dartmouth '45). Although the driving spirits behind the founding of the College of California were men with church affiliations, the college was never ecclesiastically controlled: it was to be Christian but nonsectarian. In 1860, the College of California set forth its foundational principles:

> The College of California is an institution designed by its founders to furnish the means of a thorough and comprehensive education under the pervading spirit and influence of the Christian religion. The bond which unites its friends and patrons are a catholic Christianity, a common interest in securing the highest educational privileges for youth, the common sympathy of educated men, and a common interest in the promotion of the highest welfare of the State, as fostered and secured by the diffusion of sound and liberal learning.

It was declared that a majority of the trustees of the College "should be members of evangelical Christian churches, and not more than one fourth of the actual members be of one and the same Christian denomination." Finally, the statement specified that "in the election of professors, preference shall always be given to men of Christian character, and the president and a majority of the faculty shall be members of evangelical Christian churches" [Fer 1937, p. 181].

The college was quartered temporarily in downtown Oakland while the founders began a search for a permanent site. The Rev. Horace Bushnell (Yale '27), a noted theologian and pastor of the North Congregational Church, in Hartford, Connecticut, had come to California in 1856 in an attempt to regain his health. At the behest of his classmate, Henry Durant, Bushnell spent several months traveling through Northern California evaluating potential sites for the college, after which he presented a detailed report to the board of trustees. The board also tried, without success, to persuade Bushnell to accept the presidency of the college. Shortly thereafter, the board obtained a parcel of land north of Oakland overlooking the Golden Gate as the permanent site for the college.

In a ceremony at Founders Rock at the northeast corner of the site on April 16, 1860, the site was consecrated to learning by the trustees and friends of the college. The *Pacific* editorialized a few days later, "There is not another such college site in America, if indeed anywhere in the world. It is the spot above all others we have yet seen or heard of where a man may look in the face of the nineteenth century and realize the glories that are coming on" [Fer 1937, p. 182]. As they consecrated the site, the participants were unaware that they were standing only yards away from a major earthquake fault: the Hayward Fault. In 1866, the college named these lands and its environs Berkeley, in honor of the Irish philosopher Bishop George Berkeley. Bishop Berkeley (1685–1753) was a patron of education and in 1728 had come across the Atlantic to Rhode Island with the goal of promoting education in the British colonies in the New World. Although he failed in his goal of establishing a college in the colonies and returned to Britain after three years, he was influential in aiding the formation of Kings College (now Columbia University) and the College of Philadelphia (now the University of Pennsylvania). He also endowed the Berkeley Scholarship Fund at Yale and made a gift of books to both Yale and Harvard. He had written a poem about his goals of promoting education in the colonies, which concluded as follows:

> Westward the course of empire takes its way:
> The first four acts already past,
> A fifth shall close the drama of the day;
> Time's noblest offspring is the last.

These lines were known to some of the founders of the College of California and seemed to capture the spirit of their new institution; hence the name Berkeley [Fer 1930, pp. 244–255]. It is perhaps of interest to note that Bishop Berkeley is also known in the history of mathematics for his 1737 attack on fluxions, the basis of Newton's calculus.

The college had been operating since 1855, but only as an academy providing college preparatory work, because there were no students yet prepared for college-level instruction. Five years later, the college's first class of four students began their studies under Durant, as professor of Greek, and Kellogg, as professor of Latin, who together constituted the entire faculty. Samuel Willey accepted appointment in 1862 as vice president, in effect serving as acting President.

The entrance requirements for the college were formidable and were stated as follows:

> No one shall be admitted without sustaining a satisfactory examination in the following studies, or their equivalents:
>
> Latin Grammar; Latin Reader; Caesar's Commentaries, first five books; Cicero's Select Orations; Virgil's *Bucolics*, and the first six books of the *Aeneid*; Latin Prosody and Composition; Greek Grammar; Greek Exercises; Xenophon's *Anabasis*, first five books; Greek Testament, the first two Gospels, Luke and John; the Greek Accents; English Grammar; Elements of Rhetoric; Geography; Higher Arithmetic; Algebra to Quadratic Equations; and the Rudiments of French and Spanish. [Wil, p. 60]

The college curriculum was a highly structured classical curriculum with no electives. The first two years consisted of Latin, Greek, and mathematics. In the final two years student were exposed to a variety of other subjects:

> Freshman year: Latin (Livy), Greek (the *Iliad*), algebra.
>
> Sophomore year: Latin (*Tusculan Disputations*), Greek (*Prometheus*), trigonometry.
>
> Junior and senior years: German, natural philosophy, logic, rhetoric, English, history, astronomy, chemistry, physiology, moral philosophy. [Wil, p. 89]

Kellogg taught the mathematics course in the college's first year of operation, after which the Rev. Francis Hodgson, MA, was engaged as instructor in mathematics and natural philosophy to teach the mathematics and science courses. The content of the mathematics courses as of 1865 was the following: freshmen year—Robinson's *University Algebra* from quadratics to Section 8 on properties of equations; sophomore year—plane trigonometry, spherical geometry, and trigonometry, and commencement of the study of surveying [Wil, p. 142].

In 1864, the college graduated its first class of four students, and when it ceased operation in 1869 to become part of the University of California, the college had graduated six classes, totaling 23 students. The college was perennially short of money, and found itself seriously in debt [Sh, p. 353]. Attempts to raise funds in the East produced meager results, with many potential donors asking, "With all the gold in California, why do you come to us?" The board of trustees also sought new members in an attempt to broaden its base of support. In the 1860s, the college added a number of individuals to its board who would later play major roles in the life of the university, diversifying the board denominationally. Of particular note are John W. Dwinelle (Hamilton '34), an Oakland lawyer and mayor of Oakland; and the Rev. Horatio Stebbins (Harvard '48, DD '51), pastor from 1864 to 1899 of the First Unitarian Church of San Francisco. Stebbins replaced on the board of the college his legendary predecessor in the pulpit of this church, the Rev. Thomas Starr King, who had died in 1864. In February 1867, the Rev. James Eells, the newly arrived pastor of the First Presbyterian Church (an Old School Presbyterian Church) joined the board. As Millicent Shinn [Sh, p. 353] points out, the founder and builders of the college were almost exclusively Republicans.

Thus in 1868, this parent of the University of California had been in operation for 13 years and had offered college-level instruction since 1860. It had already produced several graduating classes. It had a facility in Oakland, including a library. It had support from a broad array of community leaders and a magnificent permanent site in what was to become the city of Berkeley. It also had debts.

The second parent of the university was the Agricultural, Mining, and Mechanical Arts College, which was chartered by the California Legislature in 1866 pursuant to the Morrill Act. Under this act, receipts from the sale of 150,000 acres of federal land were to provide an endowment to support this college. Its board of directors—chaired by Governor Low, another Yale graduate and a Republican—set out as its first order of business to find a site for this land-grant college. This parent was no more than a paper institution with a governing board, but it had what the other parent lacked: money.

The leaders of the College of California as well as Governor Low, a friend and financial supporter of the college, had conceived the idea of merging the College of California, freed of its religious connections, with the Agricultural, Mining, and Mechanical Arts College in order to form a comprehensive university that would provide training in technical subjects as well the liberal arts. This new institution would fulfill an unmet provision in the state constitution of 1849, calling for the creation of a state university. At the June 1867 graduation ceremonies of the College of California, which were attended by Governor Low, the governor remarked in substance to Acting President Willey, "You have here in your college scholarship, organization, enthusiasm, and reputation, but no money. We in undertaking the state institution have none of these things, but we have money. What a pity they could not be joined together" [Wil, p. 204]. The commencement speaker on that day was the noted chemist and Yale faculty member Benjamin Silliman, Jr. (Yale '37), whose address was appropriately titled, "The Truly Practical Man, Necessarily an Educated Man." The fact that Yale University had been able to capture Connecticut's appropriated land-grant funds for its Sheffield Scientific School may have directed thinking in California along similar lines.

In fall of 1867, the trustees of the College of California voted to donate their assets, including the Berkeley site, to the state for the creation of a state university that they wished to see composed of a "College of Mines, College of Civil Engineering, College of Mechanics, College of Agriculture, and Academic College, all of the same grade, and with courses of instruction equal to those of Eastern Colleges." The College of California would then be dissolved. The trustees made their decision in spite of fears that the new institution would be politically controlled and manipulated. The following spring, John Dwinelle, then a member of the legislature, introduced a bill to organize the University of California, which quickly passed both houses and was signed by the new governor, Henry Haight, on March 23, 1868 (Charter Day). The university was to be governed by a board of regents, sixteen of whom would be appointed, with an additional six serving ex-officio. These sixteen regents were to be appointed by the governor with the advice and consent of the state senate.

The new governor, Henry Haight, was a Democrat, who had won the governorship in a recent election, replacing the Republican governor Low. Although a Yale graduate himself (class of '44), Haight had not been close to the founders of the College of California; nor had he been an active supporter of the college. The appointed regents were balanced by political party, with eight Democrats and eight Republicans, but very few of the old College of California men were appointed to the new board. Shinn explains that

> in spite of the most faithful effort to be non-political and non-sectarian, the College of California had by sheer force of gravitation become mainly Congregational, and almost exclusively of New England. This Governor Haight intended to break up; he meant to make the University as broad-based as the population of the State. He planned to represent North and South, Catholic, and Protestant, and Jew on the board.

Stebbins, who was the chairman of the board of the College at the time, and Dwinelle, who had carried the bill to charter the university, were the only two members of the board of the College of California appointed to the board of regents. Because of the Democratic victory in the recent elections, the first ex-officio regents were mostly Democrats. The board as constituted in June 1868 was composed of 13 Democrats and 9 Republicans.

Stebbins was a very learned man of considerable oratorical skills, and from all surviving evidence, he appears to have been a man of not inconsiderable ego. He was a prominent Unitarian minister and a Harvard graduate, with many connections to Cambridge. Andrew Hallidie, as president of the Mechanics Institute of San Francisco, became an ex-officio regent. Born in Great Britain, Hallidie joined the gold rush at age 16, and then studied and practiced engineering. He designed and built bridges and established a firm in San Francisco to produce wire cable. He subsequently conceived the idea of a cable railway, and by 1873 had built and operated the first (and famous) cable railway in San Francisco. Both Hallidie and Stebbins were to serve as regents for many years, into the 1890s, and were to exercise great power and influence on university affairs. Hallidie served for a long time as chair of the finance committee, and Stebbins as the chair of a number of committees overseeing educational policy. Stebbins and Hallidie, together with the very powerful secretary of the board, a full-time employee of the regents, were so dominant and intent on managing the university that for many years presidents of the university found that there was not very much in the way of governance left for them. Another influential regent was San Francisco lawyer and judge John Hager, who as state senator served in 1868 as the chair of the joint senate and assembly committee on the bill to establish the University of California. He was a Democrat and served as a regent from the founding of the university. He also later served as a US senator, and in subsequent debates in the regents that are relevant for our story, he was a key opponent of Stebbins.

The regents determined that the university would not open until September 1869, since it would take some time to appoint a president and a faculty, and so the regents asked the College of California to continue in operation for the 1968–1969 academic year. The first task of the regents was to appoint a president, and in November 1868, the presidency was offered to General George B. McClelland, the Civil War general and unsuccessful Democratic candidate in the presidential election of 1864. The Republicans on the board favored Stebbins for the office, but the final vote split along party lines, 12 to 5 for McClelland. Many observers in the state were appalled by the choice of McClelland, including ex-governor Low, who promptly resigned from the board. To the great relief of friends and supporters of the University, McClelland declined the offer. Instead of naming another candidate, the regents themselves turned to the task of appointing a faculty. Over the following ten months the board appointed a faculty of ten, including eight professors, one assistant professor, and one instructor.

The University of California opened in September 1869 in the College of California facilities in Oakland with 38 students enrolled, a number that included new students as well as the continuing students from the College of California. The land owned by the college in Berkeley had become the property of the new university, and construction of university buildings began there shortly. The university made the move to its permanent site in Berkeley in 1873. Also in 1873, blue and gold were chosen as the school colors. The choice of gold is easy enough to understand, since the campus is situated facing the Golden Gate Bridge and is located in the "Golden State." It is a plausible conjecture, supported by evidence, that the choice of blue was at least in part a heritage from the school color of Yale, given the large role that Yale graduates had played in the founding of the university [Fer 1930, pp. 631–633]. As the university developed, Harvard graduates would play a more dominant role in the mathematics department than graduates of Yale or other American universities. Indeed, all department chairs of mathematics from 1882 to 1954 held Harvard degrees, with the exception of an interim chair who served for one year, 1933–1934.

American Mathematics in the Nineteenth Century

The appointment of the initial faculty, and in particular the appointment of the mathematics faculty, as well as the subsequent events involving the mathematics faculty, must be viewed in the context of the state of mathematics in United States at that time. The preeminent mathematician in the United States at the time was without doubt Benjamin Peirce (1809–1880, Harvard '29), who had been appointed professor of mathematics and natural philosophy at Harvard in 1833. In 1842, he became the Perkins professor of astronomy and mathematics. As a young man he had assisted the astronomer and mathematician Nathaniel Bowditch with his translation and commentary on Laplace's *Mécanique Céleste*. Astronomy and celestial mechanics remained a major focus of Peirce's work throughout his life. He published a number of papers, mostly connected with astronomy and celestial mechanics, and was also the author a number of textbooks. Perhaps his most original research accomplishment was a long paper on linear associative algebra, which he circulated in 1870 in what would today be called preprint form. It was formally published only posthumously, with editorial work by his son Charles Sanders Peirce.

Peirce influenced and trained directly or indirectly many students who went on to considerable success, including most notably Simon Newcomb (1835–1909), George W. Hill (1838–1914), and his own son Charles Sanders Peirce (1839–1914). Charles W. Eliot, who assumed the presidency of Harvard in 1869 and who proceeded to shake up the institution during his 40 years in office, was also a student of Peirce and briefly served as a tutor of mathematics at Harvard under Peirce. Peirce knew and worked closely with Alexander Dallas Bache (1806–1867), who was professor of natural philosophy at the University of Pennsylvania before he became superintendent of the US Coast Survey in 1843. Peirce also knew and worked with Joseph Henry (1797–1878), the distinguished physicist who did pioneering work on electromagnetic induction and who, in 1846, became the first secretary of the Smithsonian Institution. These three men plus the naturalist Louis Agassiz, at Harvard, and a few others formed the early leadership in science in the United States. Bache was influential in lobbying Congress to pass legislation to create the

National Academy of Sciences and was its first president. Peirce himself headed the Coast Survey from 1867 to 1874 following his friend Bache's death.

Bache had located the Longitude Office of the Coast Survey in Cambridge so as to be under the guidance of Peirce, and the navy located the Naval Almanac in Cambridge. Charles Henry Davis, the naval officer and later admiral who started the Naval Almanac, was Peirce's brother-in-law, so Peirce enjoyed very good relations with the federal government. Peirce's influence on mathematics was pervasive, and in assessing Peirce many years later, the Harvard mathematician J. L. Coolidge opined [Coolidge 1925] that before the time of Benjamin Peirce it had never occurred to anyone that "mathematical research was one of the things for which a mathematics department existed."

Mathematics was represented early on at Yale, where Jeremiah Day was appointed professor of mathematics and natural philosophy in 1801. Day set about writing a number of elementary textbooks that were very widely used for decades. He assumed the presidency of Yale in 1817 and served for 29 years in that office. A son, Sherman Day (Yale '26), became a civil engineer and was an early immigrant to California, where he surveyed railroad routes, served in the state senate, and was an influential founding member of the board of the College of California (yet another example of Yale influence on the College of California). Elias Loomis (Yale '30) studied mathematics and astronomy, and after serving in several professorships of mathematics and natural philosophy, returned to Yale in 1860 as the chair of natural philosophy and astronomy.

In a younger generation, Hubert Anson Newton (Yale '50) studied mathematics as an undergraduate and then continued his studies for two years on his own before being appointed tutor in mathematics at Yale. In 1855, at the age of 25, he was appointed professor of mathematics and given a year's leave of absence to study in Europe. He went to Paris, where he studied under the geometer Michel Chasles. Soon after his return, his interests shifted to astronomy, and he became well known for his work on meteorites and meteor showers. Nevertheless, he mentored and trained students interested in mathematics at Yale until his death in 1896. He was a mentor of Josiah Willard Gibbs (1839–1903, Yale '58) and no doubt was influential in arranging Gibbs's appointment as professor of mathematical physics at Yale in 1871. It was during the 1870s that Gibbs published his pioneering work on thermodynamics, in which he made effective use of geometrical techniques. Gibbs also wrote a foundational book on vector analysis. Newton is also known to the mathematical community as the dissertation advisor of Eliakim Hastings Moore, who completed his PhD degree at Yale in 1885.

Yale began an organized doctoral program in 1861, and Harvard, under President C. W. Eliot, started one in 1871 with James Mills Peirce, another son of Benjamin, who was also a professor of mathematics at Harvard, serving in the role of what would become dean of the graduate school. Initially, neither program granted many degrees. It was Johns Hopkins University, founded in 1876, that became the first research university to emphasize doctoral education systematically. Daniel Coit Gilman, the founding president and recently departed president of the University of California, was able to attract James Joseph Sylvester from Great Britain as the professor of mathematics. Sylvester, in turn, attracted many excellent graduate students and supervised a string of doctoral students during his time at Hopkins. Simon Newcomb, who had become professor of mathematics in the US Navy in 1861 at the Naval Observatory, and who was later superintendent of the Nautical Almanac, served as Gilman's chief scientific adviser. Newcomb subsequently succeeded Sylvester at Johns Hopkins, serving for a number of years as part-time professor of mathematics and astronomy while keeping his Navy position. It should be added that George W. Hill was also employed by the navy at the Nautical Almanac from 1861 to 1892 while he was conducting his research in celestial mechanics. Charles Sanders Peirce was employed by the US Coast Survey for over 30 years, so one can see that the federal government was providing important support for mathematical research even at this early time.

In 1880, these four near contemporaries, Gibbs, Newcomb, Hill, and C.S. Peirce, emerged as the leading mathematicians in the United States of the generation following that of Benjamin Peirce; Gibbs's work was greatly admired worldwide; Newcomb did influential work in celestial mechanics, and in later years Einstein paid tribute to his work on planetary orbits; Poincaré paid great tribute to Hill's work as inspiring his own research; and C.S. Peirce was well known for his work in algebra, logic, and philosophy, and outside of mathematics he is probably better known than his mathematically famous father. Newcomb went on to gather many honors, including numerous medals and honorary degrees. Hill and Newcomb were respectively the third and fourth presidents of the American Mathematical Society.

Celestial mechanics and astronomy were clearly a major theme of mathematics in the United States for most of the nineteenth century. The theme of algebra and geometry appeared in the work of Sylvester and his students, in Gibbs's work, as well as in the later work of Peirce. A final theme of American mathematics, complex analysis, did not appear until the 1880s, when a number of American students went to Germany for doctoral or postdoctoral studies. They were attracted to work under Felix Klein, first at Leipzig, and then at Göttingen. Klein had a long string of American students who collectively had an enormous influence on US mathematics beginning in the 1890s.

The United States Military Academy at West Point (USMA) played an important role in mathematics and technical education in the United States in the first half of the nineteenth century that is perhaps not widely recognized. The USMA was created in 1802 in the Jefferson administration, but it was Sylvanus Thayer, the superintendent of the academy from 1817 to 1833, who is known as the father of the military academy. Under his leadership it became an institution devoted not only to training military officers with a solid grounding in mathematics, science, and engineering, but also to training engineers and scientists for the country. In fact, it was the country's first engineering school. Graduates entered the military, but many soon resigned and followed civilian pursuits. Such men formed a reserve of highly trained officers who were available for military service in a national emergency, but who were leading engineers in civilian life.

The curriculum was modeled on the curriculum at the École polytechnique and had a strong emphasis on the primacy of mathematics. Its first mathematics professor of note was Charles Davies (USMA '15), who wrote a series of textbooks that went through many editions and were widely used throughout the country until the 1870s. A number of his books were translations of French textbooks. His successor, Albert E. Church (USMA '28), was a disciple of Thayer and served as professor of mathematics from 1837 to 1878. He also wrote a series of textbooks that went through many editions and were likewise widely used through the 1870s. Indeed, the names of Davies and Church were well known to generations of college students in this country. Along with Church, two other disciples of Thayer dominated the West Point curriculum for decades: William Bartlett served as professor of natural philosophy from 1837 to 1871, and Dennis Mahan served as professor of civil engineering (and the science of warfare) from 1832 to 1871. All three graduated first in their class at West Point in the classes of '28, '26, and '24, respectively. Although all the evidence suggests that Dennis Mahan was a revered and famous man in the US Army, his fame is readily eclipsed by that of his son (Admiral) Alfred Thayer Mahan, whose middle name no doubt was chosen to honor Sylvanus Thayer.

Indeed, in the antebellum period, or at least through 1850, the USMA was regarded as a paragon of scientific and technical training in the United States. Florian Cajori, in his definitive 1890 study of mathematics education in the United States, stated that in these years, "West Point was the only school of mathematics and physical sciences where the rigid requirements and high standards now deemed essential were even attempted" [Caj, p. 123]. The Academy provided many engineers who contributed to the country's westward growth and many faculty members in American colleges and universities. In 1850, Francis Weyland, president of Brown University, wrote that West Point had done more to build up the system of internal improvements in the United States than all of the colleges in the United States combined [Forman, p. 88]. The academy often invited back some of its top graduates to teach there. In addition to those already mentioned, Alexander Dallas

Bache graduated first in his class at West Point in 1825 at the age of 19 and taught mathematics there for several years before assuming a professorship at the University of Pennsylvania. Ulysses S. Grant (USMA '43), according to his memoir, enrolled in West Point to prepare to be a professor of mathematics at a "respectable college." Around 1850, when Harvard and Yale were looking for professors of engineering for the new Lawrence Scientific School and the Sheffield Scientific School, respectively, they turned to USMA graduates [Forman, p. 88]. However, by 1860, the emphasis at West Point was turning away from science, mathematics, and engineering, and many other universities had developed programs in these areas. After that time, West Point no longer occupied a leading position in science and engineering education.

Appointing a Professor of Mathematics

It was against this backdrop that the regents began hiring the faculty for the state university that was planned to rival the eastern colleges. The first faculty member to be appointed was John LeConte, as professor of physics, on November 17, 1868. His brother Joseph LeConte was appointed professor of geology, botany, and natural history two weeks later, on December 1. The LeContes were natives of Georgia, and after receiving postgraduate education in the North, had settled into faculty positions in Georgia and South Carolina prior to the Civil War. Both made numerous contributions to the scientific literature. During the Civil War, John LeConte was superintendent of the Confederate Nitre and Mining Bureau, while his brother Joseph served as chemist at the same agency. Seeing few opportunities in the Reconstruction South, and hearing of openings in California, they applied for faculty positions after being assured that their "politics" would not be a factor [St, p. 51].

The scientific establishment in the East offered its strong support for both their candidacies. Joseph Henry, secretary of the Smithsonian, wrote a letter of recommendation for both LeContes, and Louis Agassiz, under whom Joseph LeConte had studied earlier in his career, provided a letter of recommendation for Joseph. Benjamin Peirce, in a letter directed personally to regent Horatio Stebbins, offered warm and enthusiastic support for John LeConte. Peirce addressed the issue of "politics" as follows: "He [John LeConte] has been so far removed from politics that it would be very cruel to permit his dilemma in a disloyal state to cloud his prospects for advancement now that peace is here—and we have too few men of his ability and devotion to science to afford [not] to use them—except on unquestionable grounds of offense" [UA, CU 1, Box 2, folder 16].

Also on December 1, the regents appointed Martin Kellogg from the College of California as professor of ancient languages, and R.A. Fisher as professor of chemistry. On January 16, 1869, they appointed a three-person executive committee, consisting of regents Ralston, Stebbins, and Butterworth, to act as the executive head of the university, and John LeConte was asked to come to California as soon as possible to cooperate with the regents in planning the new university. LeConte was subsequently appointed acting president, on June 14.

However, the regents had difficulty making further faculty appointments and had particular difficulty selecting a professor of mathematics. During the spring and early summer, the regents were considering three applicants for the professorship of mathematics: George Bates, William H. Parker, and William Thomas Welcker. It is evident from Welcker's letter of application that he believed that regent Hallidie would be his patron or sponsor for the appointment. Welcker, a native of Tennessee, had graduated from West Point in 1851, ranking fourth in a class of 41. At the outset of the Civil War he resigned his commission and served in the Confederate Army. Welcker had, however, never taught mathematics in any institution. Albert Church, professor of mathematics, and William Bartlett, professor of natural philosophy, both at West Point, wrote letters of recommendation for him. As an added incentive to the regents to hire him, Welcker stated that he would be willing to offer any military instruction that might be required. (The Morrill Act required land grant Institutions to offer military instruction) [UA, CU 1, Box 2, folder 16].

Both Bates and Parker had mathematics teaching experience, but Parker had never taught calculus, according to the records of the executive committee. At the regents' meeting of June 14, it was regent Hallidie who moved that Welcker be appointed professor of mathematics. Hallidie, of whom it was said that he practiced his Republicanism with a heroic fervor [St, p. 48], was apparently not bothered by Welcker's obviously Democratic "politics." However, Hallidie's motion was postponed [RM June 14, 1869]. The matter of the professorship of mathematics was again on the agenda at the regents' meeting of July 6, and it was postponed once more. At that meeting, the regents appointed a professor of agriculture (Ezra Carr), a professor of English language and literature (William Swinton), and a professor of modern languages (Paul Pioda). The only unfilled position was the chair of mathematics.

At the July 19 meeting of the regents, the executive committee reported to the board that although they were satisfied that any of the three candidates mentioned "will be found qualified to discharge the duties appertaining to the Chair of Mathematics," they were clearly not pleased with any of the candidates, and wrote that "in view of the reasonable prospect of being able, in a short period of time, to secure an experienced and thorough instructor from the East, the Committee respectfully recommend that the Chair of Mathematics be not filled at this time" [RM July 19, 1869]. Time was running out, though, since instruction was supposed to begin shortly after the first of September.

At the next regents' meeting, on August 10, the executive committee reported that through correspondence conducted by Professor John LeConte, two additional candidates had been identified and their names submitted for consideration. These were William Woolsey Johnson, assistant professor at the US Naval Academy, and Simon Newcomb (see above), of the Naval Observatory. Johnson was a graduate of Yale ('62) and had gone to work at the Nautical Almanac, then located in Cambridge, where he came under the influence of Peirce. Johnson was recommended by Professor Elias Loomis, at Yale, and by John Runkle, professor of mathematics and later president of MIT (and another student of Peirce). Newcomb was recommended by Joseph Henry, secretary of the Smithsonian. The executive committee wrote that they had assurances that each of the five candidates would accept the position if elected, and that in the committee's view each is "competent to fill the Chair with credit to the University." The committee declined, however, to express a preference, and asked the board to make the selection. The regents then voted, and as reported in the minutes of the meeting, the results were as follows: Johnson, 3 votes; Welcker, 9 votes; Bates, 1 vote; Parker, 1 vote. The minutes do not mention Newcomb's name, so it is assumed that he received no votes. Welcker was therefore declared elected, and he assumed duties three weeks later, on September 1, with a salary of $300 gold per month [RM, August 10, 1869].

This would appear to be a rather handsome salary at the time and to give proof of the University's goal to compete with the best institutions in the east. The only firm comparison data come from a salary comparison survey the regents conducted in 1882—some 13 years later. The salary for a UC full professor was still $3600 per year ($300 per month). Harvard paid $4000 per year, and Yale was not included in the survey. No other institution included in the survey paid more than $2500. Cornell and Michigan were $2250 and $2200 per year, respectively, and other public universities were at about $2000 or less [UA, CU 1, Box 7, folder 11].

Johnson went on to serve with distinction for 40 years on the faculty at the Naval Academy, contributing a modest number of scholarly articles to the mathematical literature and writing several textbooks. Newcomb's name, achievements, and reputation have already been mentioned, and it is evident—as may even have been apparent at that time—that as a mathematician, he was an order of magnitude above the other four candidates. However, whether the regents would have been able to discern this or wished to discern this is unknowable. Nevertheless, it is interesting to speculate what the future of mathematics at Berkeley might have been had the regents selected Newcomb in 1869 instead of Welcker. The three other candidates have left no trace in the mathematical literature, other than a textbook on elementary algebra by Welcker.

The contrast between the qualifications of Welcker and, for instance, the two LeContes (both of whom were elected to the National Academy of Sciences) is striking. The comment of the executive committee that all candidates for the professorship of mathematics would be acceptable is strange, especially given the high aspirations of the new university. Perhaps, in Coolidge's words concerning Benjamin Peirce quoted above, it did not occur to anyone that mathematical research was one of the things for which a mathematics department existed. No doubt, West Point still had considerable *caché* as a model of technical and civil engineering education built on a basis of training in mathematics, but by this time its reputation as a leading center for mathematics and science education had faded. Regent Hallidie, who appears to have been influential in this appointment, was a self-trained civil engineer, was president of the Mechanics Institute, and maintained a lifelong interest in the technical and science programs at the university. One may speculate that he wanted the mathematics faculty and curriculum at the university to be built along the lines of West Point, an institution that had been famous for training civil engineers. At all events, that is certainly what the regents got. This decision was congruent with the legacy of one of the two parents of the university and represented a choice that, as we shall see in the next chapter, would be reversed a decade later.

Immediately upon electing Welcker as professor of mathematics, the regents approved a further recommendation of their executive committee to appoint Frank Soulé, Jr., as assistant professor of mathematics at a salary of $200 gold per month [RM August 10, 1869]. Soulé was also a graduate of West Point ('66), and had served as assistant instructor of ordnance in 1867–1868 and as acting assistant professor of mathematics in 1868–1869 at West Point. The Soulé family came from Maine and could trace its origins to the *Mayflower*. Frank Soulé, senior, moved with his family to California in 1849 and was a journalist, newspaper editor, and author of some note in San Francisco. The record contains no clear indication of why the regents decided to make two professorial appointments in mathematics, the only discipline in which they did so. It is easy to speculate, though, that the board may have been more comfortable with the Soulé appointment than that of Welcker, since Soulé came from a local family of New England origins and actually had some teaching experience.

Chapter 2

Professor Welcker and the Department of Mathematics

Instruction at the University of California began in September 1869 with 38 students and 10 faculty members. With several colleges and a choice of tracks within some of them, students had more degree options than at the old College of California. The Greek and Latin curriculum was preserved as the curriculum for the BA degree in the College of Letters, but students who did not study Latin and Greek could earn a bachelor of Philosophy degree (PhB) in the College of Letters.

It is evident from contemporary documents that Welcker, assisted by Soulé, imported to Berkeley the West Point curriculum in toto. The first two years of the UC mathematics curriculum consisted of algebra, plane and solid geometry, plane and spherical trigonometry, analytic geometry, descriptive geometry, shades, shadows, and perspectives. The third-year course consisted of differential and integral calculus. This was the entire mathematics curriculum, and it differed from that of West Point only in that the latter was more intensive and completed these topics in two years instead of three. The standard West Point textbooks were used: Davies' *Algebra* and *Geometry* and Church's *Calculus*. One of these, Davies' translation and rewriting of Bourdon's algebra text *Elements of Algebra: On the Basis of M. Bourdon*, achieved an usual degree of notoriety among Berkeley students, on which we shall elaborate a bit later.

Perusal of these books, especially the calculus text, indicates, to begin with, that a great deal of facility and dexterity in formal algebraic manipulation was required of the student. Geometry is missing or at best an afterthought at key points; the notion of continuity of a function is confused, as is the convergence of infinite series. There is no indication that an integral is defined as a limit of approximating sums. The integral calculus is just an exercise in finding antiderivatives, and (amazingly) there are almost no applications. Presumably, at West Point the cadets saw the applications in their junior-year course in natural philosophy (physics) and the senior-year course in civil engineering. Cajori, in his 1890 book *The Teaching and History of Mathematics in the United States* [Caj], takes the "West Point method" to task, citing evidence of a drill-room routine that taught students procedures and processes but left them without a deeper understanding of the structure and purposes of calculus.

What is missing in the curriculum, clearly, is any opportunity for more advanced study either in additional courses or individual study with a faculty member, which was certainly available at Harvard years earlier. According to Cajori's description, this curriculum was not all that different from the array of courses that was available in most colleges in 1870, with the notable exception of Harvard and perhaps a few others. As for the possibility of individual study at Berkeley, there is evidence (see below) that Welcker's command even of topics in beginning calculus was shaky, not to mention more advanced topics in mathematics.

As enrollments at the university grew, and as Soulé went on to different professorial duties as already described, additional teaching staff in mathematics was required. Samuel Jones, another graduate of West Point, was appointed adjunct professor of mathematics for the year 1872–1873. Then in 1873, two members of the first graduating class (which consisted of twelve students, the so-called Twelve Apostles), George C. Edwards and Leander Hawkins, were appointed as instructors of mathematics. Hawkins served until 1879, and then left the university. In 1877, a Mr. Clark, an honors graduate of the class of 1877, was appointed as instructor of mathematics. He was slowly promoted up the ladder to associate professor, but left the university in 1890.

Edwards, on the other hand, served his entire career at Berkeley and became something of a campus legend. He was a successful teacher and was promoted to assistant, associate, and finally to full professor. His life work was the teaching of mathematics, and he was a revered teacher of generations of students. In particular, he began his teaching career as instructor of the freshman course in college algebra, using the Davies text *Elements of Algebra: On the Basis of M. Bourdon*—or Davies' *Bourdon* or simply *Bourdon*, as it came to be called. In June 1875, immediately after Edwards's final examination in the course in his second year of teaching, his students expressed their vehement dislike of the text in a manner that began a Berkeley tradition. They conducted a nighttime funeral procession through Berkeley city streets toward the campus carrying a coffin filled with copies of the offending book. Then they conducted a cremation on campus to the accompaniment of appropriate song and oratory. Many subsequent freshman classes repeated the ceremony with embellishments even long after the book had been replaced. Several years later, the text in the required freshman English course, William Minto's *A Manual of English Prose Literature*, was added to the funeral ceremony.

Professor Edwards's desk copy of *Bourdon* in a handsome leather binding, and with copious hand-written marginalia and some 90 pages of supplementary notes, in Edwards's own hand, that he employed when teaching the course, resides in the Bancroft Library. It is accompanied by an article describing the student tradition that appeared in the November 1947 issue of the *California Monthly* [Kyte] entitled "The Bourdon That Never Burned."

Edwards also served as the commandant of the ROTC cadet corps on campus with the rank of colonel, and was therefore often referred to as Colonel Edwards. He had a lifelong interest in track and field, and there is an athletic prize in his honor—the Colonel George C. Edwards Medal—a gold medal that is presented on the Berkeley campus to each California man who breaks the California track and field record in the 440-yard run. Edwards Stadium, the site of track and field events and a campus landmark, was also named in his honor. He retired in 1918 after 45 years of service, and in 1923 was awarded an honorary doctorate (LLD) by the university.

For all of Edwards's legendary status at Berkeley, it should be noted that for almost the entire period from 1869 to 1890, the mathematics faculty of the university consisted of professor of mathematics Edwards and at most two junior faculty members whose training consisted of bachelor's degrees from the University of California.

Other Faculty Appointments

The events that were to take place with respect to mathematics, especially the shape of the curriculum and faculty appointments, occurred within the context of the university at large, with faculty appointments in related disciplines and issues of university governance playing an important role. John LeConte continued as acting president until August 1870, at which time the regents settled on their first permanent president, the Rev. Henry Durant, one of the founders of the College of California. He served for two years and retired on reaching his 70th birthday. The regents then approached Daniel Coit Gilman (Yale '52) a rising young star at Yale serving as professor of political and physical geography in the Sheffield Scientific School. He had already declined the University of California presidency two years earlier, as well as offers of other university presidencies, but now he accepted the invitation to come to California. He was also an ordained Congregational minister, and thus both of the first two presidents of the university were New Englanders, graduates of Yale, and Congregational ministers. (It may be observed that the seventh president of the university, Martin Kellogg, also shared these three characteristics.) The regents established a standing advisory committee, as a successor to the executive committee, which had been chaired by Regent Stebbins, to work with and "advise" the president.

The two LeContes, as well as Kellogg, went on to give many years of distinguished service to the university, and indeed a building on campus is named in the LeContes' honor. The fourth among the original faculty appointees, Professor Fisher, chair of the chemistry department, met an unfortunate fate when on October 3, 1870, the regents, without warning, abolished his position of professor of chemistry, mining, and metallurgy, as it was then called (but graciously allowed him to be paid through the end of the month), and turned the duties over to Professor Carr [RM October 3, 1870]. The following spring, Joseph Henry, secretary of the Smithsonian, wrote to the regents strongly recommending a young chemist, Willard Rising, who was finishing his doctoral work under Professor Robert Bunsen in Heidelberg. Rising had taught briefly in the College of California in the 1860s before going to Germany, and on November 22, 1871, he was appointed to the re-created professorship that had been abolished a year earlier [RM November 22, 1871]. He served the university with distinction for many years.

Ezra Carr, the professor of agriculture was dismissed by the Regents on August 11, 1874, for reasons discussed below. The Regents began a search for a replacement and on October 28, 1874, appointed Eugene Hilgard, from the University of Michigan, as professor of agriculture. This was a particularly wise choice, since Hilgard went on to a long and distinguished career of research and teaching in agricultural science at Berkeley. A campus building is named after him.

In 1872, the regents decided to create a professorship of astronomy and a professorship of civil engineering. Not having funds for both positions, they merged them into a professorship of astronomy and civil engineering at their meeting of July 16. Assistant Professor Frank Soulé, Jr., applied for the position and was appointed to it at the regents' meeting of July 23 [RM July 23, 1872]. Professor Soulé went on to a long and distinguished career at the university. He was the first dean of engineering when a college of engineering was organized in 1898. Soulé Road, which runs from Northgate through the College of Engineering, is named in his honor.

In 1875, the regents created a professorship of industrial mechanics, and appointed Frederick G. Hesse (1825–1911) to the position. A native of Prussia, Hesse had attended the Royal Polytechnic School in Berlin, served briefly in the Prussian Army, and subsequently worked as an engineer for the Prussian government. He came to the United States in 1848, and after serving in a variety of positions, ended up as professor of mathematics in the US Navy assigned to the Naval Observatory. Hesse likewise had a distinguished career at Berkeley, as the founder of the department of mechanical engineering. A building on campus is named after him.

The regents made two additional professorial appointments that will enter into the story about Welcker and the department of mathematics. On November 2, 1870, the regents appointed George Davidson as nonresident (honorary) professor of geodesy and astronomy to serve without compensation [RM November 2, 1870]. Davidson (1825–1914) was born in England, and his family emigrated to Philadelphia shortly thereafter. A talented youth, Davidson became a protégé of Alexander Dallas Bache, and according to his biographical sketch [Dav], Davidson had the ambition to become a professor of mathematics and classics at an Eastern college. Bache, who was serving as the director of the US Coast Survey, convinced him to join the survey. In 1850, Bache sent Davidson west with the assignment to survey and map the Pacific Coast in the newly acquired state of California.

Davidson remained head of the Pacific branch of the survey until his retirement in 1895, and became very well known as a geodesist, astronomer, and scientist. He was the long-time president of the California Academy of Sciences, and was elected to the National Academy of Sciences in 1874. He negotiated with James Lick concerning the establishment of the Lick Observatory, and had a wide-ranging correspondence [Dav] with the leading scientists and mathematicians on the East Coast. He certainly knew Benjamin Peirce, who in fact was his supervisor in the US Coast Survey for seven years (1867–1874). Davidson was also a close personal friend of the powerful secretary of the regents, Robert Stearns, and it was Stearns who was influential in lobbying Governor Irwin to have Professor Davidson appointed as a regent of the university in June 1877 [Dav].

On July 22, 1874, the regents appointed the noted California mining engineer William Ashburner as professor of mining to serve part time at a salary of $50 per month [RM July 24, 1874]. It transpired that Ashburner could not devote even this level of commitment to the post, and he subsequently requested that his appointment, like Davidson's, be made an honorary one, without compensation. On April 2, 1880, he was appointed by Governor Perkins as a regent of the university, making him the second honorary professor on the Board of Regents. It was alleged in subsequent newspaper articles in the *Wasp* that the secretary of the regents, Stearns, had arranged not only this regental appointment behind the scenes, but also the reappointment of Regent Stebbins when his first term expired in 1878, as well as the appointment of George Davidson as regent in 1877. As noted, Stearns's role in the Davidson appointment is confirmed by documents in the Davidson papers. The correspondence makes clear that among the many duties of the secretary of the regents was that of chief lobbyist for the university in Sacramento.

Political Turbulence

In the period 1881–1882, the regents made decisions that were to change the shape of the mathematics department. These decisions flowed from some deep political conflicts in the state that directly affected the university. It is evident that the university had been struggling during its first twenty years of existence. It had been beset by political turbulence and a fierce debate over its mission and structure, and over who would control it. Moreover, there were no regular state appropriations to the university, and the land sales from the land grant under the Morrill Act had not generated as much endowment as expected. Hence, the university was constantly short of funds. Enrollments stagnated, never rising above 400 until 1890. With the notable exception of the brief tenure of Daniel Coit Gilman as president (1872–1875), the leadership of the university was weak, and the regents assumed a predominant managerial role. "Micromanagement" is probably too weak a term to describe the regents' mode of governance at this time. Such a high degree of control at the regental level is generally not a good thing, but the situation was made worse by the regents' periodic internecine battles, with factions often forming along political lines.

The debate over mission, structure, and control began almost immediately after the creation of the university. In 1870, the Mechanics' Institute, under Hallidie as president, petitioned the legislature to have instruction in practical and applied sciences in the university located in San Francisco. It was argued that such instruction would be more readily available to the "common man" than it would be at Berkeley, which was seen as accessible mostly to the sons and daughters of the wealthy. The university beat back this proposal, which would have in effect split the university. However, soon thereafter, a coalition of farmers and laborers began a drumbeat of criticism of the university as furthering the interests of the elite and privileged classes. The California State Grange, an organization representing small farmers that had been nursing a long list of populist grievances against many institutions including the railroads, the banks, and the legislature, was joined by the Mechanics Deliberative Assembly in criticizing the university. They claimed that insufficient resources had been devoted to the agricultural college, and that the regents were violating the Morrill Act by using land-grant funds to support classical and liberal education, thereby neglecting practical education. In a memorial to the legislature, the California State Grange and the Mechanics Assembly asked the legislature for legislative redress and urged that the independent board of regents be abolished and that the university be placed under an elected state board of education that would control the common schools and the state normal school.

Ezra Carr, the professor of agriculture, and his wife supported the Grange, as did Professor Swinton, who was a friend of the populist agitator and newspaper editor Henry George. Swinton published lengthy broadsides against the university, and Carr was using this forum to grind an axe about internal resource-allocation decisions that in his view disadvantaged agriculture. Carr was finally fired by the regents in 1974, and Swinton was also eased out. In 1874, the legislature conducted an investigation into charges that the university was paying insufficient attention to its agricultural and mechanical mission. These investigations put Gilman and the university on the defensive, but in the end, no adverse legislation was proposed at that time. But the battle lines were drawn—there were those who wanted the university to become a polytechnic institution, eliminating all instruction except that in the agricultural and mechanical arts; and there were those who wanted the university to remain true to its founding mission as a comprehensive institution embracing the liberal arts and sciences. Many newspapers disparaged and ridiculed the university, writing about "Gilman and the kid-glove junta" and that the "university was teaching rich lawyers' boys Greek with the farmers' money" [Sh, p. 857]. The latter quotation appears in an article by Millicent Shinn in the *Overland Monthly*, in which the writer opined that the people of the state were not in general sympathetic to the university.

Gilman complained that although the university was nominally administered by the regents, it was in fact administered by the legislature. But the university also had enemies on the other side; religious groups complained about the lack of religion in the state university, and these critics did not feel that classical education could be entrusted to the legislature. Gilman testified to the legislature that there were religious bodies that would like to control the university or see it die in order that separate denominational colleges might arise in its stead. It is not surprising that Gilman, having spent his life at Yale, was frustrated by California politics and that when the offer came to head the new Johns Hopkins University, he accepted it, leaving California in 1875.

In 1876, the speaker of the assembly, Gideon Carpenter, introduced a bill implementing the populist reforms of disbanding the regents and reforming the university's curriculum, making it a polytechnic institution. This bill passed the assembly, but was killed in the state senate on March 29. In an interesting sidelight, on March 31 a senior executive of the Central Pacific Railroad began a letter to Robert Stearns, secretary of the regents, by congratulating him on the "death of the serpent" [UA, CU 1, Box 3, folder 19]. But the serpent was far from dead, for a similar bill was introduced in the senate two years later by Senator Curtis. Continuing populist agitation, inflamed by widespread economic distress and unhappiness with the 1849 constitution, led to the establishment of a constitutional convention in 1878, at which university governance would be a significant issue.

Supporters of the university at the convention proposed a provision that would establish the university in the constitution as a public trust, ensuring independence from the reach of the legislature. Opponents ridiculed the proposal, claiming that the university "now seeks, by a clause in the Constitution of the State, to forever secure for itself the million or more that it gobbled several years since, despite the protests of farmers and mechanics," as quoted in [Doug, p. 66]. Interests opposed to the university proposed various versions of the Grange, Carpenter, and Curtis bills to be included as provisions in the new constitution. In the end, by persistence and good fortune, the university prevailed, and the provision for designating the university as a public trust was adopted by the convention. See Douglass's recent book [Doug] for a fascinating description and analysis of these events. To this day, the university remains a public trust in accordance with Article 9, Section IX of the California Constitution, and this constitutional provision has ever since been the single most significant fact concerning the governance of the university.

The new constitution, which was narrowly approved by the voters in 1879, had many populist features and contained a provision concerning public schools that was to harm the university over the following two decades. The populist forces had inserted a provision into the constitution that state education funds could be used only for the common schools and not for high schools. It was the view that a common-school education was all that farmers or laborers needed, and they saw no reason to spend public money on high-school education that would prepare only the children of the well-off for college. The state was therefore left with free common schools and a free university, but without adequate provision for the link between them. Many high schools had to close, and high-school enrollment in the state dropped steeply. This was a contributing factor to the decline in university enrollments in the early 1880s and in their general stagnation for another decade [Doug, pp. 70–71]. Economic factors contributed as well, but university enrollments in 1881 fell to under 200, about half of what they had been a few years earlier, and they did not rise above 400 until the 1890s. These circumstances brought calls for retrenchment in the university in the short term, and prevented faculty growth and diversification of the curriculum in the longer term. The high-school funding problem was partially fixed by legislation passed in 1893, but was not fully fixed until 1903 [Doug, p. 138].

The 1879 constitution had resolved the governance issues that the university had faced, and agitation for a polytechnic university subsided. Among the faculty of the university, Professors Carr and Swinton, as already noted, had played a substantial role in the debate, and both had involuntarily departed the university. Professor Welcker appears to have kept his head down in this debate, since there is no evidence that he was directly involved in the debates and conflict about the mission of the university: a polytechnic institution versus a more broadly based university. However, his own background and the support he enjoyed from Regent Hallidie suggest that he likely would have sided with the critics of the university who were urging that it become a polytechnic school.

Internal Conflicts within the University

After the departure of President Gilman in 1875, the regents recalled Professor John LeConte to the presidency. But over time there was growing dissatisfaction with his leadership and with the general state of the university among some regents. For instance, in a July 24, 1879, letter to Regent George Davidson, Horatio Stebbins, chair of the advisory committee of the regents, wrote,

> We are weak, Mr. Davidson, in the general moral force of the Faculty, and much of the trouble we have got into with the Class that has been suspended in my opinion is to be attributed to that cause. Of course, this is to be spoken of with some delicacy, but it seems to me that in all our appointments it is to be steadily and firmly kept in view. We have from the beginning been weak in the general tone and spirit of the Faculty, and I am so impressed by it now that almost the only view I have taken of the matter has been in the direction of its influence in giving some new force to the internal administration of the University. [Dav]

Very much in line with these thoughts, in May 1880, the regents reconstituted their advisory committee and renamed it the Committee on Instruction and Visitation. The duties "shall be to visit the University to learn what are the methods of instruction and government; to make themselves personally acquainted with the mode in which daily recitations and public examinations are conducted; and to inquire and report upon the condition, the prospect and the wants of the Institution." There is considerable evidence that Stebbins had formed a very low opinion of John LeConte's leadership and administrative abilities as president. It is not unreasonable to conclude that once the university had won political independence of the legislature, Stebbins and his friends on the board of regents would move to consolidate their victory, to establish and upgrade the scholarly credentials of the faculty, and to eliminate any remnants of the polytechnic alternative.

The board was now heavily Republican. The Republican governor Perkins did not reappoint the three Democratic regents whose terms were expiring in early 1880, instead replacing them with three Republicans. The political affiliations of the appointed regents were 11 to 4 Republican with one vacancy. As previously noted, one of these new regents was (honorary) Professor Ashburner. The Committee on Instruction and Visitation, unlike other committees, was to be selected by ballot of the regents, and the five members selected were Stebbins, Davidson, Ashburner, Campbell, and Redding. Fred M. Campbell was an Oakland schoolteacher who had been elected as state superintendent of public instruction and was therefore an ex-officio regent. Redding was a member of the California Academy of Sciences and was the land agent for the Central Pacific Railroad. Stebbins chaired the committee.

This committee took seriously its expansive charge, which in effect was to conduct a review of the university similar in nature to reviews done today of departments or entire institutions by external review committees. The committee sought comment externally, as indicated by a note Stebbins sent to Davidson on March 17, 1881: "I send you a note from Pres. Eliot. I think we ought to have meeting of the Committee on Instruction and Visitation and make up a statement of the condition of the University and of the changes necessary to be made, and how to make them" [Dav].

It is not clear at what point Professor Welcker knew that he was a target of the Committee on Instruction and Visitation, but he was, and of regent Davidson in particular. On March 28, 1881, Welcker wrote to his "patron," regent Hallidie, that he (Welcker) had heard that Davidson had said some derogatory things about the department of mathematics, among other things that its tone was very low, far inferior to Yale or Harvard. Welcker then offered the best defense that he could to these criticisms [UA, CU 1, Box 11, folder 4]. (The letter is reproduced in its entirety in Appendix 1.) Given his background, Davidson clearly knew a fair amount of mathematics, and it is reasonable to assume that he was also aware through his many contacts on the East Coast of current developments in mathematics, especially the dramatic changes that had taken place during the decade of the 1870s. He must also have been aware of how limiting the West Point curriculum was and of the lack of research activity in mathematics at the University of California. What Cajori wrote a few years later, in 1890, was likely in the air already:

> It is our opinion that under Professors Davies and Church, the philosophy of mathematics was neglected at West Point. If this criticism be true of West Point, which was for several decennia unquestionably the most influential mathematical school in the United States, how much more it must be true of the thousands of institutes throughout the country which came under its influence. [Caj, p. 121]

Events moved rapidly at this point, and the Committee on Instruction and Visitation produced a report with ten recommendations that were presented to the regents at their meeting on May 31. The highly controversial report was ordered printed and was slated for discussion at the annual meeting of the regents, scheduled for June 7. The report had become public, and a number of the press were at the meeting. One of the first items of business was a motion to allow the press into the meeting. This was defeated by a wide margin; a later motion offered just before the committee's report was to be taken up to allow the press into the meeting was also defeated, but more narrowly, by a vote of 9 to 8.

The report was wide-ranging and very critical of many aspects of the university's operation and programs, and of the president in particular; see Appendix 2 for the full text. In one key passage the committee writes, "The faculty, not well united, needs direction, studies require to be relatively adjusted; and the whole intellectual and moral force of the University brought to bear upon one point: the increase of intellectual activity and scholarly manners." The committee also writes that "the university does not have any vital sympathy or connection with the progress of education and in some of the departments, outgrown methods are still adhered to." The report concludes with a list of ten recommendations. The first three recommendations were (1) to separate the office of the president from any professorship so that the duties of the president would be administrative and executive, (2) to declare the presidential office vacant, and (3) to declare the chair of physics vacant.

At the outset of the meeting LeConte tendered his resignation as president. The regents then took up the recommendations of the committee. Recommendation (1) passed; recommendation (2) was moot in light of LeConte's resignation; and the committee then withdrew recommendation (3), expressing admiration and kind sentiments for LeConte's work as professor of physics. Five of the other recommendations, which involved dismissals or reassignment of instructors, were approved without discussion, and the tenth recommendation was for a 17% salary cut for everyone as an austerity measure. This recommendation was tabled to a future meeting, at which it was passed.

The remaining one of the ten recommendations, number 5, was the major focus of the meeting. It called for the dismissal of Welcker as professor of mathematics, or as it was put, to declare the chair of mathematics vacant. This recommendation provoked an extended discussion, which was reported in the press in considerable detail, in spite of the press being excluded from the meeting. The second of the two quoted passages from the committee report cited in the previous paragraph, especially the point about "outgrown methods," was clearly aimed directly at Welcker, and while the first quoted passage had a wider range of targets, it is reasonable to assume one of them was Welcker. In response to questions from the board concerning the rationale for the recommendation to dismiss Welcker, Davidson responded on behalf of the committee that he did not like the West Point style of teaching, that it was an old-fashioned textbook manner with no attempt to draw out the mind of the student and that the standing of the department was nothing approaching that of Harvard and Yale. Davidson and Judge Hager, who had been a regent from 1869, apparently had exchanged heated remarks during the discussion as to Welcker's competence at the time he was hired. According to press reports of the meeting, Hager said that at the time Welcker was hired, he had asked Welcker whether he could work out Taylor's theorem. Hager went on to say that he honored Welcker for his honest answer of "no," but Welcker added that he could work it up in a short time, since he was only rusty. This is no doubt the first and last time that Taylor's theorem was a topic of discussion at a regents' meeting. The tide was clearly against

Welcker, and when the motion to dismiss Welcker was put to a vote, it passed 11 to 6 [RM June 7, 1881].

The next day's *San Francisco Chronicle* played the story in a way that was not particularly sympathetic to the regents, headlining the article "Results of the Star Chamber Proceeding—Five Scholastic Heads in the Basket—LeConte Resigns, Welcker Walks the Plank." Some press reports criticized Stebbins and his committee, alleging that Stebbins and Ashburner had a conflict of interest, since both sought the presidency of the university. The committee was also criticized for bringing these recommendations to the board for action at a time when Hallidie, who allegedly would have opposed them, was out of town and unable to attend. Some reporters thought the action politically motivated, since LeConte and Welcker were Democrats and the board was heavily Republican. Welcker was asked about this, to which he replied, "It may or may not be a political one, but I think not," and added that his appointment had been supported by a prominent Republican [Hallidie]. Other press coverage was more muted and favorable to the regents. The *Daily Evening Bulletin* called the controversy a "tempest in a teapot" and noted that "in the early appointments some mistakes were made and had to be corrected."

The *Daily Evening Bulletin* comment on the early appointments was probably more on target than the overheated coverage of other newspapers. Of the eight professors and one assistant professor among the original appointments, only four— the two LeContes, Martin Kellogg, and Frank Soulé—survived. The others had been fired outright or eased out. The regents clearly regarded the faculty as "at will" employees, with no idea of academic tenure. The faculty was being steadily upgraded, and as we shall see in the next chapter, the same "upgrading" happened in mathematics.

Welcker did not go silently; within days, the regents received lengthy petitions signed by many students and alumni who protested Welcker's dismissal and defended his West Point mode of instruction. The originals of these petitions are preserved in the Bancroft Library and are quite impressive. Welcker did not file suit as a dismissed faculty member might do today, but rather sought a rather more innovative method of revenge. With the support of some university alumni, he won the Democratic nomination for the office of state superintendent of public instruction, and was elected to this office in the general election in the fall of 1882, defeating the incumbent Fred M. Campbell, who had been a member of the Committee on Instruction and Visitation that had recommended his dismissal. It should be added that the idea of a defrocked university professor running for this statewide office was not original with him. After being fired as professor of agriculture by the regents in 1874, Ezra Carr also ran and was elected to the office of state superintendent of public instruction.

Because then, as today, the state superintendent of public instruction was an ex-officio regent of the university, Welcker assumed that office, and he then proceeded to lead a fight to disband the Committee on Instruction and Visitation. The issue came to a head with a confrontation with Stebbins at the regents' meeting of May 19, 1883. According to press coverage, regents Wallace and Hager joined Welcker and argued for elimination of the committee on the grounds that it had assumed dictatorial power, had displaced the president as the chief executive, had usurped the functions of the Academic Senate, and had kept the regents as a whole from becoming acquainted with the university. The press reported that Stebbins rose to defend the committee with an impassioned argument that he [Stebbins], "under the blessing of God had been the founder of the University, a fact recognized by Governor Haight, and the Board could not dispense with the Committee." This comment has the ring of "après moi, le deluge."

The motion to disband the committee was finally put to a vote, with 11 votes in favor and 8 opposed. However, the chair of the meeting declared the motion defeated because in his view, it represented a change in procedures and thus required a two-thirds majority. This ruling of the chair was appealed to the full board, which overruled the chair, with the result that the motion to disband the committee carried [RM May 19, 1883]. The *Chronicle* the next day again provided florid coverage, headlining its account appropriately, "Welcker's Revenge." Welcker served out his four-year term as state superintendent successfully, and then practiced law in San Francisco. Although information concerning Welcker and his life is scant, he seems to have been an intelligent and able man, but clearly miscast by events as professor of mathematics at the University of California. In an unusual twist, in 1898 the regents granted Welcker, on the occasion of his 67th birthday, the title Professor Emeritus, and thus restored him to the Academic Senate. He died shortly thereafter, in 1900. It is perhaps relevant to note that Stebbins had left the board of regents in 1894.

Chapter 3

Professor Stringham and the Department of Mathematics

But we have jumped ahead a bit in the narrative, so let us return to 1881. When Welcker was dismissed, his teaching duties were assigned to an instructor in physics, and a search was undertaken for a new professor of mathematics. The Committee on Instruction and Visitation undertook the search itself, bypassing the faculty and the new president, who appeared to be mere bystanders in the process. The committee wrote to various notables in the East seeking nominations and recommendations. On May 21, 1882, the Committee on Instruction and Visitation presented their report to the regents and recommended the appointment of W. I. Stringham as professor of mathematics. They reported that Stringham had been recommended by President Gilman, of Johns Hopkins, and by President Eliot and Professor James M. Peirce, of Harvard. The full text of the report is included in Appendix 3. It is interesting to note that the committee report is signed by only four members of the committee, not including Regent Davidson. There is no indication whether he was unavailable, had withdrawn from this deliberation, or may have disagreed with the recommendation.

Washington Irving Stringham was born December 10, 1847, in the town of Yorkshire, Cattaraugus County, in western New York State. After the Civil War, his family moved to Topeka, Kansas, where the young Stringham established a house- and sign-painting business and worked in a drugstore while attending Washburn College part-time. He also served as librarian and teacher of penmanship at Washburn College. With this unusual background, Stringham was admitted to Harvard College as a freshman in 1873 at the age of 25. He studied mathematics under the influence and guidance of Benjamin Peirce and graduated with highest honors in mathematics in 1877. His senior thesis concerned applications of quaternions and was subsequently published in the *Proceedings of the American Academy of Arts and Sciences* in 1878.

Stringham remained at Harvard for a semester as a PhD candidate, but he was then attracted to Johns Hopkins to work under J. J. Sylvester. He completed his doctoral degree in 1880 as Sylvester's fourth student at Hopkins with a dissertation entitled "Regular Figures in n-Dimensional Space," which was published in the *American Journal of Mathematics* in 1880. He also published a paper determining the finite subgroups of the quaternions. He was influenced by William Story (1850–1930), a junior faculty member in mathematics at Hopkins, from whom he took courses in elliptic functions. Stringham noted his indebtedness to Story in the published version of his dissertation. Like Stringham, Story had studied mathematics under Peirce at Harvard, graduating in 1871. He then went to Germany for graduate study at Berlin and Leipzig, receiving his doctoral degree from the University of Leipzig in 1875. Story subsequently left Hopkins in 1889 to found the Mathematics Department at Clark University, where he had a long string of doctoral students, the most famous of whom was Solomon Lefschetz.

After receiving his degree from Hopkins, Stringham, quite likely at the suggestion of Story, then went to Germany for two years of postdoctoral study at Leipzig under Felix Klein, who had just come to Leipzig. Stringham was Klein's first (of many) American students. Stringham, as one can tell by his later work, was deeply influenced by Klein's famous lectures in 1881–1882 on geometric function theory and on elliptic functions. Thus, Stringham brought to Berkeley in 1882 comprehensive training in the forefront of research mathematics of the day encompassing algebra, geometry, and complex analysis, having studied under Benjamin Peirce, J. J. Sylvester, and Felix Klein. There is no evidence, however, that he had any background in celestial mechanics, one of the important themes of mathematics at the time. Finally, as noted already, he had strong personal recommendations from Presidents Eliot, of Harvard, and Gilman, of Hopkins.

Let us pause for a moment to reflect on these events of 1880–1883. One can view the dismissal of Welcker and his replacement by Stringham as one of the last chapters (or even the last chapter) in the decade-long battle between the competing visions of a polytechnic versus a comprehensive university. If one wanted a polytechnic, then someone like Welcker might be a reasonable choice for a professorship in mathematics. After all, West Point, especially in the 1830s and 1840s was the best example of a polytechnic around. If, on the other hand, one wanted a comprehensive university that would challenge the best institutions in the East in 1880, Welcker was clearly not the individual to lead the Mathematics Department. Stringham, on the other hand, clearly had sterling credentials for the job. And so the change was made.

Another observation is the extent to which the regents, and indeed small groups of regents, exercised day-to-day management of the academic affairs of the university, usurping the role of the president and the (almost invisible) Academic Senate. Stebbins and his committee repeated their performance in 1882–1883 and acted as the search committee for the professorship of English language and literature, recommending a candidate to the regents as well as the appointment or reappointment of nine other faculty members. One can see in these reports the basis of complaints aired in 1883 by some regents in the debate on the motion to eliminate the Stebbins Committee on Instruction and Visitation. One can also see in these events deep divides among the regents.

When Stringham arrived in Berkeley, he quickly moved to revamp, expand, and modernize the curriculum. Byerley's calculus replaced the West Point book of Church, and from inspection of these books today, it was a considerable improvement. Expansion of the curriculum was limited by the small enrollments and the small faculty, which consisted of Stringham plus two others whose training consisted of a bachelor's degree at Berkeley. More advanced courses in calculus were added to the curriculum as well as a course in quaternions, a topic of particular interest to the new chair. Obviously, Stringham was mathematically isolated in Berkeley and did not publish very extensively in the research literature. His primary interest appeared to focus on elliptic functions. In a letter to his mentor Felix Klein in 1888, he wrote that "[t]he plants of intellectual culture grow but slowly, and on new raw ground like that of California they can hardly flourish without very great efforts. I impatiently await the time when it will be possible to bring some important researchers in the field of mathematics to the California coast," as quoted in [Pa-Ro, p. 266].

In 1888 Stringham married Martha Sherman Day, a great-granddaughter of Jeremiah Day, President of Yale, and granddaughter of Sherman Day, one of the founders of the College of California. The wedding took place in New Haven, with the Rev. Timothy Dwight, President of Yale, performing the ceremony. The couple had three children.

Although Stringham was deeply versed in Klein's theory of elliptic, Abelian, and algebraic functions and published some well-received expository articles, he did not contribute to advancing the frontiers of research in these areas. He was nonetheless recognized in the national mathematics community, as indicated by his election to the Council of the American Mathematical Society for a three-year term in 1903 and as vice president of the society for a one-year term in 1906.

Stringham became involved in university administration, serving in several decanal positions starting in 1886. He ultimately became dean of the faculties under President Wheeler, which appears to be a position similar to that of executive vice president, for he was officially designated to serve as acting president when Wheeler went on a sabbatical leave in the fall of 1909 as Roosevelt professor at the University of Berlin, and to attend regents' meetings in the absence of the president. He was a deeply revered and respected senior member of the faculty. Throughout his tenure he was regarded as the leader of the mathematics department and spoke for it on issues of resources and curriculum. Although the term "department chair(man)" was not formally used in the university until some years later, he acted in that role up until shortly before his death, in 1909. His service in significant campus-wide administrative roles in his final years no doubt meant that he had to leave some day-to-day matters concerning the department to others, most notably to Mellen Haskell, who succeeded him as chair.

Stringham soon had an opportunity to bring more mathematical researchers to California when the fortunes of the University took a turn for the better. In 1887, the legislature passed the Vroom Act, which for the first time provided state funds to the university on a regular annual basis; the amount was equal to one cent per $100 of taxable property, the so-called mill-tax. This was increased to two cents in 1897, but it was not until the emergence to power of the Progressive Party in 1911 that state finances were rationalized and the university received an appropriation based on enrollment rather than property values. In any case, the mill-tax revenues made growth possible. By the 1890s, new high schools were being established in spite of provisions of the 1879 constitution, and the population was growing through immigration. In addition, the university began a program of accrediting high schools under which a student who graduated from an accredited high school would be admitted to the university on the recommendation of the principal. This admissions policy, which was to remain in effect for many decades, smoothed the transition from high school to the university and facilitated enrollment growth. The program started in 1888 with six accredited high schools, and the list grew to 116 by 1900.

All of these factors combined to make enrollment growth at the university possible. Indeed, undergraduate enrollment on the Berkeley campus, which stood at about 400 in 1889, which was only marginally more than it had been a decade earlier, grew to 2000 in 1900 and then to almost 4000 by 1910 and to nearly 10,000 by 1920. Up until 1890, the campus had only a handful of graduate students (9 in 1886 and 21 in 1889), and this number also grew rapidly, to 220 by 1900 and to over 500 by 1910 and 1200 by 1920. Thus, the period starting in 1890 was one of rapid enrollment growth for the university and of correspondingly rapid growth in the faculty. The mathematics faculty stood at three (one of whom had a PhD) in 1889, and by 1902 it had grown to ten (eight of whom had PhDs), and to 13 in 1910. It was still 13 in 1920, with the result that the growth in the mathematics faculty did not keep pace with the overall growth of the university after 1902.

University governance became more stable with the appointment of the respected senior faculty leader Martin Kellogg, first as acting president in 1890, and then as president in 1893. He served until his 70th birthday, in 1899, and stepped down voluntarily at that time. His nine-year tenure (and voluntary departure) set something of a record. The regents also began a process of stepping back from participation in the day-to-day affairs of the university, and as previously noted, Stebbins left the board in 1894. Hallidie left in 1900, and the regents disbanded their Committee on Internal Administration (the successor to the advisory committee and the Committee on Instruction and Visitation) in 1897. The process of withdrawal was not complete until the accession to the presidency of Benjamin Ide Wheeler in 1899, who insisted on firm understandings about the relationship between the president and regents as a condition of accepting the post. Wheeler served for 20 years and presided over the growth and maturation of the university.

The first new professorial appointment in mathematics that was made since the appointment of Stringham was that of Mellen Woodman Haskell as assistant professor in 1890. Haskell had received his doctoral degree under Felix Klein in 1889 and had spent one year as instructor at the University of Michigan before coming to California. He was born in Salem, Massachusetts, on March 17, 1863, the son of a Unitarian clergyman, and had graduated from Harvard in 1883 in mathematics. He remained at Harvard, receiving a master's degree in 1885 before going to Germany for more advanced study at the University of Göttingen, to which Klein had just moved. Parshall and Rowe [Pa-Ro, p. 210] write that Haskell in his dissertation gave

> an elaborate construction of the projective Riemann surface associated with a certain normal curve Klein had constructed in conjunction with his work on the geometric Galois theory of the modular equation for the prime $n = 7$. Thus Haskell's work represented a blend of some of Klein's fondest ideas and interests. It came closer than the work of any of Klein's American students to capturing that peculiar mix of ideas from group theory, algebraic geometry, and complex function theory that lay at the heart of Klein's own mathematical research. In fact, Haskell's effort ultimately represented one of the last concerted efforts to push through this particular set of ideas.

Haskell prepared an English translation of Klein's Erlanger program that was published by the American Mathematical Society upon his return to the United States. Evidently, Haskell and Stringham had many mathematical interests in common, including of course the common experience of having studied under Klein, which no doubt served to temper the isolation both may have felt in Berkeley. The curriculum was further expanded to include courses in algebraic forms, algebraic curves, elliptic functions, what we would today call Galois theory, and spherical harmonics, among others.

Haskell was promoted to associate professor in 1894 and to full professor in 1906. He published a modest number of papers focusing on algebraic geometry and especially a classical kind of enumerative geometry of plane curves. He supervised the doctoral work of 14 students, who for the most part wrote on these same topics. One exception was Benjamin Bernstein, whose dissertation concerned issues in mathematical logic and who became a long-time Berkeley faculty member. He also influenced Bing Wong, whose dissertation advisor was Derrick Lehmer, and who also served on the Berkeley faculty. He was said to be a fine teacher, and he especially valued good teaching in his colleagues. Haskell served as acting chair of the department in Stringham's absence, and became the chair in 1909 after Stringham's death. He served in this capacity until his retirement at age 70 in 1933. He also served on a number of Academic Senate committees during his career and was a highly respected member of the campus community. On the national scene, he was elected vice president of the American Mathematical Society in 1914.

During the next few years, following Haskell's arrival in Berkeley, Stringham hired several former graduate students at Berkeley, although in fact, the only two of these who remained on the faculty ended up receiving their doctoral degrees in Germany. In these early days of graduate study, a doctoral candidate at Berkeley could specify three fields of study. One early graduate student connected with the mathematics department was Armin O. Leuschner, who listed astronomy, physics, and mathematics as his three fields. Leuschner received an AB from the University of Michigan in 1888 and came to the University of California as a graduate student based at Lick Observatory. In 1890, he became an instructor in the Department of Mathematics and in 1892 was promoted to assistant professor. He taught both in mathematics and in astronomy, and in 1894 his title was changed to assistant professor of astronomy and geodesy. In 1895, all of his teaching was in the astronomy department. He worked on methods for determining orbits of minor planets and comets, and subsequently was awarded a doctoral degree in 1897 at the Frederick Wilhelm University in Berlin with a dissertation *"Beiträge zur Kometenbahnbestimmung"* ("Contributions to the Determination of Orbits of Comets"). Thus, Leuschner represented an important theme of nineteenth-century mathematics that had been missing from Berkeley, namely celestial mechanics. He went on to a long and distinguished professorial career at Berkeley and was elected to the National Academy of Sciences. The Leuschner Observatory on campus is named after him. He also played an important role in the events of 1933 and the recruitment of Evans as described in Chapter 5.

Louis Theodore Hengstler was a graduate student associated with the Department of Mathematics who listed political science, mathematics, and German literature as his three disciplines, and wrote a dissertation entitled "The Antecedents of English Individualism." While a graduate student he served as instructor of mathematics, and after receiving his degree in 1894, he served briefly as assistant professor of mathematics, but then left Berkeley.

Charles A. Noble graduated with a BS in mathematics from Berkeley in 1889 and taught mathematics in secondary schools in the San Francisco Bay area for four years. He spent three years in Göttingen pursuing a doctoral degree, and then returned to Berkeley as a graduate student in 1896. He taught for four years as a fellow and as an instructor in mathematics. Returning to Göttingen in 1900, he completed his doctoral degree under Hilbert in 1901 with a dissertation entitled *"Eine neue Methode in der Variationsrechnung"* ("A New Method in the Calculus of Variations"). He was then reappointed as instructor in 1901 and advanced to assistant professor in 1903. Noble subsequently advanced to associate and then to full professor, remaining at Berkeley for the remainder of his career, until his retirement in 1937. He served as acting chair of the department for the 1933–1934 academic year as a transition between Haskell and Evans. As a faculty member, Noble focused his energies on mathematics education, and along with E. R. Hedrick, who also received his doctoral degree under Hilbert in 1901, he translated Klein's famous book *Elementary Mathematics from an Advanced Point of View*. Hedrick went on much later to become the founding member of the UCLA Department of Mathematics.

As graduate enrollment grew and the faculty expanded, the interdisciplinary classification of graduate studies apparently waned, and graduate studies became departmentalized. In 1897, the Academic Senate created the Graduate Council as a committee of the senate to oversee graduate instruction. In 1897–1898, the campus had 113 graduate students, 9 of whom were listed as in the Department of Mathematics. Graduate education was still in its youth at Berkeley; for instance, in the years prior to 1900, the entire campus had awarded only 12 doctorates. The first PhD in the mathematics department was granted in 1901 to Frank Elmore Ross, who wrote a dissertation "On Differential Equations Belonging to a Ternary Linearoid Group" under the direction of Irving Stringham. There was, however, a considerable gap until the second doctoral degree in mathematics was awarded in 1909.

The department continued to expand as enrollments grew and university funding stabilized. Beginning in 1897, the department reached outside for new faculty, now to a new source in Chicago. The University of Chicago had been founded in 1892, and Eliakim Hastings Moore, who received his PhD from Yale in 1885, was brought to Chicago to lead the mathematics department. He recruited Oskar Bolza and Heinrich Maschke to Chicago, and under Moore's leadership, Chicago very quickly became a leading department and produced a number of doctoral students who have through many generations of students played a major role in shaping American mathematics.

Moore's first student was Leonard Eugene Dickson, who after finishing his doctoral degree in 1896 spent a year of postdoctoral study split between the Universities of Paris and Leipzig with Camille Jordan and Sophus Lie, respectively. He then came to Berkeley in 1897 as an instructor in the Department of Mathematics. Berkeley, however, had difficulty retaining this young star, and there is correspondence between the department and the president of the university with a decidedly modern ring to it. In 1899, the University of Texas offered Dickson an associate professorship at a significantly higher salary, and Associate Professor Haskell, who was acting chair in Stringham's absence, wrote an urgent letter to the president, enclosing a glowing appraisal from an outside reviewer of Dickson's work, and begging him to make haste and obtain the regents' approval of Dickson's (accelerated) promotion to assistant professor to entice him to remain at Berkeley. The promotion was ultimately approved, but Dickson, a native of Texas, ended up leaving for Texas [UA, CU 1, Box 24, folder 11]. However, Texas could not hang onto him either, and Moore brought him back to Chicago as an assistant professor in 1900. Dickson remained at Chicago for the rest of his career and was a major figure in American mathematics, with his work in group theory, number theory, and algebra making him the country's leading algebraist for many years.

In 1898, Berkeley hired Ernest Julius Wilczynski, who had received his doctoral degree from the University of Berlin in 1897 and had spent a year at the Nautical Almanac Office. Wilczynski worked in projective differential geometry, and Berkeley also had some difficulty in retaining him. He left in 1907 for the University of Illinois, and then he was brought to the University of Chicago as a replacement for Maschke, who had died in 1908. According to Saunders Mac Lane in his history of the Chicago department, Dickson, Bliss (Bolza's student), and Wilczynski were the triumvirate leading the Chicago department in the generation that followed Moore [AMSC, vol 2, 124]. Two of the three were former Berkeley faculty members, and their losses, ultimately to Chicago, were serious blows to Berkeley. It is interesting to speculate on how mathematics at Berkeley might have developed had Dickson and Wilczynski remained at Berkeley.

In 1899, Albert W. Whitney was hired as an instructor. He was born June 20, 1870, in Geneva, Illinois, and graduated from Beloit College. Subsequently, he was a fellow in mathematics and physics at the University of Chicago, and taught at the Universities of Nebraska and Michigan but received no graduate degrees. His expertise was in the mathematics of insurance, and at Berkeley he regularly taught courses in probability theory and prepared mathematics students for work in the insurance area. He was promoted to assistant professor in 1904 and to associate professor in 1910, but resigned from the university in 1913.

Berkeley continued to grow, and in 1900 the mathematics department turned to another Moore student, Derrick Norman Lehmer, who came to Berkeley as an instructor immediately following the receipt of his doctoral degree. Lehmer was born in 1867 in Somerset, Indiana, and his family moved to Nebraska when he was a child. He graduated from the University of Nebraska in 1893 and went on to receive a master's degree there in 1896. His first publication, "Proof of a Theorem in Continued Fractions," appeared in the *Annals of Mathematics* in 1896. He then enrolled in the doctoral program at the University of Chicago, obtaining his doctoral degree for a dissertation entitled "Asymptotic Evaluation of Certain Totient Sums." Lehmer was Moore's third doctoral student, and he was in good company, for Moore's fifth, seventh, and eighth students were Oswald Veblen, R. L. Moore, and G. D. Birkhoff, and as noted earlier, his first student was L. E. Dickson. Lehmer married Clara Eunice Mitchell on July 12, 1900, in Decatur, Illinois, and arrived with his bride in Berkeley that summer. His salary was the princely sum of $1000 per year. Berkeley was able to retain Lehmer for the remainder of his career. He was promoted to assistant professor in 1904, to associate professor in 1910, and to professor in 1918.

Lehmer worked in number theory and was perhaps best known for his books *Factor Tables for the First Ten Millions (containing the smallest factor of every number not divisible by 2, 3, 5, or 7 between 0 and 10,017,000)* (1909) and *List of Prime Numbers from 1 to 10,006,721* (1914), the largest accurate table of primes at the time (though, of course, minuscule compared to what can be done today). He published actively, with many articles in the *Bulletin of the American Mathematical Society* on various topics in number theory and on plane algebraic curves. His publication list includes nearly 70 items, rather more than any of the pre-Evans faculty. Together with his son, Derrick Henry Lehmer, he developed electrooptical and mechanical sieves that could be used for factoring large numbers and that were precursors of the computers that would be developed subsequently. He had a strong avocation in music and poetry, and he composed and collected many songs. He also wrote two operas and a number of poems. These interests apparently became stronger in his later years. During his career, he supervised the dissertations of 18 doctoral students, more than any other mathematics faculty member at the time. His students wrote dissertations either in number theory or in classical enumerative algebraic geometry. One of his students was Bing Wong, who subsequently served for many years on the faculty of the department. Lehmer retired at age 70 in 1937 and died the following year on September 8.

In 1900, 1901, and 1902 three more recent recipients of doctoral degrees from outside Berkeley were hired. First, Edwin M. Blake, who received his doctoral degree from Columbia in 1893 with a dissertation entitled "The Method of Indeterminate Coefficients Applied to Differential Equations," was hired as an instructor in 1900. He left Berkeley in 1904.

The second, Thomas M. Putnam, was born in Petaluma, California, on May 22, 1875, and came to Berkeley as an undergraduate majoring in mathematics. He graduated in 1897 and entered graduate school at Berkeley. He then came under the influence of L. E. Dickson, who had just come that year as an instructor, and followed Dickson first to Texas in 1899 and then to the University of Chicago in 1900, where he received his doctoral degree in 1901 as Dickson's first (of 62) doctoral students. His dissertation was entitled "Concerning the Linear Fractional Group on Three Variables with Coefficients in the Galois Field of p^n Elements." Putnam was hired as an instructor at Berkeley in 1901 and was subsequently promoted up the ladder to assistant professor in 1907, to associate professor in 1915, and to full professor in 1919. In 1914 he became dean of the lower division, and in 1919 became dean of the undergraduate division (a title later changed to dean of undergraduates). He remained in administrative posts until 1940, when he resigned to return to full-time duties in the mathematics department. He supervised the work of one doctoral student during his career and also served as associate secretary of the American Mathematical Society for a number of years. He died prior to retirement in 1942.

In 1902 John H. McDonald was hired as an instructor. McDonald was born in Toronto on December 11, 1874. He received a BA from the University of Toronto in 1895 and came to the University of Chicago for graduate study, receiving his doctoral degree under Oskar Bolza with a dissertation entitled "On a System of a Binary Cubic and Quadratic and the Reduction of Hyperelliptic Integrals to Elliptic by a Transformation of the Fourth Kind." McDonald was promoted up the ladder, reaching the rank of professor in 1927. His memorial article [IM] states that McDonald had a profound and broad knowledge of the entire field of mathematics, and that his graduate courses were characterized by perennial freshness as he developed his subject from a new and independent point of view. The authors go on to write about McDonald's research as follows:

> McDonald's publications were mainly in the theory of numbers, Bessel's functions, transformation of elliptic integrals, and on differential equations. They appeared in *Transactions of the Royal Society of Canada*, *Transactions of the American Mathematical Society*, and the *Bulletin of the American Mathematical Society*. This gifted mathematician withheld much of his research from publication, but his varied interests found expression in many doctoral dissertations.

McDonald supervised the work of 14 doctoral students, who wrote dissertations on a remarkable variety of topics, including algebraic geometry, ordinary differential equations, partial differential equations, differential geometry, and complex variables (schlicht functions). Raphael Robinson, who went on to serve with distinction on the faculty for many years, was one of his students. Every year, McDonald visited his teacher Oskar Bolza, to whom he was especially devoted and who had returned to Germany, and spent time discussing mathematical problems. McDonald remained at Berkeley for his entire career, retiring in 1945. He died in 1953.

In summary, starting in 1888, the department enjoyed rapid growth and maturation and by 1902 had reached a complement of ten faculty members in regular ranks, which at the time included instructor and assistant, associate, and full professor. All but two of the regular faculty had doctoral degrees. As noted, the department awarded its first doctoral degree in 1901. The department had come a long way between 1888, when the department consisted only of Stringham and Edwards and some part-time assistants, and 1902. It was now beginning to receive a modicum of recognition. For instance, on March 14, 1899, E. H. Moore, who was president of the American Mathematical Society, wrote to Stringham that the society was going to start a new journal devoted to the publication of research, to be called the *Transactions of the American Mathematical Society*. He went on to say said the society needed $1000 as a startup fund for the new journal and asked each of ten departments to contribute $100 to defray these costs [UA, CU 1, Box 24, folder 11]. Stringham took this as signal that Berkeley (or California, as it was known in those days) was viewed as among the top ten departments in the country.

This growth had started under Kellogg's presidency and continued in the first years of Benjamin Ide Wheeler's, which began in 1899 and continued through 1919. After 1902, the department grew rather slowly for many years, reaching 11 faculty members in regular ranks by 1920, and 15 by 1928. Total campus enrollments grew by a much greater percentage during this period. The regular faculty was supplemented by a number of other teaching staff, including lecturers, associates, assistants, and teaching fellows.

According to departmental documents, in 1902, a typical member of the mathematics faculty taught four or five classes per semester and spent 13 to 15 hours per week in the classroom. Lower-division classes ranged in size between 30 and 50 students, with upper-division classes somewhat smaller, in the 20–30 range. Graduate courses drew 5 to 10 students, and there were only about four or five graduate courses offered, as opposed to about 35 at the undergraduate level. Stringham wrote to President Wheeler on August 10, 1902, complaining about a lack of resources and citing the statistic that the mathematics department at California handled 115 course enrollments per semester per faculty member, while the mathematics department at Harvard had only 54 course enrollments per semester per faculty member [UA, CU 1, Box 24, folder 11]. Stringham's pleas evidently fell on deaf ears, given the lack of growth in the faculty in subsequent years.

Even though there was very little net growth, there were separations from the faculty that had to be replaced by new appointments. New appointees to the faculty were new or recent PhDs and were normally brought in as instructors. Some of these instructors departed after a year or two, and were replaced by new appointees. Service as instructor could be for a considerable length of time, as many as five to eight years or more. Those who were promoted to assistant professor generally remained with the university for the rest of their careers, so it appears that the rank of instructor was used de facto for the probationary period.

Irving Stringham became one of the leaders of the entire campus faculty, and in the process moved even further away from research interests. He held several decanal posts in addition to his position as the leader of the mathematics department. (Formal designation of individuals as department chair or head was several years in the future.) President Wheeler evidently placed considerable confidence in Stringham, for he asked Stringham to represent him at the May 1903 regents' meeting to present faculty recommendations on degrees and scholarships. As mentioned earlier, in the summer of 1909, Wheeler formally appointed Stringham to the new position of dean of the faculty. Professor Alexis Lange, of the education department, was designated dean of the graduate school, a position that was regarded as next in the chain of authority to the dean of the faculties.

It was during this period of service as acting president of the university that Stringham was stricken ill on October 4 and was taken to the hospital. He died the following day at age 63 after 27 years as leader of the mathematics faculty. The accounts of the memorial service that was held in his honor gave further proof of the high regard in which Professor Stringham was held by the entire campus community [Str]. Alexis Lange took over his duties as dean of the faculties, and Professor Mellen Woodman Haskell assumed the position as leader of the mathematics faculty, a position that he held for 24 years until his retirement in 1933. During this period, he presided over appointments and promotions in the department.

Chapter 4

Lagging Behind: 1909–1933

This chapter deals with the period from 1909, when Stringham died and Haskell took over leadership of the Department of Mathematics, up until Haskell's retirement in 1933. Under the presidency of Benjamin Ide Wheeler, 1899–1919, and his successors, the University of California began to achieve real distinction as a research university, at least in certain fields. As the number of such fields grew, these strong departments exerted influence on weaker ones and began to pull them up as well. Mathematics during this period lagged behind other science departments in developing excellence in research. Under Haskell's leadership, far more energy and effort went into developing the teaching mission of the department than to developing its research excellence. In the latter half of this period, every new appointee to the mathematics faculty was a Berkeley doctoral student. This practice of inbreeding weakened the department at a time when it should have been reaching out to make strong appointments from the outside. Consequently, mathematics began to lag behind other departments, and the lag became greater and greater until the realization in 1932 as Haskell's retirement neared of how far the department had fallen behind in research. This realization led to the revolution in 1932–1934 that brought Griffith Evans to Berkeley to reshape the department. This chapter is then the prelude to the events of 1932–1934 that will be described in the next chapter.

After the rapid buildup in size of the department between 1898 and 1902, there was a hiatus in establishing new positions, and the department remained at about the same size through 1919. The first new appointments from outside came in 1911. One of these was Frank Irwin. Irwin was born on June 5, 1868, in New Jersey and graduated from Harvard College in 1890. He undertook a number of business ventures before returning to complete his doctoral studies in 1908 with a dissertation entitled "Invariants of Linear Differential Expressions" written under the direction of Maxime Bôcher. Irwin spent three years as an instructor at Princeton and then came to Berkeley as an instructor in 1911. He was promoted to assistant professor in 1918 and to associate professor in 1923, retiring at that rank in 1938. According to his memorial article [IM], he read contemplatively and extensively, but did not publish original mathematical work.

The second appointment in 1911 was Thomas Buck. Buck was born in Maine on December 25, 1881, and graduated from the University of Maine in 1901 at the age of 19. He remained at the University of Maine as an instructor until 1906, when he undertook graduate studies in mathematics and astronomy at the University of Chicago. He completed a dissertation in 1909 entitled "Oscillating Satellites near the Lagrangian Equilateral Points Periodic Solutions" under the mathematical astronomer F. R. Moulton. He then served as an instructor of mathematics at the University of Illinois from 1909 to 1911 before coming to Berkeley as an instructor. He was promoted to assistant professor in 1918 after seven years of service, to associate professor in 1923, and then to full professor in 1931. Buck was on war leave for part of 1918 as first lieutenant in the United States Army calculating range tables for new high-velocity artillery shells. His appointment at Berkeley added strength in celestial mechanics to the department, and complemented work in this area in the astronomy department led by Armin Leuschner.

Buck was an excellent teacher. His upper-division course in differential equations, Fourier series, and hyperbolic functions was taken by generations of engineering students. His graduate courses on special functions and partial differential equations were taken also by generations of graduate students in the physical sciences and engineering, and both courses were widely admired and appreciated. He retired in 1952 after 41 years of service, and died in 1969 at age 87.

In 1907, Benjamin Bernstein was appointed as instructor even though he was still a graduate student in the department, and in fact would not receive his degree until 1913. Bernstein was born in Lithuania on May 20, 1881, and came to the United States at an early age. He received his bachelor's degree at Johns Hopkins in 1905 and continued as a graduate student for two years at Hopkins, but did not receive a graduate degree. While an instructor at Berkeley he continued his graduate studies, receiving his doctoral degree in 1913 from Berkeley. He was an early worker in mathematical logic and was very much influenced by the *Principia Mathematica* of Russell and Whitehead. His dissertation, completed under Mellen Haskell, was entitled "A Set of Postulates for the Logic of Classes." It would seem that he worked mostly on his own, since there was no one at Berkeley with scholarly interests in this field. He was an intensive and critical student of the *Principia* and discussed it in many subsequent research papers. In later years he collaborated with Alfred Foster in work on the foundations of commutative rings, and he published actively throughout his career. His publication record in mathematical research ranks him among the top two or three of the pre-Evans faculty in the department. He was promoted to the rank of assistant professor in 1918 after eleven years as instructor, to associate professor in 1923, and to full professor in 1928.

Bernstein was one of a group of American mathematicians who have been termed the American postulate theorists [Sca]. Their work, generally in the period 1900 through 1930/40, was inspired by the work of Hilbert and Peano, and was devoted to analyzing axiomatic systems for various mathematical structures. They studied independence of axioms and also developed the notion of categoricity of postulate sets, a key notion in the later development of logic. Prominent workers included E. V. Huntington, Oswald Veblen in his early years, E. H. Moore, R. L. Moore, and Bernstein, among others. This school of mathematics more or less died off as the new developments in logic by Gödel and Tarski came to the forefront. Bernstein was particularly interested in the axiom systems for Boolean algebras, which complemented his lifelong engagement with the *Principia Mathematica* (PM), of which he was quite critical. On June 7, 1932, he wrote to G. H. Hardy, "Whitehead and Russell have intended in the PM to clarify the nature of logic and mathematics, but the PM I feel strongly greatly obscures both logic and mathematics, and my papers on the PM are merely attempts to bring these obscurities to light" [Ber, Box 2]. In the same year, correspondence with Huntington highlighted their differing views on the *Principia*. Lukesiewicz and Tarski had written to Bernstein in 1929, requesting reprints of his papers for their seminar in Poland [Ber, Box 2]. Bernstein's work was thus clearly known and recognized in a wider community. When Tarski arrived in Berkeley in 1942, it was not long before the two clashed, as we recount later on, in Chapter 6. Bernstein retired in 1951 after 44 years of service on the faculty, and died in 1964.

The next appointment was that of Pauline Sperry, who joined the faculty in 1917. Sperry was born in Massachusetts on March 5, 1885, and graduated from Smith College in 1906. She taught mathematics at Smith from 1908 to 1912 and then decided to pursue graduate study in mathematics at the University of Chicago. There she worked under Professor Ernest Wilczynski, formerly of Berkeley, in projective differential geometry. Her dissertation was entitled "Properties of a Certain Projectively Defined Two Parameter Family of Curves on a General Surface," and a paper based on it was subsequently published in the *American Journal of Mathematics*. She returned to Smith for a year as assistant professor before coming to Berkeley as an instructor in 1917. She was promoted to assistant professor in 1923 and to associate professor in 1931. Although she published infrequently, she supervised five doctoral dissertations during the 1930s and 1940s. One publication of note was her 1931 *Bibliography of Projective Differential Geometry*. She was a devoted teacher and prepared two textbooks for freshman mathematics. In 1945 she was selected as chair of the Northern California section of the *Mathematical Association of America*.

Sperry was raised as a Quaker and throughout her life practiced her ethical beliefs. In 1950 her belief that a required loyalty oath encroached on the political freedom of the university led her to refuse to sign the oath as a matter of high principle. She was fired from the university, but after litigation, she was reinstated just prior to her retirement in 1952 (see Chapter 9). After retirement she moved to Carmel and devoted herself to her causes and to charitable projects. On the occasion of her 80th birthday, she published an article in the Smith *Alumnae Quarterly* [Sp], "Formula for Happiness at Eighty," which begins,

> Everybody knows that unless you personally do something about it, you will feel needed less and less as you grow older. At eighty, I feel needed more and more, and I am eager to tell the secret. Oscar Wilde once said, "Men who are trying to do something for the world are always insufferable; when the world has done something for them, they are charming." The world has done so much for me that I do not mind being insufferable, and I let the charm fall where it may.... I have always burned for causes.

One of her causes was an orphanage in Haiti, which she writes about in the article. This article provides insight into the generosity and spirit of this remarkable woman who had not been treated at all well by the university. She died in 1967, shortly after writing this article.

In 1918, Professor George Edwards retired after 45 years of service as a teacher and citizen of the campus. In connection with his retirement, the department ventured into the field of the history of mathematics through the appointment of Florian Cajori. Cajori came with quite a different background from those of others who had recently been appointed. He was born on February 28, 1859, in Switzerland and came to the United States in 1875. He studied at the University of Wisconsin, Johns Hopkins, and then at Tulane, where he received his doctoral degree. He served on the faculty at Tulane and then at Colorado College. He had also served, before coming to Berkeley, on some national curriculum commissions of the National Education Association and was the author of ten books on a variety of topics in the history of mathematics and the teaching of mathematics. He was the author of many journal articles and was regarded as a leading scholar in the history of mathematics and mathematical pedagogy. In 1918 he was appointed professor of the history of mathematics at the age of 59. He continued his publication rate unabated, and was the first member of the department designated a faculty research lecturer (1926). He also served, in 1917, as one of the very early presidents of the Mathematical Association of America. He retired from the university in 1929 and died a few months later.

The appointments to the faculty over the 20-year period 1898–1918 covered a range of areas of mathematics. Only one, Bernstein, was a Berkeley PhD, and it could be argued that he was in fact intellectually independent of the faculty in the department, and supervision of his dissertation was pro forma. A number of those recruited to Berkeley during this period received their degrees at the University of Chicago. Many of the new appointees ended up devoting their major efforts to teaching. Among all the faculty during this period, Lehmer was the most active in research.

The graduate program provides a useful indicator of the mathematical life of the department during this period. As already noted, the first doctoral degree in mathematics was awarded in 1901, and there was a gap until the next one was awarded in 1909. Graduate enrollments throughout the university expanded rapidly in this period. The graduate curriculum in the mathematics department was not fully fleshed out in anything resembling the curriculum of today until 1910 or later. In the 24-year period from 1909 to 1933, when Professor Haskell served as department chair, the department awarded 41 doctoral degrees. Somewhat surprisingly, supervision of doctoral dissertations was completely dominated by three faculty members: Lehmer supervised 18; Haskell, 13; and McDonald, 8. Two other faculty members supervised one each. Haskell and McDonald did not publish actively, but it appears that their mathematical ideas appeared in the dissertations of their students.

Starting in about 1919, the department focused its hiring almost entirely on candidates who had received their doctoral training at Berkeley. Of the nine appointments made from this time through 1933, all were at the beginning level of instructor, and all nine were Berkeley PhDs. Why this happened is not clear. One factor may have been that until this time, the department had not produced that many doctorates, so previously there had not been many internal candidates. As noted, the first doctoral degree was granted in 1901 and the second not until 1909. The department may have tried outside recruitment as the junior level and been unsuccessful, but there is no record of any such efforts. The leader of the department, Mellen Haskell, placed great emphasis on the teaching mission of the department, and he had the opportunity to observe closely the teaching of the local graduate students, and he could be assured that he was hiring good teachers by this practice of internal hiring. This policy choice led to considerable inbreeding and the lack of new ideas and enthusiasm coming into the department. Initially, there appeared to be no administrative or campus oversight of the situation, and this lack perhaps may have been connected to the retirement in 1919 of Wheeler and the faculty revolt against his autocratic manner. Although Wheeler's successors were perhaps not as vigorous in their oversight, by 1927 the campus leadership realized that things were not going as well as they should in mathematics. The administration led the way in trying to bring in new blood at the senior level. These efforts failed, and the inbreeding continued until 1933.

The first such appointment was that of Arthur Robinson Williams. Williams was born in 1887 in Westport, Connecticut, and graduated from Yale College in 1907. He came to Berkeley in 1910 as a graduate student, first in economics and then in mathematics, receiving his degree under Lehmer in 1916 with a dissertation entitled "A Birational Transformation Connected with a Pencil of Cubics." He was engaged in war-related work until 1920, at which time he was appointed an instructor at Berkeley. After ten years of service, he was advanced to the rank of assistant professor, and he retired in 1950 as assistant professor emeritus. He had very nearly been terminated in the restructuring of the department in 1933–1935, when Evans arrived to take over the department. The decision at the time was to keep Williams on as assistant professor but with the assumption that he would never be promoted. His teaching was valued and well received. Williams worked in classical algebraic geometry, notably on the geometry of ruled surfaces. Over the course of many years, he published about ten papers on these topics in the *Bulletin of the American Mathematical Society*. He frequented the department in the years following his retirement, and was a forlorn presence. He died in Berkeley in 1961.

In 1921, Sophia Levy was appointed as instructor in mathematics. Levy was born in Alameda, California, in 1888 of parents who were also native Californians. She attended the University of California and graduated with a major in astronomy. Continuing with graduate studies in astronomy, she completed a dissertation under Armin Leuschner in 1920 on the motion of comets and minor planets. During her time as a graduate student, she also served for four years as assistant to the dean of the graduate division and for two years as secretary of the Commission on Credentials of the State Board of Education. After appointment as instructor in 1921, she was advanced to assistant professor in 1924, associate professor in 1940, and full professor in 1949.

She contributed scholarly papers to the literature concerning the motions of comets and minor planets. The memorial article on her life states that

> Since her work in astronomy required handling of extensive numerical data, she quite naturally directed herself to the field of numerical analysis, including such subjects as interpolation methods, mechanical quadratures, the numerical solution of algebraic and transcendental equations, Fourier analysis and periodogram analysis. [IM]

During World War II she taught courses in antiaircraft gunnery to armed services personnel at Berkeley and later published a text on the subject with the University of California Press. She was also deeply engaged in the preparation of secondary-school mathematics teachers and served as the departmental advisor for prospective teachers and served on a number of regional and statewide committees on mathematics education. With A. L. McCarty, of San Francisco City College, she founded in 1939 the Northern California Section of the Mathematical Association of America, serving first as secretary of the section, then as vice-president, and then president and sectional governor.

At some point in her career, at a time lost in the mists of history, she formed a close personal relationship with her departmental colleague John Hector McDonald, who was 13 years her senior. They had hoped to marry, but marriage was precluded by the university's strict nepotism rules, which did not allow close relatives to be employed in the same department. One or the other would have had to resign. In any case they waited until John reached mandatory retirement age and in 1945 they married. She was subsequently known as Sophia Levy McDonald. John died in 1953 and his In Memoriam article states, "The colleagues who were privileged to have insight into his character, his intellectual power, and his artistic sensitiveness are grateful to Mrs. McDonald for the comfort and happiness which her devotion brought to his later years."

She retired from active duty in the department in 1954 and died in 1963. Her memorial article concludes as follows:

> The daughter of pioneer parents in California, Sophia Levy McDonald viewed herself as somewhat of a pioneer for women in study and research in the exact sciences. She contributed to the fame that the astronomy department had enjoyed under the leadership of Professor Leuschner in the field of celestial mechanics, and she contributed significantly to the teaching of mathematics in the schools and colleges of California. [IM]

In 1922, another Berkeley PhD graduate, Bing Chin Wong, was appointed as instructor. He received his degree under Lehmer with a dissertation entitled "A Study and Classification of Ruled Quartic Surfaces by Point-to-Line Transformations." Bing Wong was born on September 1, 1890, of Chinese-American parents who were farmers in the Sacramento Valley and who had come to this country in the 1850s. He spent several years as a boy in China, and after returning to the United States, graduated from Sacramento High School in 1913. He came to Berkeley as a freshman and remained in Berkeley for the remainder of his life. He was appointed as an instructor in 1922, and then promoted to assistant professor in 1927 and to associate professor in 1935. In 1942 he suffered a major stroke during a class that left him paralyzed and speechless until his death in 1947. He played a significant role on campus as one of the first Asian-Americans to be appointed to the faculty of the university.

Over his professional career of nearly 20 years, he published frequently on topics in algebraic geometry in the *Bulletin of the American Mathematical Society* and in other journals. His work was devoted to a type of enumerative algebraic geometry based on work of Francesco Severi and Giuseppe Veronese of the Italian school. Some typical titles of papers are "On the Number of Apparent Double Points of an r-Space Curve" and "On Loci of $r - 2$ Spaces Incident with Curves in r-Space." His work is noted and remarked on in the *Encyklopädie der mathematischen Wissenschaften*. During the 1920s and 1930s, Wong and Bernstein were the two most active faculty members in research and publication among the pre-Evans appointees.

In 1923, Raymond Sciobereti was hired as an instructor in mathematics. Sciobereti was born on April 7, 1886, in Basel, Switzerland. He developed interests in navigation, climatology, and oceanography, and served as navigator on various sailing vessels. After a varied early career, he enrolled at Berkeley in 1919, earning a BA in 1920 and a PhD in 1923 in astronomy. According to his memorial article, his early work concerned Leuschner's method for computing orbits, and in later years his work turned to solutions of ordinary and partial differential equations and to problems in geometry and continued fractions. He was advanced to assistant professor in 1927 and to associate professor in 1948. He retired from the university in 1954.

In 1924, Annie Dale Biddle Andrews was hired as an instructor in mathematics. Under her maiden name of Biddle, she had received a doctoral degree from Berkeley in 1911 under the joint supervision of Lehmer and Haskell with a dissertation entitled "Constructive Theory of the Unicursal Plane Quartic." She was the third doctoral student to graduate from the department and the first woman doctoral graduate. Andrews taught in the department on and off for a number of years in various temporary titles that were ordinarily used for graduate students before her appointment as an instructor in 1924. She was dismissed from her position in 1933 as a part of the reorganization of the department that will be described subsequently. She was evidently a successful teacher, but she was not engaged in any research activities.

It is helpful at this point to review the academic progress and growth at Berkeley in the other science departments in order to provide a context for events in the mathematics department. The first department to gain national recognition was the astronomy department. Actually, there were two astronomy departments: the Lick astronomy department and the Berkeley astronomy department. After 14 years in gestation, the Lick Observatory on Mount Hamilton was completed in 1888 and turned over to the university. It was constructed through the generosity of James Lick, a San Francisco merchant with an interest in astronomy. The 36-inch refractor was the largest telescope in the world at the time of its construction, and was surpassed only in 1894 by a 40-inch instrument at the Yerkes Observatory at the newly created University of Chicago.

The Lick facility attracted prominent astronomers, among them the young William W. Campbell, who succeeded to the directorship of Lick in 1901. Under his leadership, Lick was regarded as one of the premier observatories of the world. Campbell was elected to the National Academy of Sciences and received many other honors. In 1923 he became president of the University of California, and on his retirement from that position in 1930, he was elected president of the United States National Academy of Sciences.

Instruction in astronomy was offered from the very inception of the Berkeley campus. George Davidson, mentioned earlier in Chapter 2, offered such courses, and in 1872, when Frank Soulé, Jr., who had been appointed as assistant professor of mathematics in 1869, was appointed professor of astronomy and civil engineering in 1872, he offered astronomy courses until 1892. At that time, Armin Leuschner, who had been an instructor in mathematics, took over instruction in astronomy and became in 1894 an assistant professor in astronomy. In 1898, a doctoral program in astronomy was created, and Leuschner was appointed chair of astronomy in 1900. He held this position for 38 years until his retirement in 1938. Leuschner was a distinguished scientist who was elected to the National Academy of Sciences and was a major and influential figure on the campus for decades. Under the dual leadership of Campbell and Leuschner, astronomy flourished at the University of California.

Chemistry was the next unit to grow to distinction. Willard Rising, who studied under Bunsen in Heidelberg, was appointed to the chair of chemistry in 1872 and served in this position for 36 years until his retirement. A doctoral program had been established, but when Rising retired in 1908 only four doctoral degrees had been conferred. In 1912, Wheeler in a bold stroke recruited a group of chemists, led by Gilbert Newton Lewis, from MIT. Lewis was appointed dean of the College of Chemistry, a position he held until 1941. Coming to Berkeley at the same time were Richard Tolman, Merle Randall, and William Bray. Joel Hildebrand joined the faculty a year later and was followed by others, including Wendall Latimer in 1917 and William Giauque in 1922. This was an extraordinarily distinguished group of scholars, and it remained so even after Tolman left after four years.

It is interesting to note that, as Jolly points out in his book on the history of chemistry at Berkeley [Jolly], Lewis almost exclusively relied on Berkeley PhDs to build his department. Of the 16 faculty appointments that he made between 1915 through 1941, all but two (one of whom was the Nobelist Melvin Calvin) were Berkeley PhDs. The list of faculty includes Nobelists William Giauque, Willard Libby, and Glen Seaborg, as well as Wendell Latimer and Kenneth Pitzer, among others. Thus, Lewis succeeded in building an outstanding department using his own doctorates, whereas Haskell failed in a similar task in mathematics. One clear difference in chemistry was a stronger infusion of senior faculty leadership early on and the ability of the department then to attract strong graduate students into the program.

Lewis was famous for the Lewis electron-pair theory, the Lewis acid-base theory, and the discovery of deuterium. Berkeley became known as a center for thermodynamics with Lewis's work as well as Latimer's work on entropy of aqueous ions and Giauque's low-temperature work that won him a Nobel Prize. Chemistry thus had reached a high level of distinction.

The Physics department was led by John LeConte from 1869 until his death in 1891. LeConte published actively and was an early member of the National Academy of Sciences. Frederick Slate then served as head of the department until his death in 1918. A doctoral program had been established, and the first doctoral degree was awarded in 1903. E. P. Lewis, who joined the faculty in 1895, was the only active researcher in the department until Raymond P. Birge arrived in 1918 as an instructor. Birge moved up through the ranks and was appointed chair in 1932, the same year that he was elected to the National Academy of Sciences. The appointments of J. Robert Oppenheimer and E. O. Lawrence in 1929 set the department on a course to become the premier department in the country. Many elections to the National Academy of Sciences and Nobel Prizes followed, although physics on balance was a few years behind chemistry in its climb to excellence.

Joseph LeConte was the early leader in the Department of Geology, and like his brother was an early member of the National Academy of Sciences. The growing demand for mining engineers and geologists led to the expansion of the program. Andrew Lawson was appointed to the faculty in 1890 and became a legendary figure on campus for decades. George Louderback was another leader in the department from his appointment in 1906 until 1944. Hence, the earth sciences were on their way to distinction as well.

By comparison with these other science departments, it is evident that the mathematics department was not doing as well during the 1920s and the early 1930s. It did not have the high level of distinction that other science departments had developed and were developing. Under Haskell's leadership, it had become primarily a teaching department serving the needs of students in other departments and preparing future secondary-school and community-college mathematics instructors. Graduate students served as teaching fellows and assistants during their term of studies, and so their abilities as teachers were well known to the department. Selection of those with excellent teaching records to be junior faculty was a natural course of action. But, in general, such a policy created a department that was not really at the forefront of research. Bing Wong was something of an exception to this and was doing very credible research, as were Benjamin Bernstein and Derrick Norman Lehmer from an earlier era of appointments. For instance, in the 15-year period 1919–1934, a total of 17 articles by the entire Berkeley mathematics faculty appeared in the three primary American mathematics journals of the day (*American*

Journal of Mathematics, Annals of Mathematics, and the *Transactions of the American Mathematical Society*). Eight of these were by Wong, seven by Bernstein, and one each by Lehmer and McDonald. Florian Cajori, as already noted, published actively, but his work was in another direction altogether.

In 1927, a consensus developed that a senior appointment of a leading scholar in mathematics was needed. This may have been related to the impending retirement of Cajori in 1929. Feelers were sent out to E. B. Wilson, professor of vital statistics at Harvard, but nothing came of this initiative. Attention turned to Griffith C. Evans, of Rice University. Evans was indeed a leading scholar, who had recently declined a call to Harvard. An ad hoc committee on a professorship of mathematics, consisting of Professors Haskell and Putnam, of Mathematics, and G. N. Lewis, A. Leuschner, and a Professor Williams, from outside the department, was formed by the budget committee to consider the appointment. Vice President Walter Morris Hart visited Evans in Houston in November 1927, and Hart wrote to Evans on December 5, thanking him for his hospitality and for suggestions of names of young mathematicians who might be hired at Berkeley [UA, CU 5, series 2, 1927]. Some background information on Griffith Evans will be helpful before the efforts to bring him to Berkeley are recounted.

Griffith Conrad Evans was born May 11, 1887, in Boston, Massachusetts. His forbears had emigrated from Wales, first to Ireland and then to the United States, settling in Boston. Evans's father, George W. Evans, was a graduate of Harvard College (1883) and became a mathematics teacher and administrator in the Boston schools. He was a mathematics teacher at Boston Public Latin School and later was principal of the Charlestown School. He was the author of high-school textbooks on algebra and on geometry, and was a frequent contributor to the *American Mathematical Monthly* and to the *Mathematics Teacher*, the journal of the National Council of Teachers of Mathematics [Evans, Box 6].

Griffith Evans graduated from the English High School of Boston in 1903 at age 16 and entered Harvard College the following fall. After graduating summa cum laude in 1907, he continued at Harvard in the graduate school. He received his PhD in 1910 under Maxime Bôcher with a dissertation entitled "Volterra's Integral Equation of the Second Kind with Discontinuous Kernel." On receiving his degree, he was awarded a Sheldon traveling fellowship and spent two years as a postdoctoral scholar working in Rome under Professor Vito Volterra [Evans, Box 6].

According to his own notes, Evans's employment opportunities at the end of his postdoctoral studies in 1912 were faculty positions at Yale, MIT, Berkeley, and Rice. He selected Rice in part because it was a new institution that was just opening for students in 1912. Indeed, Evans was the first faculty member hired at Rice, and his initial appointment was as assistant professor. He evidently prospered there, being promoted to full professor in 1916 at the age of 29. He met his future wife, Isabel Mary John, a native of Houston, when she was a student in one of his classes at Rice. Mary John was a descendant of Sam Houston. They were married on June 20, 1917, and had three boys, Griffith Conrad, Jr., George William, and Robert John. Evans volunteered for military service as a scientist in World War I and was commissioned as a captain in the US Army Signal Corps, serving in France and Italy, where he did scientific and experimental work [Evans, Box 6].

Following the war, Harvard began a courtship of Evans. In 1919 Harvard offered Evans a position as assistant professor of mathematics (not withstanding his rank of full professor at Rice). Evans declined the offer. The following year, Harvard offered him a position as associate professor of mathematics, which he again declined. In 1926, Harvard finally offered a full professorship, which Evans again declined, much to the surprise and consternation of his many friends at Harvard, as one can tell by the correspondence. A letter from departmental chair G. D. Birkhoff of December 17, 1925, mentions a salary of $6000, with an expected increase to $7000 five years hence, and another expected increase to $8000 after another five years, an increase that was reserved "for men of preeminence." After Evans declined this call to Harvard, Harvard offered the position to Marston Morse, who accepted the appointment [Evans, Carton 1].

As already noted, soon thereafter, Berkeley began its courtship of Evans, and on December 14, 1927, UC President William W. Campbell telegraphed Evans an offer of a professorship as follows:

Professor G.C. Evans, Rice Institute Houston

I take unusual pleasure in offering you a Professorship of Mathematics here effective next July salary nine thousand dollars per annum plus non recurrent removal item seven hundred fifty dollars stop This is our maximum salary and is larger than any salary in department of mathematics stop I would hope and reasonably confident maximum will go up to ten thousand eighteen months hence from which advance I would expect you to profit stop Within few years Professor Haskell will retire and I should then prefer you assume leadership of Department stop. Kindly telegraph answer. W.W. Campbell [Evans, Box 6]

Evans responded three days later on December 17, "very much appreciate generous offer and magnificent opportunity. Decline regretfully Griffith C. Evans." However, President Campbell and others, especially Leuschner, were persistent and persuaded Evans to reconsider his decision, and they continued to discuss and negotiate with him about coming to Berkeley. Evans did come to Berkeley to teach in the summer session 1928 and so had a firsthand experience of Berkeley. Discussions about the offer continued into the fall of 1928 with correspondence between Leuschner and

Evans. Evans was quite frank with Leuschner, telling him that he had concerns about the future of the department at Rice and that he intended to raise these issues with President Lovett of Rice and wrote, "If he [Lovett] can give the proper answers without having to coerce the Trustees under the bludgeon of a present emergency, I shall be willing and even happy to stay; because this is my battlefield, just as yours is the University of California, and anything else is, in a very definite sense, a retreat" [Evans, Box 6].

On October 25, Chair Haskell wrote Evans a letter stating that Evans "would be warmly welcomed by the whole department." Haskell then continued, "As you know, we have an excellent staff of teachers, but they have never been strong in research. I think we should have the approval of the administration in building up in that direction, and I should value your assistance and cooperation in doing this." This was certainly an honest assessment of the department, but it also indicated a lack of comprehension about the direction in which the university was moving. In addition, the telegraphed offer from Campbell was curious in that it did not offer Evans the chairmanship right away, but only after Haskell retired, which was not scheduled to occur until 1933. In the end, Evans declined the offer finally on January 30, 1929, in a letter to Leuschner, indicating that President Lovett had responded at least in part to his concerns. Evans closed the letter with, "I can only say that I very much envy the man who will be so fortunate as to take advantage of the opportunity which you offer" [Evans, Box 6].

When Evans, after the year-long-plus courtship finally declined the offer to come to Berkeley, the department did not turn to seeking another senior person to take advantage of the opportunity the administration had provided. Indeed, when Florian Cajori retired on July 1, 1929, thus potentially freeing up more resources for senior appointments, Chair Haskell then proceeded instead to engage in a sudden faculty hiring spree focused on Berkeley graduate students. That spring, Haskell hired four Berkeley doctoral graduates as instructors starting July 1, 1929. Moreover, all four were his own students, and their dissertations topics were rather narrowly focused in one area or even one subarea. These were as follows:

1. Dewey Duncan, PhD 1928 under Haskell; dissertation title: "Rational Quintics Autopolar with Respect to a Finite Number of Conics."

2. Elmer Goldsworthy, PhD 1929 under Haskell; dissertation title: "Curves Autopolar with respect to a Cyclic Set of Conics."

3. Edward Roessler, PhD 1929 under Haskell; dissertation title: "A Family of Autopolar Sextic Curves."

4. Lee Swinford, PhD 1929 under Haskell; dissertation title: "On Autopolar Rational Quintic Curves."

These appointments pushed the department even further in the direction of a teaching department and away from any path to preeminence in research and scholarship. The stock market crash that fall and the ensuing depression resulted in a decline of resources that were available to the university. The opportunities for new appointments were very limited, and as the depression deepened, the university adopted a policy of generally not replacing faculty who had retired or separated. The only impending retirement was that of Haskell himself, which was scheduled for July 1, 1933. By then there was a new president, Robert Gordon Sproul, who had assumed office in 1930, and a new academic vice president/provost, Monroe K. Deutsch.

Chapter 5

The Beginning of the Evans Years: 1932–1936

By August 1932, if not earlier, Provost Deutsch was aware of serious problems in the mathematics department. On September 1, 1932, he wrote to President Sproul,

> The Department, in the eyes of the University, has not been maintained at the proper standard, and undoubtedly a large part of that responsibility devolves upon Professor Haskell. The Department seems to most men rather dead, and it is to be noted that for a considerable period of time the only persons added to the staff are those who have been on the ground here, have either been undergraduate students at this University, or graduate students, or both.... In view of the importance of mathematics, it seems to me that we should use this opportunity to bring in a strong man as professor. It might be desirable to consider dropping one of the men further down in order that there might be some saving. [UA, CU 5, series 2, 1932, 100-Mathematics]

Deutsch, as a scholar of Greek and Latin, likely had little personal knowledge of the situation in mathematics, but was rather probably relying on the judgment and evaluations of senior faculty in related disciplines: astronomy, physics, chemistry, and engineering.

Deutsch then sought the advice and guidance of the Academic Senate's Committee on Budget and Interdepartmental Relations, and after a few iterations, the Budget Committee proposed that a committee be formed to "[c]onsider the personnel of the Mathematics Department, including the elimination of such member or members as may be desirable for the welfare of the Department" and to "[m]ake an independent canvass of the field for the purpose of selecting a professor of mathematics to succeed Professor Haskell." The Budget Committee nominated the following individuals to serve on the committee: G. N. Lewis, chair; R. T. Birge; C. Derlath, Jr.; A. O. Leuschner; B. M. Woods; and J. H. Hildebrand. This was indeed a distinguished committee. Lewis was dean of chemistry, Birge was chair of physics, Derlath was dean of engineering, Leuschner was chair of astronomy, Woods was chair of mechanical engineering, and Hildebrand was a senior faculty member in chemistry and a widely admired and respected leader of the Berkeley faculty. Four of these individuals have campus buildings named after them. It was a deliberate decision not to include any members of the mathematics department among the members of committee. However, as noted earlier, Leuschner had served briefly in the mathematics department some 35 years earlier, and Woods had also served briefly as a member of the department 20 years earlier [ibid.].

One aspect of what was happening is the process of excellence in an institution being propagated outward from departments that have achieved a high level of accomplishment and stature to other departments that have lagged behind. As noted already, astronomy and chemistry had achieved this kind of excellence early on, and physics somewhat more recently. Birge was a strong scientist himself, and in 1929 Berkeley had made two junior appointments in physics that would shape the future of the department: Ernest O. Lawrence, an experimentalist, and J. Robert Oppenheimer, a theoretician. For a number of years, Oppenheimer had a split appointment between Berkeley and Caltech, and split his time between the two. Oppenheimer, who had studied and absorbed quantum mechanics in Göttingen, brought this new field to Berkeley and attracted a swarm of brilliant students, almost all of whom graduated from Berkeley, not Caltech.

It is interesting to speculate whether these leaders of other science departments may have had another agenda beyond just improving the mathematics department. Quantum mechanics had arrived on the scene in the late 1920s, revolutionizing physics with important impacts on chemistry, such as in molecular structure and spectra, and on astronomy and astrophysics, for example in stellar structure. It was apparent that whole new areas of mathematics were now relevant to the physical sciences: Hilbert spaces, operator theory, matrices, and group theory and group representations, especially with the publication in 1929 of Hermann Weyl's book *The Theory of Group and Quantum Mechanics* and of Eugene Wigner's book *Group Theory and Applications of Quantum Mechanics to Atomic Spectra*. None of these areas was represented in the mathematics department at the time, and it is possible that Birge, Hildebrand, and others had some realization that the growing importance of mathematics meant that for their departments to be successful at the highest level, the campus needed to have a first-rate mathematics department and that the department needed to have scholars who could interact scientifically with colleagues in other departments. However, there is no record that such thoughts were reduced to paper at the time, nor is there evidence, for instance, of substantial interaction of Oppenheimer and his group of students with the (revitalized) mathematics department during the 1930s.

However, another important way in which mathematics would affect the sciences was through statistics, and this discipline, too, was missing from the mathematics department. Several years later, Birge was especially influential as an advocate for an appointment in statistics, and of Neyman in particular. Also, later (see Chapter 7), Professor Trumpler, in astronomy, realized the importance of statistical techniques in astronomy especially as a way to build on the pioneering work of Hubble in the 1920s on the distribution of galaxies. It is not known definitely whether ideas for developing statistics on campus were a force in the move to revitalize mathematics. At any event, when Neyman was appointed in 1938, there was an immediate impact on the other sciences.

While it is interesting to speculate what thoughts were going through the minds of the members of the committee, there was clearly a desire to make over the mathematics department so that its stature was more in line with that of other departments and to promote interaction between mathematics and the natural sciences at levels higher than merely providing introductory instruction in mathematics. The dual charge to the committee to seek a new chair and to review current personnel was seen as serving two functions: first, the review of current personnel leading to eventual dismissal would free up funds for a senior appointment; and second, by making these personnel decisions in advance of the new chair's arrival, it would spare the new chair the burden and possible odium associated with such dismissals. The committee was appointed and set to work in mid October 1932 [ibid.].

The Committee filed its report on December 7, and its conclusions were very tough. The Committee concluded that

> the future welfare of the Department of Mathematics depends upon securing at an early date the addition of several new members who, through their interest in the development of mathematics, through their productivity, and through their ability to attract graduate students, will introduce a new spirit into the Department. In order to make room for these additions to the personnel, some immediate elimination of present members of the Department should occur.

Specifically, the committee recommended that Instructors Andrews, Duncan, and Roessler be notified at once that they would not be appointed beyond July 1, 1933, and that Instructor Swinford be notified that his reappointment is doubtful. In addition, the committee recommended that Assistant Professors Levy and Williams be immediately notified of the termination of their appointments, allowing perhaps a leave of absence for them before separation in recognition of their long service [ibid.].

The committee also stated that the problem of finding replacement staff and a new chair was so urgent and so important that it was recommending that Professor Hildebrand be asked to survey the situation and visit the eastern United States to obtain advice and interview candidates. The committee stated that "Professor Hildebrand has a keen appreciation of the importance to the University of some radical improvements in the Department of Mathematics." These recommendations for Professor Hildebrand to seek new faculty were accepted with alacrity, and then the administration confronted the harsh task of dismissal of several junior faculty members in the depth of the Great Depression with very few jobs available. In the end, some of the recommended actions for non-reappointment were modified, and some were reversed. Meanwhile, Hildebrand set off on his journey, to find a new chair and to identify strong young mathematicians who could be recruited to Berkeley. Just as Hildebrand was about to depart Berkeley for the East, Leuschner wrote to Evans on February 8, 1933, to ask whether he had any interest in a renewal of the offer from 1928, and if so, he suggested that Hildebrand include Houston on his itinerary. Evans responded in the affirmative a few days later, and Hildebrand met with Evans in Houston as part of his journey [Evans, Box 6].

Chair Haskell was beside himself with anger when he was informed of the review committee's recommendations and that the president and provost were to begin implementing them immediately. All these deliberations had taken place without consultation with him. The committee chair, G. N. Lewis, interviewed some senior members of the department—Lehmer, McDonald, Putnam, Bernstein, and Buck—with Haskell (and Noble) notably absent from the list. On February 19, Haskell wrote a four-page letter to President Sproul, defending the department on the basis of its teaching mission. "If members of the present staff are now summarily dismissed," he wrote, "it can only appear, to them and also to those who remain, that good teaching is a matter of slight importance, to properly be neglected for a feverish rush to get into print. The whole department will feel that a great injustice has been done." He continued, "My own suggestion would be, not to dismiss any of the present staff, but to increase it gradually by the addition of promising mathematicians from other institutions." He cautioned the president that "it should be borne in mind that men who are absorbed in research will not have the time to devote to the proper teaching of fundamental courses, nor will they be interested in such teaching." Haskell had developed a mindset about the proper mission of the department that was increasingly out of step with the evolving mission of the campus [CU 5, series 2, 1933, 1-Mathematics].

Hildebrand reported on the results of his canvass and interviews, and on March 7, 1933, Chair Lewis wrote to Sproul recommending on behalf of the committee the appointment of Griffith Evans as professor and chair at a salary of $9,000. The committee noted that Evans had been offered a professorship four years earlier at the same salary. In addition, the committee recommended that Evans be informed that as an added inducement, there would be several additional positions in the department to be filled after consultation with the new chair. These recommendations were quickly approved, and on March 30, President Sproul communicated the formal offer to Evans, which had been endorsed by the budget committee and approved by the regents [ibid.]. However, there is no evidence that anyone in the mathematics department was consulted or even informed of these recommendations and actions.

What followed was a period of intense negotiation that extended over a period of seven weeks, during which time Evans twice declined the offer, only to reconsider twice before finally reaching a final decision to accept. The primary issue was salary, and specifically salary cuts that were on the table throughout academia at this point in the depression. The offer to Evans stipulated that his salary would be subject to the same percentage cut that would be imposed on all faculty salaries in the university. The most likely outcome anticipated was a 10% cut, but it was recognized that, depending on legislative deliberations that were occurring at that time, the cuts could be larger, and in the worst case, salaries might be cut by as much as one-third. Evans's salary at Rice was also $9,000, but a 10% decrease in all salaries at Rice was expected for the following year, so he was satisfied with a prospective salary of $8100, but the possibility of a salary of $6000 caused him grave concern [ibid.].

In response, Sproul and Deutsch, after conferring with Leuschner and Hildebrand and the budget committee, proposed to offer Evans a firm salary of no less than $8100, even in the unlikely event that no regular professorial salary would exceed $6000. Leuschner told his colleagues that Evans's election to the National Academy of Sciences was about to be announced, and this fact, coupled with a feeling of great urgency to land Evans and put the department on a new track, contributed to this unusual decision. The salary proposal was finally approved, and the modified offer was tendered to Evans in early May. But it was May 23 before Evans finally accepted. By this time, the budget issues had been resolved, and the salary cut was only 10%, a fact that in the end reassured Evans. He had the competing concerns of his family's financial welfare on the one hand, and on the other hand, concerns about the prospect that his salary would be much higher than that of any other faculty member. But at this point it was now so late in the year that Evans did not feel that he could honorably tender his resignation from Rice effective July 1. What was agreed was that his "acceptance" would be kept secret and not be publicly announced, and that in the coming autumn, another formal offer with an effective date of July 1, 1934, would be tendered, which he would accept and have the acceptance made public at that time [ibid.].

It is worth speculating why, after refusing a very generous offer at Harvard in 1925 and a very generous offer from Berkeley in 1927 to remain at Rice, Evans would now accept an offer from Berkeley. There are some hints in his correspondence: Evans wrote to his good friend George Birkhoff at Harvard on February 20, 1933, just after being approached by Hildebrand, that "The situation at Rice is not what it used to be and some who have been here the longest are suddenly ready to leave" [Evans, Box 6]. Then, at the conclusion of the negotiations with Berkeley, he wrote to Sproul in his acceptance letter on May 23, 1933, "I should have persisted in my refusal if certain matters, not concerning me personally, could have been arranged here. But in discussing them, I did not wish to use the threat of leaving" [Evans, Box 6]. It seems that unspecified institutional difficulties at Rice had finally overcome the attachment that Evans and his wife held for Rice and Houston.

The entire scenario shows some remarkable similarities with the events 60 years earlier, in 1881, when Professor Welcker was dismissed out of a desire to upgrade the tone and spirit of the department and reorient it toward scholarship and research under the leadership of a new chair, Irving Stringham. In the first instance, in 1881 it was a committee of the regents who did the review of the department, made recommendations for change, and recruited a new chair. In the second instance, it was faculty members in related disciplines who first persuaded the provost and the president that there was a problem in the department, and then it was a faculty review committee nominated by the budget committee of the Academic Senate that performed the same set of functions. These faculty members were motivated by self-interest, for they saw a thriving mathematics department as important for the success of their own departments and for the university as a whole. Leuschner and Hildebrand appear to have been in the lead in this effort. Deutsch, in the context of the salary guarantee to Evans, wrote that "Leuschner and Hildebrand emphasized the key importance of the work in mathematics for the fields of Chemistry, Physics, Astronomy, etc. We have, they say, long looked forward to the opportunity to place the work in Mathematics on a high plane, and most sincerely hope that we shall not lose this chance." Baldwin Woods, chair of mechanical engineering and a member of the review committee, wrote to Evans in early May stressing the opportunity to build a great mathematics department at Berkeley and that departments interested in the applications of mathematics, such as physics, chemistry, and the several engineering departments, would offer him a high degree of cooperation [ibid].

Evans's election to the National Academy of Sciences in 1933 just as he was deciding to move to Berkeley was recognition of his many achievements in research, which up to that time had lain in functional analysis, especially the theory of integral equations, and potential theory. His interest in mathematical economics developed at Berkeley. The description of his work in his memorial article [IM] cannot be improved on, and we simply quote it:

> Professor Evans' scientific interests and publications concerned three fields: functional analysis, potential theory, and mathematical economics. His first paper was on the first of these subjects and was published in 1909. During the ensuing ten years he contributed a great deal to this field. His principal results concerned certain integro-differential equations and integral equations with singular kernels. Evans received early

recognition of his work in that field when he was invited to deliver the Colloquium Lectures before the American Mathematical Society on the subject "Functionals and Their Applications." These lectures were published in the American Mathematical Society Colloquium Publications. In 1920 he published the first of his famous research papers on potential theory, a field in which he was certainly the foremost authority in this country for many years. By using the then new general notions of integration and certain classes of functions (now known as Sobolev spaces), he was able to obtain basic results on all the important problems in the field, such as the Dirichlet and Neumann problems, and to discuss the differentiability properties and boundary behavior of the solution functions. One of his most beautiful results is his proof of the existence of a surface of minimum (electric) capacity among all surfaces spanning a given curve; this was shown to be the equipotential surface of a certain harmonic function defined on a two-leaved three-dimensional space having the curve as branch-curve. Such a space is a three-dimensional analog of a Riemann surface in the complex plane. This led him into his extensive research on multiple-valued harmonic functions, his principal interest during his later years.

Evans' work in mathematical economics was that of a pioneer. At a time when economists disdained to give mathematical treatments of economic questions, he boldly formulated a model of the total economy in terms of a few macro-economic variables and proposed the related index number problem of how to define the aggregate variable in terms of these micro-economic components. Today such models are commonly used and the index number problem is still important. Evans' study of the treatment of production in terms of cost functions, after Cournot, has led to the discovery of a duality between these functions. In Berkeley, Evans held a seminar in mathematical economics which soon became internationally known, providing an inspiring educational activity and establishing a tradition of mathematical economics on the Berkeley campus. [IM]

It is clear from this summary that Evans had very broad interests in mathematics, but what is perhaps somewhat less clear immediately is that he had a strong view that mathematics, properly defined, included a broad range of applied topics, an intellectual position that was far more common in Europe than in the United States. Evans saw no boundary line between pure and applied mathematics but rather saw it as a continuum. It was also his view that these applied areas should all be included organizationally within a mathematics department as properly conceived. This belief will become very evident in Evans's sustained efforts to keep statistics within the mathematics department despite Neyman's sustained campaign to split off a separate Department of Statistics, a campaign, as will be recounted in Chapter 11, that was ultimately successful, but only after 18 years and Evans's retirement. Evans's views are also reflected in projects to develop applied mathematics after World War II under the umbrella of a comprehensive mathematics department. Evans was also supportive of Courant's efforts to build just such a mathematical enterprise at New York University, and he shared Courant's view of the breadth of mathematics. Evans may have been influenced by his war work during World War I and in the years after World War II by his work with the National Defense Research Council. His views are expressed in a speech he gave in 1953 at Berkeley on "Applied Mathematics in the Traditional Departmental Structure" [Evans, Box 6].

Evans received many awards and honors and was a deeply respected senior faculty member on the Berkeley campus for many decades as well as a major figure in American mathematics. He served as president of the American Mathematical Society for a two-year term in 1939–1940. He served as department chair at Berkeley for 15 years, from 1934 to 1949, during which time he brilliantly performed the task of reshaping the department. He supervised the doctoral work of 27 students, 11 at Rice and 16 at Berkeley. He retired in 1954, but continued to publish after retirement. He died in 1973 at the age of 86, and Evans Hall is named in his honor; see Chapter 15.

With Evans's acceptance assured, but with an effective date a year later than had been hoped, attention turned to the situation of the junior faculty. The appointment of Instructor Annie Andrews was terminated as of July 1, 1933, by recommendation of the review committee. The file notes state that she was married to a practicing attorney and would not therefore become destitute if she lost her job, and this fact apparently provided a level of comfort to those making the decision. Instructor Duncan, who was informed that he would be terminated as of July 1, 1933, made a plea for an extension based on the fact that he had a wife and small children to support. His appointment was ultimately extended through July 1, 1935, in recognition of the hardship early termination would cause. Instructor Edward Roessler was informed that his reappointment was unlikely, but a transfer to the Davis campus, which had been in the works prior to the review committee report, was arranged effective July 1, 1933. Instructor Lee Swinford was also informed that his appointment was unlikely to be renewed. It was decided to defer final action on the two assistant professors recommended for termination, Sophia Levy and Arthur Williams, although they were also informed that their reappointments were in doubt. It was noted that Sophia Levy was the sole support for her ailing mother. After much discussion over the next two years, it was decided that Swinford, Levy, and Williams were to be retained, but more about that later [UA, CU 5, series 2, 1933, 1-Mathematics].

During his trip to the East in February, Hildebrand had accumulated the names of a number of promising young mathematicians who might be candidates for hiring as a part of the reorganization. Since the appointments of Andrews and Roessler had been terminated, there was the possibility of two new instructor appointments. On June 7, 1933, Hildebrand presented a list of eleven names to President Sproul for consideration [ibid.]. On the basis of the subsequent careers of these mathematicians, one can say only that Hildebrand got some very good advice and had very good taste in selecting young mathematicians. Among those at the top of the list were Charles B. Morrey, Jr., and Alfred L Foster. Morrey received his degree at Harvard in 1931 under George Birkhoff and was on a postdoctoral fellowship at Rice University working with Evans in the area of partial differential equations. Foster had received his doctoral degree in 1930 from Princeton under Alonzo Church and was an instructor at Princeton. His interests were in the foundations of mathematics and in certain mathematical problems related to physics. Both were highly recommended by senior mathematicians and by Hildebrand himself on the basis of interviews. After consultation with Evans, both were offered instructorships at Berkeley, and they accepted, effective July 1, 1933. Thus, two of Evans's new appointments actually preceded him, arriving at Berkeley a year before he did [ibid.].

Charles B. Morrey, Jr., was born July 23, 1907, in Columbus Ohio. His father was a professor of bacteriology at Ohio State University (a personal note: the author's parents were undergraduates at Ohio State University and first met each other when they were both taking a course from Professor Charles B. Morrey, Sr.). Morrey's mother was a concert pianist and headed the Morrey School of Music, in Columbus, and was regarded as the principal pianist and teacher of piano in central Ohio. Morrey's lifelong interest in the piano was clearly a result of his mother's influence. He entered Ohio State University, receiving a BA in 1927 and then an MA in 1928. He then went to Harvard for doctoral studies, where he prospered. He is said to have written the best PhD examination on the subject of real variables ever held at Harvard [Hildebrand's report to Sproul cited above]. Morrey received his doctoral degree in 1931 with a dissertation entitled "Invariant Functions of Conservative Surface Transformations" under the direction of George Birkhoff.

Morrey flourished at Berkeley, and his research in partial differential equations won wide recognition. He was given accelerated promotion to assistant professor in 1935 to ward off an offer from outside. Promotion to associate professor came in 1938 and to full professor in 1945. In her biography of Courant, Constance Reid tells of Kurt Friedrichs' reaction to his first meeting with Morrey [Reid 1976, p. 272]:

> It was one of the New York meetings of the American Mathematical Society about 1938; and this inconspicuous-looking boy came up to me and said modestly that he wanted to tell he had been working on partial differential equations and he had solved such and such a problem. I said that was very nice, and went on. Then—wait a minute!—I suddenly turned and went back to him and said, "Would you tell me once more which problem you said you had solved?" I couldn't believe it. It was one of those problems many of us had worked on for years and years—I couldn't believe it. Oh yes, Morrey is powerful.

Friedrichs may have been referring to Morrey's important 1938 paper on nonlinear elliptic equations, but more likely was referring to Morrey's results on existence and differentiability of solutions of variational problems for multiple integrals. These results were of major importance and have been very powerful tools in nonlinear analysis, which provided the key to the solution of Hilbert's 19th and 20th problems. Morrey had announced these results at conferences in 1939 and in a *Bulletin of the American Mathematical Society* research announcement in 1940, but the results were not published in full until 1943 and in a journal that was not widely circulated. In this paper, he introduced certain function spaces that were widely used subsequently and were in effect Sobolev spaces. Morrey's memorial article argues that there is a case to be made that these spaces should be called Morrey spaces or at least Morrey–Sobolev spaces instead of Sobolev spaces. Morrey subsequently published important results on the Plateau problem, and on the regularity of solutions of nonlinear elliptic systems. A 1952 paper on quasiconvexity and lower semicontinuity of multiple integrals, perhaps not noticed so much at the time, has turned out to be of fundamental importance in subsequent work in the area. This paper was 30 years ahead of its time. Morrey also served as a mathematician at the Aberdeen Proving Grounds in the years 1942–1945 during the Second World War.

On June 28, 1937, Morrey married Frances Eleanor Moss, who had graduated in 1933 from UC Berkeley with a BA in mathematics. She wanted to be a mathematics teacher and continued her study of mathematics, receiving an MA in 1935. She and Charles met shortly after he arrived in 1933. She taught in the Oakland City schools until their first of three children was born in 1941. After 18 years she returned to teaching, at first part-time at Mills College, and then full-time at Merritt Community College.

Morrey's work was recognized by election to the National Academy of Sciences and the American Academy of Arts and Sciences. He served as president of the American Mathematical Society and was selected to give the AMS colloquium lectures in 1964. Morrey was a devoted citizen of the department, serving as chair for five years, 1949–1954, during very difficult times, and as acting chair a few years later. He supervised the doctoral work of 15 students at Berkeley, and retired in 1973. He died in 1984.

Alfred Foster was born in New York City on July 13, 1904. He attended Caltech as an undergraduate, receiving a BS degree in 1926 and an MS in 1927. He then went to Princeton for doctoral studies, where he came under the influence of Alonzo Church. He received his doctoral degree under Church in 1931 with a dissertation entitled "Formal Logic in Finite Terms." He was Alonzo Church's first doctoral student, and indeed, Church was only one year older than Foster. In 1930, Foster married Else Wagner, and they had four children. After receiving his doctoral degree he was awarded a Rockefeller International Research Fellowship, which he took in Göttingen; the following year he returned to Princeton. Then, in 1933 he was appointed as an instructor at Berkeley.

Foster's research developed slowly and centered on questions in abstract algebra. He developed with Benjamin Bernstein a duality principle according to which basic theorems in commutative algebra should have dual theorems. He also worked on Boolean algebras, and his departmental file makes several references to important work of Foster that was virtually complete when he was scooped by Marshall Stone. Given that the only overlap in the work of Stone and Foster was on Boolean algebras, it is a reasonable supposition that Foster had independently discovered some portion of the results that appeared in Stone's two long and very influential papers on Boolean algebras in 1936 and 1937. Foster then put his own work on this subject aside. Foster's subsequent work moved toward universal algebra, and he published at a regular rate. Promotion to assistant professor came in 1937, to associate professor in 1945, and to full professor in 1951. His memorial article discusses his work in universal algebra:

> In the course of this work Foster realized that the more general setting of the new area of universal algebra was more appropriate for continued development of his ideas, and in this context he developed the theory of primal algebras and in 1953 showed that the variety generated by a primal algebra has the same essential structure as the variety of Boolean algebras. The mathematician Bjarni Jonsson, in a recent paper dedicated to the memory of Alfred Foster, states: "This result is an acorn from which a mighty oak has grown," and Foster himself devoted the rest of his life to the development of this important work.

Foster supervised the work of 17 doctoral students during his career. He retired in 1972 and died in 1995 at the age of 90.

With the retirement of Haskell on July 1, 1933, and with Evans's arrival delayed until the following year, there was a need for an acting chair for a year. Charles Noble was selected for this post and was informed that Joel Hildebrand would serve as an "advisor" to him. This must have been an uncomfortable position for Noble, for he no doubt agreed with the direction Haskell had been taking and had not been consulted by the review committee. He knew that Evans would be arriving in a year with an assignment to remake the department in a new direction, but that this fact was supposed to be kept secret for several months. In addition, Hildebrand would be looking over his shoulder, representing Evans. With respect to the status of the junior faculty members whose future reappointments were still in doubt, Noble received a pledge from Provost Deutsch that no final action would occur until Evans had been involved.

The appointment of Evans was formally announced on November 1, 1933, and at that point he could be more openly involved with the affairs of the department.

Noble wrote to him at about this time, recommending that Duncan, Swinford, Levy, and Williams all be retained on the basis of their teaching and in order to "restore the morale of the department" as he saw it [UA, CU 5, series 2, 1933, 1-Mathematics]. In January 1934, Deutsch requested that Evans visit Berkeley and consult with him, President Sproul, and members of the department. Prior to his arrival, Evans wrote concerning curricular matters:

> The schedule of undergraduate classes seems to me to be too cumbersome, and that of graduate courses neither sufficiently advanced nor of adequate range. This question is less delicate [than personnel issues]; but it is a decided advantage to make changes with the cooperation of members of the Department rather than over their heads. Presumably there are also special details in the relations with the Department of Education, which I have not had to consider at the Rice Institute. Moreover I am hoping that the schedules of those young members of the Department whom we regard as having decided promise of productive scholarship will not be excessive—say, not over nine hours per week. [UA, CU 5, series 2, 1934, 1-Mathematics]

During his visit, there was clearly discussion of the personnel cases of those junior faculty members whose reappointment was in question. One possibility under consideration was the transfer of Sophia Levy to the Department of Astronomy, but Leuschner wrote on March 7 as chair of astronomy that Levy was not well matched to the needs of astronomy, since the department had to expand in the area of astrophysics, not theoretical astronomy (i.e., celestial mechanics). He added that Levy was well suited to mathematics because of her role in teacher education. Soon thereafter, Deutsch prepared a draft budget for mathematics that deleted the provisions for Duncan, Swinford, Levy, and Williams. On March 29, Chair Noble protested this action, stating that there was an agreement that there would be no changes in the status of these four faculty members until Evans had arrived and had "experience on the ground." Deutsch pointed out that Evans had been consulted during his visit in February, which met the agreed-upon condition of Evans's involvement. Noble protested vigorously that this was not his understanding, and Deutsch eventually backed down, agreeing that here had been a genuine misunderstanding between himself and Noble [ibid.].

On February 20, Evans had written to Hildebrand recommending the appointment of Ralph D. James, another of the young mathematicians that was on Hildebrand's list from the previous spring. James was born on February 8, 1909, in Liverpool, England. His family emigrated to Canada, and he attended high school in Vancouver from 1921 to 1924. He then entered the University of British Columbia, earning a BA in 1928 and an MA in Mathematics in 1930. He then went to the University of Chicago for doctoral study and received his doctoral degree there in 1932 working under Dickson with a dissertation entitled "Analytic Investigation in

Waring's Theorem." James won a National Research Council fellowship, which he took at Caltech, where he worked under E. T. Bell, and he won high praise from both Dickson and Bell. Evans noted that G. H. Hardy had also spoken highly of James and had mentioned his important work on Waring's problem. The campus responded favorably to Evans's recommendation, and James was offered an instructorship effective July 1, 1934, a post that he accepted. However, Chair Noble had not been consulted and expressed his great dismay when he discovered that a new faculty member had been added to the department without his knowledge. He asked who would be displaced in order to make this appointment, but he never received an answer, except by implication that it would be one of the four current faculty members that the administration was planning on dropping [UA, CU 5, series 2, 1934, 1-Mathematics]. James prospered at Berkeley, publishing research and working with students, but resigned from the university in 1939 to accept a professorship in Canada, but more about that later.

Starting in the autumn of 1933, the university began negotiations with Richard Courant, who had been suspended from his professorship at Göttingen by the Nazi government soon after their rise to power, and who had determined to leave Germany. Abraham Flexner, of the Rockefeller Foundation, was an intermediary. In September, Courant was offered a visiting professorship for the spring 1934 semester with a stipend of $3000. But, Courant had already accepted a position at Cambridge University, and in October he asked whether the Berkeley offer could be postponed six months. The university agreed, and the offer was renewed for the fall semester. Courant was definitely interested, especially since he had visited Berkeley in the recent past. However, he temporized in responding because he was naturally also seeking something more permanent. In January 1934 Courant was offered a two-year appointment at New York University with the possibility of permanence. Berkeley was not able to promise a more permanent arrangement that would be competitive and attractive to Courant. In particular, Evans was opposed on principle in spite of the great eminence of Courant and wrote in January 1934, "I cannot think that it is to the advantage of a university to fill major appointments with foreign professors when there are no reciprocal appointments abroad. I fear that the result is to discourage the legitimate aspirations of our young scholars, and that the process has already been carried too far." Courant went to New York University, and the rest is history. Courant retained an interest in Berkeley, and that interest resurfaced in 1949, as we recount in Chapter 8 [ibid.]. Eventually, Evans softened his stance somewhat against hiring senior foreign professors, as will be recounted.

When Evans arrived at Berkeley, a good start on rebuilding had begun: Morrey and Foster had already been in residence for a year, and James was arriving from

Caltech. Very shortly thereafter, Evans received word that the University of Virginia was about to make an offer to Morrey. Evans undertook a review of the junior faculty, and after his "time on the ground" presented his recommendations to Sproul on December 20. He recommended that the appointment of Instructor Dewey Duncan be extended for at most one more year if necessary. (In fact, Duncan departed the university on July 1, 1935.) Evans further recommended that the appointment of Instructor Lee Swinford end no later than July 1, 1937. He then recommended that assistant professor Sophia Levy be retained and that assistant professor Arthur Williams be retained but informed that promotion would be unlikely. In addition, Evans recommended that Raymond Scioberetti, who was not mentioned in the review committee report, be transferred to the Department of Astronomy either at Berkeley or at UCLA. With respect to the new junior faculty, Evans recommended the accelerated promotion of Morrey to assistant professor in light of the impending outside offer and recommended that instructors Foster and James be given salary increases and informed that their work was appreciated [UA, CU 5, series 2, 1935, 1-Mathematics].

The recommendations for the advancement of the new faculty members were acted on favorably, but no action was taken on his recommendation in regard to Sciobereti, who ultimately remained in the department until his retirement in 1954. He was promoted to associate professor in 1948. For reasons that are not entirely clear, Swinford's appointment was continued in spite of the recommendation of Evans. After 16 years as instructor, Swinford was promoted to assistant professor in 1945, and he retired in 1954. Williams remained an assistant professor until his retirement in 1950. The contributions to the department of these three faculty members lay entirely in their teaching, in line with the pre-1933 orientation of the department. The department realized the value that Sophia Levy provided beyond her teaching in her leadership in teacher education and her work with the School of Education and the state board of education. Her research on the celestial mechanics of orbits of comets and minor planets reached publication and was appreciated, even though it fell more into astronomy than mathematics, given the current division between these disciplines. She was promoted to associate professor in 1940 and to full professor in 1949.

Elmer Goldsworthy, one of the four Haskell doctoral students hired as instructor in 1929, was promoted to assistant professor in 1931 and was not mentioned in the 1932 review committee report. He was a very successful teacher and undertook part-time service as assistant dean of undergraduates in 1935. His decanal work was much appreciated. He took military leave from the university shortly after the United States entered World War II to serve as lieutenant colonel in the air force in charge of instruction at the School of Applied Tactics, at Orlando. It should be noted that he had served in the British Army in World War I and had been wounded. After World War II, he went to Caltech, where he served as assistant professor of mathematics and as master of the student houses. He remained on extended leave from Berkeley during this period, since he was undecided whether he wanted to return to Berkeley. However, his health declined during this period, and he died on April 30, 1949.

The situation at the end of the 1934–1935 academic year was that Evans had replaced Haskell as chair, the appointments of three instructors had been ended, and these had been replaced by three very promising and productive young mathematicians: Morrey, Foster, and James. These were the first three of the 21 appointments that Evans would preside over before he stepped down as chair after 15 years in 1949, and that would remake the department. It was decided that several faculty members in the junior ranks whose contributions would be limited to teaching would be continued, but with at most a scant chance of advancement. There is not much direct evidence suggesting the reasons that Evans did not make a complete clean sweep of the staff along the lines suggested by the 1932 review committee. Perhaps part of it was concern for the morale of the department, part of it compassion and reluctance to dismiss faculty members in the midst of the depression when there were few opportunities, and perhaps a realization that their efforts were needed to staff classes, particularly because enrollments were starting to increase.

The faculty in the department in the ranks from professor to instructor totaled 19 in 1934–1935. This represented only a very modest increase over the faculty level of 17 that the department had enjoyed ten years earlier. The department had three retirements approaching: Lehmer and Noble in 1937 and Irwin in 1938. While the department did not have reasonable expectations of growth, it was expected that these upcoming retirements would permit replacements to be hired.

During his first year at Berkeley, Evans proposed the establishment of a new mathematics journal, the *Pacific Journal of Mathematics*. He proposed that it be published and financed by the UC Press, and he sought joint sponsorship from the mathematics departments at other institutions, each of whom would provide a yearly subvention of $100 to help underwrite costs. He cited the establishment of the *American Journal of Mathematics* early on at Johns Hopkins and the more recent establishment of the *Duke Mathematical Journal*, arguing that creation of such a journal would give a boost and provide visibility to mathematics departments on the West Coast. The UC Press turned him down for budgetary reasons, and he never got beyond recruiting UC Berkeley and UCLA as sponsors. The idea of establishing a West Coast journal was a good one, but it was a bit early. In 1949, the *Pacific Journal of Mathematics* started publication with a roster of West Coast sponsoring institutions.

It may seem amazing by standards and expectations of today that the mathematics department did not have any budgeted clerical assistance when Evans arrived. The files indicate that all correspondence from the department chair to the administration was written in the hand of the chair. The chair personally answered the telephone and responded to any requests for information or assistance. Evans was able to obtain some temporary assistance in the spring of 1935, and he reiterated his request for permanent provision for clerical assistance. In the spring of 1936, the administration suggested that the mathematics and French departments share a stenographer. Evans resisted this idea, and finally, realizing that the department was a large one in terms of the size of the faculty and number of students, the administration agreed to provide a half-time stenographer position [UA, CU 5, series 2, 1936, 1-Mathematics].

Evans was able to recruit Miss Sarah Hallam to fill this position starting in the summer of 1936. Sarah Hallam had graduated from Reed College in 1935 with a major in mathematics, and was working on the Reed College campus. She was interested in pursuing graduate study in mathematics, and she was accepted as a graduate student in the department for the fall of 1936. Evans offered her half-time employment as stenographer at a salary of $400 per year [ibid.]. Hallam received her master's degree, and eventually Evans succeeded in convincing the campus to increase Sarah Hallam's position to a full-time one, and in 1941 she worked full-time with a yearly salary of $990. Over the years, the department's administrative support needs grew and additional staff were added. Sarah Hallam became the manager in charge of the departmental staff, and also served as budget officer and advisor to the department chair for many years until her retirement in 1975. Upon her death in 1994 the department received a bequest of $300,000 to endow the Sarah Hallam Graduate Fellowship.

In the spring of 1935, Evans became aware through correspondence with R. D. G. Richardson, of Brown University, of an opportunity to bring Hans Lewy to Berkeley. Richardson was the long-time secretary of the American Mathematical Society and served as the graduate dean at Brown. He was a very influential figure in American mathematics at the time. Lewy had been displaced from his position in Germany shortly after the Nazi regime came to power and had emigrated to the United States in 1933. He had held a temporary two-year position at Brown University that was coming to an end. Richardson had a proposal for Berkeley. Berkeley was asked to make the following commitment: if Lewy lived up to the high expectations everyone had of him during a two-year temporary appointment at Berkeley, then there would be a reasonable expectation that Lewy would be offered a permanent appointment. In that case, Richardson felt that external funding could be obtained for the two-year appointment. The plan was to obtain funding from the Rockefeller Foundation and the Emergency Committee in Aid of Displaced German Scholars. University funds were short, and Evans was still trying to free up money by arranging the transfer of assistant professor Sciobereti to astronomy either at Berkeley or UCLA. This effort was ultimately unsuccessful, but with the anticipated retirements of Professors Noble and Lehmer in 1937, Evans could look forward to having the funds required to be able to add Lewy to the regular faculty at the end of the two-year probationary period. Evans was successful in persuading President Sproul to move in the direction of making an advance commitment in anticipation of a retirement. Thus Lewy was offered appointment effective July 1, 1935, as lecturer under the terms outlined by Richardson at a salary of $3,000. Half of the salary was provided by the Rockefeller Foundation and half by Berkeley [UA, CU 5, 1935, 1-Mathematics].

A few months after Lewy's arrival in Berkeley he wrote to his sponsor Richardson. On January 17, 1936, Richardson responded in a way that reflected conceptions on the East Coast about Berkeley. Richardson wrote in part,

> We miss you here but are happy to know that you are fitting into the situation in California. There is a possibility of making that one of the great centers of mathematics in the country such as Harvard, Princeton, and Chicago already are. It would be an eye-opener to Easterners to go west for postdoctoral training. I hope that the National Research Fellowships of the new vintage will encourage this migration to different regions of the country. The undergraduate teaching is the main work of nearly everyone in academic work in mathematics in America, and this is universally true of the younger men. We have to stand or fall by our success in this particular. [Lewy, Carton 1]

Lewy's letter to Richardson is not preserved, but it is not difficult to imagine what Lewy had written that prompted Richardson's response about Berkeley and teaching responsibilities.

Hans Lewy was born October 26, 1904, in Breslau and received his PhD at Göttingen in 1926 under Courant with a dissertation entitled "*Über einen Ansatz zur numerische Lösung von Randwertproblemen.*" He served as *Privatdozent* at Göttingen from 1927 until his dismissal from the University in 1933, shortly after Hitler came to power. By the time he arrived in the United States, he had made seminal contributions to partial differential equations, including work on numerical stability, the analytic nature of solutions of elliptic equations, and nonlinear hyperbolic equations. He built on these early contributions in his subsequent work, and it was clear just a few months after his arrival at Berkeley that the department would want to add him to the regular faculty. He was appointed assistant professor in 1937 and was promoted (with great enthusiasm) to associate professor in 1939, and to full professor in 1946. He became a naturalized US citizen in 1940 and served as a mathematician at the Aberdeen Proving Grounds during the war.

His research included important estimates for the Monge–Ampère equations, waves on sloping beaches, free boundary and cavitation problems, and his display of a partial differential equation of a basic kind that did not have any solutions of any kind—a result that astonished the mathematical community and that remains important to this day. His research also included work on reflection principles, hulls of holomorphy, as well as on many other topics. The combination of Lewy and Morrey along with Evans himself ensured the preeminence of Berkeley in partial differential equations for decades to come. Lewy's work was recognized by his election to the National Academy of Sciences and the American Academy of Arts and Sciences, and by the award of the Wolf Prize in 1985. He married the former Helen Crosby, a political analyst who had been a member of the postwar mission to monitor the elections in Greece in which Jerzy Neyman had participated. She had worked for the Office of Strategic Services during the war [Reid 1982, p. 207]. The Lewys had one son, Michael. Lewy supervised the work of ten doctoral students during his career. He retired in 1972, but continued to publish papers actively until his death in 1988.

In the late fall of 1936 and early months of 1937, Evans, in a long memorandum to Sproul, outlined his ideas about a possible senior appointment that would advance the standing of the department. The discussion was somewhat connected with the three upcoming retirements of Lehmer, Noble, and Irwin in 1937 and 1938, but Evans, citing the substantial increase in mathematics enrollments, was thinking about expansion of the departmental faculty. Student credit hours in the fall semester had increased from 4981 in 1933 to 7015 in 1936. Evans analyzed the possibilities of two possible appointees from Europe in whom he was interested: Karl Menger (Vienna) and Constantine Carathéodory (Munich). But, he wondered whether bringing mathematicians from Europe would foreclose opportunities for promotion of the staff, a point he had already made in connection with Courant. He went on to discuss and analyze possible appointments of American mathematicians as well as some mathematicians who had recently emigrated from Europe. The names he mentioned are R. L. Moore (Texas), J. F. Ritt (Columbia), M. H. Stone (Harvard), A. A. Albert (Chicago), T. Rado (Ohio State), J. D. Tamarkin (Brown), G. Szegő (Washington University in St. Louis), H. A. Rademacher (University of Pennsylvania), and L. V. Ahlfors (Harvard). Evans stated that he was looking for candidates who were either "a member the National Academy or obviously destined for such membership" [UA, CU 5, series 2, 1936, 1-Mathematics]. Evans had set his sights high in this exercise, but nothing came of it, for as described in the following chapter, the subsequent appointments, with the exception of Neyman, and Tarski, were at junior levels.

Chapter 6

The Evans Years: 1936–1945

As early as 1935, Evans turned his attention to the field of statistics with the intention of recruiting a major figure in this field to Berkeley in the mathematics department and creating a center for statistics in the West. He had become convinced of the intellectual need to develop this field and wrote to President Sproul proposing a search. Many others on campus shared the view that Berkeley needed to create a center for statistics. One suspects that R. A. Fisher—formerly of Rothamsted, England, but at the time serving as the Galton Professor at University College, London—would have been Evans's top choice for this appointment. Evans knew Fisher from the summer of 1931, when they were both teaching at the University of Minnesota, and thought very highly of him [Evans, Box 1]. Evans nominated Fisher to be a Hitchcock lecturer at Berkeley as perhaps a first step in recruitment. Fisher spent three weeks on campus in September 1936 presenting his Hitchcock lectures. However, these lectures received a rather lukewarm reception from many colleagues on campus, and this unenthusiastic reception of his lectures dimmed the prospect that Evans could sell Fisher to his colleagues across the campus for this major appointment. Fisher had been offered a professorship at Iowa State University in 1936 but had turned this offer down. Evans then sought Fisher's advice on candidates, and Fisher recommended Frank Yates, who was his successor at Rothamsted. Evans also contacted Egon Pearson, as well as S. S. Wilks (an ex-Texan) and Harold Hotelling, for advice. His intent was, as he put it, to attract someone of the caliber of R. A. Fisher [ibid.]. At one point, Yates appeared to be at the top of Evans's list, but he did not remain there.

Evans then received two additional recommendations from Berkeley colleagues. Raymond Birge, chair of physics, recommended Jerzy Neyman, then at University College, London; and Carl Sauer, chair of geography, recommended Willi Feller, of Princeton. Birge had heard about Neyman from his longtime friend and coauthor W. E. Demming, who had a very high opinion of Neyman [ibid. and Reid 1982, p. 141]. In the spring of 1937 Neyman had visited the United States for six weeks on a speaking tour but did not come to California. In the fall of 1937 Evans determined that Neyman would come closest to meeting the needs for statistics as he saw them. Sproul blessed the choice, and a lengthy negotiation with Neyman commenced. After a campus review, Neyman was formally offered a full professorship with a salary of $4500. Neyman wanted a larger salary, but the major part of the negotiations involved the creation of a statistical laboratory with its own space and the provision of equipment and assistants. There was also a competing offer to Neyman from the University of Michigan that complicated matters. But the saga finally ended with Neyman arriving in Berkeley as professor of mathematics in the summer of 1938. In Evans's words, "it was quite a hunt" [ibid.].

Jerzy Neyman was born on April 16, 1894, in Bendery, Russia. His family on both sides considered themselves Polish patriots, even though there had not been a Poland for over a century, the territory having been partitioned among Russia, Germany, and Austria–Hungary. Neyman's father was a lawyer and a judge, and the family moved around quite a bit in southern Russia and the Crimea. When his father died in 1906, his mother moved to Kharkov, and Neyman subsequently enrolled in the University of Kharkov to study mathematics. When World War I began, he was rejected for military service in the Russian army because of his bad eyesight, so he was able to continue his studies without interruption. After receiving his baccalaureate degree in 1917, he continued on for graduate work in mathematics with the goal of pursuing an academic career. However, the turmoil and chaos of revolution and civil war complicated matters, and Neyman was arrested and imprisoned more than once, presumably because he was Polish, and Russia and Poland were at war. During this period he married a Russian woman, Olga Solodovnikova. Finally, in the summer of 1921, he, his wife, and other members of his family emigrated to the newly created state of Poland as part of an exchange of nationals arranged by the two governments.

Neyman immediately went to Warsaw with the intent to study for a doctoral degree under Wacław Sierpiński. He concentrated his work on probability theory and received his doctorate in 1924 under Sierpiński with a dissertation entitled "Justification of Applications of Calculus of Probabilities to the Solution of Certain Questions of Agricultural Experimentation." But he claimed that his real "mathematical father" was Lebesgue. Neyman received fellowship support for postdoctoral work at University College, London, with the intent to study statistics with Karl Pearson. He met and interacted with many scholars, including R. A. Fisher and W. S. Gosset (a.k.a. Student), and then began a long and fruitful collaboration with Egon Pearson, Karl Pearson's son. This long-distance collaboration between Pearson in England and Neyman in Poland was helped along by short visits. In 1934, Pearson was able to invite Neyman to come to University College for three months, an invitation that Neyman accepted with enthusiasm. Poland was descending into chaos, and for several reasons, Neyman's prospects for a professorship in Poland seemed dim. This short visit turned by stages to a longer visit, and finally Neyman was given a permanent faculty appointment at University College in 1935. For more information on Neyman's life, see Constance Reid's biography, *Neyman: From Life* [Reid 1982].

Neyman's goal from the outset in coming to Berkeley was to create and head an independent department of statistics. Realizing that he would have to get there in stages, he began with the statistical laboratory, but he probably did not think that it would take 17 years to achieve the final goal. The statistical laboratory was established in space separate from the mathematics department, and Neyman was provided with a secretary, funds for assistants ("computers," as they were called), and equipment. Evans was supportive of Neyman's never-ending requests, and a substantial portion of the correspondence between the department and the president that is contained in the archives concerns the statistical laboratory.

It was Neyman's work with Egon Pearson in the 1930s in which they developed the modern theory of hypothesis testing that brought him fame. His research at Berkeley was devoted more to applications of statistics to the sciences. His work on cosmology and the distribution pattern of galaxies, on many issues concerning public health, and on weather modification won him wide recognition. His work was recognized by his election to the National Academy of Sciences and the American Academy of Arts and Sciences, and in 1968 by the award of the National Medal of Science. He attracted many students, and in a sense educated a whole generation of statisticians; altogether, he supervised the work of 38 doctoral students, and his students included many future faculty members at Berkeley: Erich Lehmann, Joseph Hodges, Evelyn Fix, and Lucien Le Cam. George Dantzig was an early student of Neyman, and while he was briefly on the Berkeley faculty, he went to Stanford, and is justly famous for work in operations research as discoverer of the simplex method. According to the "Mathematics Genealogy" website, Neyman has 805 doctoral "descendants" as of this writing. He worked unceasingly, as we shall see in subsequent chapters, first to create a center for statistical research and then a department of statistics. He reached the mandatory retirement age of 67 in 1961, but he was recalled to full-time active duty every year, and he was vigorously engaged in research and teaching up to his death in 1981 at the age of 87.

In 1938, Abraham Wald had been dismissed from his position at the University of Vienna because he was Jewish, and he came to the United States with a fellowship at Columbia. He appealed to Neyman in March 1939 in the hope of obtaining a position at Berkeley. Neyman tried to persuade Evans to hire him, but without success. Wald then accepted a position at Columbia that Hotelling had arranged for him [Reid 1982, p. 165]. It is interesting to speculate how statistics at Berkeley might have developed had Neyman succeeded in having Wald appointed. We shall return to Neyman and the development of statistics at Berkeley shortly.

In 1937, Evans hired Raphael Mitchel Robinson as an instructor in the department. According to his memorial article, Robinson

> was born on November 2, 1911, in National City, California, and was the youngest of four children of Bertram H. Robinson, an atypically peripatetic lawyer who wrote poetry, gave his sons romantic names, and ultimately drifted away. His mother, Bessie Stevenson Robinson, supported the family as an elementary school teacher. Robinson attended the University of California, Berkeley, where he took a BA in 1932, an MA in 1933, and a PhD in December 1934.

Robinson was a brilliant student who completed his doctoral degree in record time. His dissertation, written under the direction of John McDonald, was entitled "Some Results in the Theory of Schlicht Functions." He was unemployed in the spring of 1935, and in the fall of that year landed a half-time instructorship at Brown University. The half-time salary forced him to live in poverty, and he contracted tuberculosis. After two years in part-time status, his appointment to a full-time position and return to California was no doubt a blessing.

Robinson was an extraordinarily gifted mathematician with very broad interests. Over his long career he made significant contributions to complex analysis, geometry, number theory, combinatorics, set theory, and logic. The span and the depth of his work are amazing. He was a very successful teacher and was promoted to assistant professor in 1941, to associate professor in 1946, and to full professor in 1949. In 1951, he was the first to code a successful computer program to test very large Mersenne numbers for primality. In fact, the program was for the SWAC computer at UCLA (see Chapter 8), and Robinson programmed from the manual without ever seeing the machine. The program was error-free and ran the first time, something that at the time was a remarkable feat. In the 1990s he gave a lecture to the department in which he described six problems he had solved or theorems he had proved, each in a different area of mathematics, and each from a different decade of his career at Berkeley. The performance was a tour de force. S. S. Chern, who joined the department in 1960, once remarked to Constance Reid that he felt that Robinson was a much underrated mathematician, adding "he could solve problems that I could not solve" [Constance Reid, personal communication]. Chern's view of Robinson is shared by many others who knew him. Robinson supervised the doctoral work of only one student during his career, and retired in 1971, several years before reaching the age of mandatory retirement. But he continued to publish actively in research journals until his death in 1995, after 65 years of affiliation with Berkeley as student and faculty member.

In 1939, one of the students in an undergraduate course in number theory taught by Robinson was Julia Bowman. Teacher and student became friends during long walks, and the friendship subsequently warmed into a courtship. Julia graduated in June 1940 and enrolled as a graduate student in mathematics in the fall. She and Robinson were married in December of 1941. Julia went on to receive her doctorate under Alfred Tarski and complied a distinguished record of research, sharing in the solution of Hilbert's tenth problem. She taught from time to time on a temporary basis in the department, and in 1976 she was appointed to a professorship at Berkeley. She was the first women mathematician to be elected the National Academy of Sciences and the first woman to serve as president of the American Mathematical Society. We will return to Julia Robinson and her career in greater detail in Chapter 18.

In 1939, Evans appointed Tony Morse as instructor of mathematics. Anthony Perry Morse was born in 1911 in Ithaca, New York, and graduated from Cornell University in 1933 with a degree in mathematics. As an undergraduate he had come under the influence on R. P. Agnew and J. R. Randolph. He then went to Brown University for his doctoral work and received his degree there in 1937 under Clarence R. Adams, who in turn was a student of George Birkhoff. Morse's dissertation was entitled "Convergence in Variation and Related Topics," and he was described to Evans as the best student at Brown in several years. It seems certain that Morse and Raphael Robinson knew each other while Robinson was an instructor at Brown. After completing his degree, Morse won a fellowship for postdoctoral work at the Institute for Advanced Study in Princeton. During this period he worked on measure theory and abstract analysis and laid the groundwork for his subsequent work on differentiation of measures and on product measures.

Morse was a successful teacher, and his work in measure theory progressed well. He was promoted from instructor to assistant professor after two years of service in 1941, to associate professor in 1946, and to full professor in 1949. Morse spent the war years as a mathematician at the Aberdeen Proving Grounds. Morse in his later years also worked on developing a new set theory different from the Zermelo–Fraenkel and the von Neumann–Bernays–Gödel theories that merged logic and set theory. His memorial article adds the following information:

> Morse's set theory is definitely stronger than each of these two systems (and so there inheres in it a somewhat greater risk of contradiction). It is not finitely axiomatizable, admits a universe that belongs to no class, and has a classification theorem scheme that says that for any proposition whatsoever there is a class that consists of precisely those members of the universe for which the proposition holds. The theory is very usable, very flexible, and with its concurrent treatment of logic and its use of formal language makes a complete framework for mathematics.

Evans saw Morse and Robinson as very much on a par with each other and made considerable effort, as is evident from correspondence, to see that their promotions up the ladder occurred congruently. Soon after the war, their work attracted attention that led to outside offers of full professor to both of them: Brown and Wisconsin tried to lure Morse in 1949, while Pennsylvania tried to lure Robinson also in 1949. These offers led to their promotion to full professor at Berkeley. Morse was more active than Robinson in doctoral training and had 11 doctoral students. His first Berkeley student was Herbert Federer, who finished in 1944 and went on to a very distinguished career. He was the second Berkeley doctoral student to be elected to the National Academy of Sciences, in 1974. (The first to be elected was George Dantzig in 1971, followed by Federer, then Julia Robinson in 1976, Erich Lehmann in 1978, and others later on.) Morse retired from active service in 1972 and died in 1984.

A third appointment from this period was that of Derrick Henry Lehmer, the son of Derrick Norman Lehmer, a long-time Berkeley faculty member in mathematics who had retired in 1937. D. H. (or Dick) Lehmer was born on February 23, 1905, in Berkeley and received his BA degree with a major in mathematics from UC Berkeley in 1927. He then left to go the University of Chicago for graduate work to study under Dickson. According to [Lehmer], Lehmer did not like working under Dickson and left after one year to go to Brown University. He received his degree there in 1930, with a dissertation entitled "An Extended Theory of Lucas Functions," which was completed under the direction of J. D. Tamarkin. Lehmer worked rather independently, since Tamarkin was not an expert in number theory. In fact Lehmer's dissertation was sent to E. T. Bell at Caltech for an evaluation before it was accepted. While already an undergraduate, Lehmer had been interested in mechanical devices for computing linear congruences that would function as sieves for factoring large numbers. With his father, he oversaw the construction of several such devices. The first one, in 1927, used bicycle chains, and is now in the Computer Museum in San Jose, California. He and his father demonstrated a more-advanced electrooptical device of this sort at the 1933 Chicago World's Fair, which is described in a 1933 publication of the Carnegie Institution.

After completing his PhD, Lehmer won a National Research Council fellowship, which he took at Stanford and then Caltech, where he worked under E. T. Bell. He was a National Research Fellow at the Carnegie Institution in 1932–1933 and then was a member of the Institute for Advanced Study in 1933–1934. He then joined the faculty at Lehigh University as instructor in 1934, and was promoted to assistant professor in 1937.

While an undergraduate at Berkeley, Lehmer had met his future wife Emma Trotskaya. She was born November 6, 1906, in Samara, a city on the Volga River in

Russia. Her family moved to Harbin, Manchuria, in 1910 where her father served as the far eastern representative of a Russian firm. They were thus spared the trauma of World War I and the revolution, and they remained there as expatriates after the war. Emma had originally hoped to return to Russia for college, but the purges and famine in Russia made that impossible. She instead looked to the US, and she applied and was admitted to UC Berkeley in 1924 as a freshman. She developed an interest in mathematics and then decided to major in it. During the summer of 1926, following her sophomore year, she got a job assisting Professor Derrick Norman Lehmer on a research project in number theory. While working on this project, she soon met and worked with Derrick Henry Lehmer, who was one year ahead of her but also a math major. Their friendship ripened into love, and Dick and Emma were married April 20, 1928, after Dick had returned to Berkeley following his year of graduate study at the University of Chicago. After a trip to Manchuria to meet her family, they returned and both went off to Brown University, where Emma enrolled in the master's program at Brown, winning her master's degree while Dick completed his doctoral degree. Emma and Dick had two children, born in 1932 and 1934.

Evans first considered Lehmer as a candidate for appointment at Berkeley in December 1937. (For documentation for the following four paragraphs, see [UA, Lehmer file].) Prior to his father's retirement a few months earlier, the university's nepotism rules would have precluded such an appointment. Evans wrote to Sproul about the possibility of appointing the younger Lehmer, but Sproul responded negatively, asking why the department needed two faculty members in number theory instead of just the one that they already had, Ralph James. It is not clear what Sproul may have had in mind in addition to the notion of limiting the department to one number theorist. Was he worried about inbreeding, as had happened at the end of Haskell's tenure in the 1920s? The record is silent. Evans responded to Sproul in turn that appointing pairs of faculty with parallel interests was an appropriate and quite effective way to build the research strength of the department. He cited the examples of Morrey and Lewy in partial differential equations and of McDonald and Wong in algebraic geometry. Evans must have sensed a level of opposition from Sproul, who was well known for forming opinions from which he was not easily dislodged. He dropped the idea of appointing Lehmer for the time being.

The possibility of appointing D. H. Lehmer next arose in the spring of 1939. Ralph James had been promoted to assistant professor in 1936, but in the spring of 1939 he was approached by the University of Saskatchewan, in Saskatoon, and was asked to consider an appointment there. The president of that university visited James in Berkeley to make the pitch that James accept appointment as full professor and chair of the department at a salary of $4200, a substantial increase over his

Berkeley salary of $2700. James, in spite of pleas from his friends and colleagues and their advice that he was making a big mistake, decided to accept this offer to return to Canada. After several years in Saskatoon, James moved to a professorship at the University of British Columbia, which was his undergraduate alma mater.

E. T. Bell, at Caltech, wrote to Evans on May 3, 1939, bemoaning the loss of James, in whom he had been interested in recruiting to Caltech, and Bell went on to suggest the appointment of Lehmer. Bell also mentioned a promising young mathematician who was receiving his degree at Caltech that year, Robert Dilworth, as another possibility. Evans decided to renew his request for the appointment of Lehmer, but now at the associate professor level, and requested outside letters from Dickson (Chicago), Rademacher (Penn), Robertson (Princeton), and Tamarkin (Brown). In spite of the powerful evidence presented by these referees in favor of the appointment and a favorable recommendation from the budget committee, Sproul again had doubts about the appointment, and said that he was reluctant to make direct appointments as associate professor. It was also quite late in the year, August, before the decision was made, so an appointment in 1939 was probably not possible anyway [UA, Lehmer file].

In the fall, Evans renewed the request for the appointment of Lehmer effective July 1, 1940, but now as assistant professor, with a general understanding that if things worked out, promotion to associate professor would follow in a year. During a trip to the East in November, Sproul asked Lehmer to meet him for an interview. The interview apparently went well, for almost immediately Sproul offered Lehmer appointment as assistant professor effective July 1, 1940. Lehmer accepted with the message, "You do not know what this means to me" [ibid.]. Evans recommended the promotion of Lehmer to associate professor for 1941, but the recommendation was deferred for a year; promotion was finally granted in 1942. Lehmer was promoted to full professor in 1947 and went on to a distinguished career at Berkeley.

He made contributions to many different areas of number theory, but one of his predominant interests was in computation. His bibliography contains more than 175 items ranging over many topics. Primality testing was a major area of interest and contribution, as were factoring of integers and various questions about the distribution of primes. His work on the partition function was widely recognized, as was his work on the zeroes of the Riemann zeta function. He was an early user of computers on this problem, and by 1956 had verified by computation that the first 25,000 zeros of the zeta function lie on the critical line. Today, of course, modern computational power has allowed workers to go many orders of magnitude beyond this result, but at the time, Lehmer was the pioneer. His approach was computational and constructive, as he wrote himself,

When I was young, mathematicians still had a slight feeling of guilt about a proof that established only the existence of a solution of a problem, without supplying a constructive procedure for finding the solution. Many times I was told that the existence proof was only the first step, and now that the solution is known to exist, very soon some, presumably less pure, mathematician would come up with a constructive solution. Meanwhile the thing to do is to generalize the existence proof to Banach space and beyond. In almost all cases I am still waiting for the second man. Often I am asked the rhetorical question "How can you even begin to solve a problem until it is known that the solution exists?" The answer to this smug question is "Easy, I have done it many times." [Lehmer]

The pamphlet [Lehmer] contains more detail on the life and achievements of all the Lehmers.

Lehmer received a number of honors for his work, including selection as Gibbs Lecturer of the American Mathematical Society, and he supervised the doctoral work of 19 students, many of whom are prominent number theorists today. As we shall see later on in the narrative, he was a determined advocate of developing computational facilities at Berkeley. Lehmer served as chair of the department from 1954 to 1957, just as Berkeley was beginning to emerge as a top-ranked department. He retired in 1972, but remained active in research and computation until his death in 1991.

His wife, Emma Lehmer, was also an important figure in the Berkeley Mathematics Department. Emma began publishing mathematics papers shortly after receiving her master's degree in 1930. Her publications included three short notes in the *Bulletin of the American Mathematical Society*, followed by a paper in the *Annals of Mathematics* in 1938 and then a *Bulletin* note jointly authored with her husband in 1941—all of this of course with small children in the house. The Lehmers spent the 1945–1946 academic year at the Ballistic Missile Research Laboratory at Aberdeen where Dick was working with the ENIAC computer that was designed and used to compute ballistic trajectories. But, on some weekends the Lehmers could use it as a numerical sieve. Emma recalls that, "When they could arrange child care, they would stay at the lab all night long while the ENIAC processed one of their problems. They would return home at the break of dawn" [Lehmer].

After her 1941 paper Emma did not publish anything until 1951, when she resumed publishing mathematical papers, and a steady stream of publications in number theory followed. The last one appeared in 1993 when she was 87. In over 60 years of married life, which combined devoted family life as well as mathematics, the Lehmers co-authored 11 papers in number theory. But, these form only a small part of Emma Lehmer's 56 total publications. Emma was also widely known for her fine translation of Pontrjagin's book on topological groups. Once Dick held a faculty appointment, the university's nepotism regulations did not permit her to hold a faculty position except for some short-term visiting positions to meet teaching needs. By the time these regulations wee eliminated in 1971, both were virtually at the age of mandatory retirement. In any case, Emma never felt excluded from the mathematical community, and indeed was a vital part of it. She traveled with her husband to mathematical conferences around the world and had many research accomplishments [Lehmer]. Quoting from the article in [Lehmer], which is based on interviews with her, "Emma Lehmer considers that she is quite fortunate in the way her career turned out. She would have liked to teach more (she taught some during World War II under special wartime exceptions to the university nepotism rules that usually prevented more than one members of a family from holding a faculty position). She considered that not having to teach freed her up to do research." Today she lives alone in the house in the Berkeley Hills that she shared with Dick, and will turn 100 as this account appears in print.

Evans's perseverance in the matter of the appointment of Lehmer over several years was rewarded with success and was of great benefit to the department in the following years. In 1940, after six years as chair, Evans had made or overseen eight continuing faculty appointments: Morrey, Foster, James, Lewy, Neyman, Robinson, Morse, and Lehmer. One, James, had been lured away, but the other seven remained at Berkeley for their entire careers. The first three had been in a sense replacements for the three junior faculty members let go in the reorganization in 1933–1934, while the remaining appointments were seen generally as replacements for the three senior members, Noble, Lehmer, and Irwin, who retired in 1937 and 1938. Evans had succeeded in expanding the continuing faculty by one position to 19, but he also had the opportunity to appoint several instructors who served on a temporary basis to replace faculty members who were on leave or who had resigned and to help with the expanding workload. These included John Hurst in 1937–1938, Joel Brenner in 1939–1940, and Ronald Shepard in 1941–1942. In addition, Francis Dresch, Evans's own doctoral student, was hired in 1938 and served until he went on leave in 1941 in the armed services for the duration of the war. He did not return to Berkeley.

Neyman's statistical laboratory continued to grow in size and independence. By the 1941–1942 academic year, it had become a separate budget line of $3850 within the department's overall budget of $96,626.21. Neyman's professorial salary was included within the departmental budget; the $3850 was the equivalent of salaries of nearly two beginning instructors. He had attracted a number of students to work in the laboratory. The payroll included Joseph L. Hodges and Julia B. Robinson as computers, Evelyn Fix as technical assistant, Elizabeth Scott as research associate, and Mark Eudey as research assistant and associate. All but Julia Robinson went on to receive their doctoral degrees under Neyman, and except for Mark Eudey, they became tenured faculty members at Berkeley. Neyman continued to agitate for an independent department of statistics, and while Evans has been generous in his support of the statistical laboratory, he remained firmly opposed to a separate department. He believed that statistics was part of mathematics and should remain in the mathematics department. The statistical laboratory was a great success, and Neyman received much praise from members of many academic departments for the statistical consulting assistance that he provided. He was attracting capable graduate students who would later earn their degrees and form a nucleus of the future faculty.

Neyman also felt a keen need at this time (1941–1942) to have someone mathematically trained to help with graduate education and the theoretical aspects of the consulting work that the laboratory was doing for other members of the faculty on campus. He thought of trying to attract Joseph Doob, but that was not possible for budgetary reasons, so his thoughts in 1942 turned to Doob's students. Doob, who was at the Institute for Advanced Study, in Princeton, wrote that he had only three students who would be suitable and that two of them were already placed. The third, David Blackwell, was with him at the institute, and Doob recommended him to Neyman warmly, describing him as the best student at Illinois the previous year. Neyman took to the idea and approached Evans. Evans consulted with Sproul and then queried Neyman to get confirmation that Blackwell was the best person available. Evans then said to Neyman, as Constance Reid writes in her biography of Neyman, "So let's have him." However, Blackwell's appointment foundered on the color of his skin. Reid, in the biography of Neyman, recounts what happened [Reid 1982, p. 182]:

> Shortly after this conversation, before a formal offer could be made to Blackwell, Neyman learned that the wife of a mathematics professor, born and bred in the south, had said that she could not invite a negro to her house or attend a department function at which one was present.

And now Reid quotes Neyman:

> Then I do not remember who started it, but someone, one of the old fellows said, "Well now, Professor Neyman, don't you think it would be a bad idea to have our department split?" And to my great regret, I said—but I was new you see, "Well, I certainly do not want to split the department."

Neyman was of course far from the only person who shared the deep regret and embarrassment that these efforts to bring Blackwell to Berkeley ended in this fashion. The faculty wife mentioned in Reid's narrative was of course Mrs. Evans. Blackwell was to join the Berkeley faculty a few years later, first as visiting professor in 1954–1955 and then as professor beginning in 1955, and was to be one of the stars of the Berkeley faculty for many decades. After Blackwell joined the faculty, the Evanses entertained the Blackwells in their home, and the two couples spent a pleasant evening together [David Blackwell, personal communication].

Although disappointed with this failure to recruit Blackwell, Neyman turned to long-term thinking and planning for the future of statistics at Berkeley and compiled a list of people he wanted to attract to Berkeley after the war. The list included Blackwell, Abraham Wald, Willi Feller, P. L. Hsu, Henry Scheffe, and George Pólya. Neyman had the highest regard for Wald, and considered him the "ideological successor" of himself and Egon Pearson [Reid 1982, p. 210].

After the appointment of Lehmer, which was consummated in December 1939, there were no faculty changes until 1942. As already noted, associate professor Bing Wong suffered a serious stroke in January 1942 that left him paralyzed and unable to speak. Also, Thomas Putnam was in declining health and was scheduled for a leave of absence in 1942–1943. However, he died on September 22, 1942. The defining event at the time was the attack on Pearl Harbor and the United States' entry into World War II in December 1941. It was expected that a number of faculty members would be on war leave. Indeed, Elmer Goldsworthy and Francis Dresch served in the armed forces, and Charles Morrey, Hans Lewy, and Anthony Morse spent much of the war on assignment as mathematicians at the Aberdeen Proving Grounds, in Maryland. Consequently, even with decreasing enrollments because of the war, the department would be shorthanded. Evans received permission from President Sproul on March 14, 1942, to recruit two temporary instructors at salaries of $2000 starting only a few months hence. There would be no guarantee of employment beyond the duration of the war, since the university wanted to be able to guarantee reemployment to anyone on war leave. Evans used this opportunity to make two fine appointments on short notice that in the end turned out to become permanent: František Wolf and Alfred Tarski. What was perhaps not foreseen at this time was the GI Bill and the resulting enormous increase in enrollment that would ensue after the war, with the need for an even larger faculty above and beyond those returning from war service.

The Wolf appointment was the more straightforward of the two, and it will be discussed first. František (or Frank, as he was known) Wolf was born November 30, 1904, in Prostějov, which was then in the Austro-Hungarian Empire but is now in the Czech Republic. He attended Masaryk University, in Brno, which was about 60 miles from his hometown. (Kurt Gödel was born in Brno on April 28, 1906, so the two are near contemporaries.) Wolf was a brilliant student and graduated with a *Rerum Naturarum Doctor* in 1928 under the mentorship of Otakar Boruvka. His dissertation was entitled "*Contributions à la théorie des séries trigonométriques généralisées et des séries à fonctions orthogonales.*" He then became a professor of mathematics in the Czech secondary-school system in Prague. His first mathematical paper was on trigonometric series and was published in 1931. Then, in 1937 he was appointed as Privatdozent and lecturer at the Czech University of Prague. Simultaneously, he was awarded a Denis fellowship to spend the calendar year 1937 studying at Cambridge University, where he came under the influence of G. H. Hardy, J. E. Littlewood, and A. S. Besicovitch.

Wolf returned to Prague in January 1938, and was promptly inducted into the army. Later that year, the German invasion of Czechoslovakia resulted in the closing of Czech universities and made teaching and research impossible. Wolf was a passionate Czech patriot, but conditions were such that he decided to try to leave the country. He had been invited to give a series of lectures at the Mittag-Leffler Institute, and the German occupation authorities granted him permission to be away for three weeks to visit Stockholm. When he arrived in Sweden, he joined the underground and was granted a scholarship from the Swedish government for study in Stockholm. He used the scholarship to study under Torsten Carleman for three years, from 1938 to 1941, and then looked into the possibility of coming to the United States. As early as February 1940, he wrote to Evans about a position at Berkeley and also explored other options. These efforts were successful and resulted in his securing a position at Macalaster College, beginning in 1941. Emigration to the United States in 1941 was quite a trek, and he came by way of the Trans-Siberian Railroad, then to Japan, and then an ocean voyage across the Pacific to San Francisco. He stopped in Berkeley to see Evans on his way to take up his appointment as an instructor at Macalaster College for 1941–1942 [UA, CU 5, series 2, 1942, 400-Mathematics] and [DM Wolf file].

With his authorization from Sproul, Evans approached Wolf in March 1942 to see whether he was interested in a temporary instructorship at Berkeley. Wolf, as it turned out, had an offer of a permanent instructorship with a promise of early promotion to assistant professor at the University of Kansas. However, Wolf declined the Kansas position in order to come to Berkeley, and he arrived in 1942 as instructor at the lowest end of the pay scale with an impressive bibliography of eight papers in highly regarded journals and considerable teaching experience. His initial research concentrated on convergence and summability of trigonometric series, but he had already extended his range of work to include studies of harmonic functions. His record at Berkeley was sufficiently impressive that in spite of the initial temporary nature of his position, he was promoted to assistant professor in 1944. Then, in response to an offer of an associate professorship at Syracuse University, Wolf was promoted to an associate professorship in 1947. Promotion to full professor came in 1952. His scientific interests had by then broadened further to include Fourier integrals, potential theory, and complex analysis, and finally he moved on to do major work on perturbation of linear operators: the behavior of spectral properties of an operator under small changes in the operator. Twenty-four doctoral students completed their work under him. He reached retirement age in 1972 and died in 1989 at the age of 84.

We turn now to the appointment of Tarski. Alfred Tarski was born in 1901 in Warsaw. His birth name was Tajtelbaum, and this is the name he used on some of his first publications in 1922. In about 1923, he changed his name to Tarski and converted to Catholicism, since he wished to become more Polish during this period of intense Polish nationalism. His brother made the same change in name and religion. Tarski published his next mathematical paper under the name Tajtelbaum-Tarski, and by 1924, he used the name Tarski on his publications. His wife, Maria, whom he married in 1929, was born a Catholic.

Tarski attended the University of Warsaw, where his mentors and teachers included Kotarbiński, Łukasiewicz, and Leśniewski. He was also deeply influenced by Banach and Sierpiński, and he received his doctoral degree in 1924 under Leśniewski. He was known for a number of notable achievements, but results of particular note include the Banach–Tarski paradox (1924), his theory of truth in formalized languages (1933), and his development of the beginnings of model theory, as well as many, many other results. He was a leading member of the Polish school of mathematics that flourished in the interwar period. His interests were very broad and included set theory, logic, universal algebra, topology, and measure theory.

The only position at the University of Warsaw that he was able to obtain was that of docent at a very low salary that was inadequate to live on. His main support came from a professorship in a lycée from 1924 to 1939. It is certainly reasonable to assume that the denial of a more senior university post to this distinguished scholar was a result of widely prevalent anti-Jewish sentiments at the time. His conversion to Catholicism and change of name apparently made little difference. In August 1939 he traveled to the United States for a lecture tour, and was in this country when Germany invaded Poland and World War II began. He remained in this country, and tried to arrange for his wife and children to join him, but without success. Through good luck and determination, his wife and children survived the war in Poland and joined him in Berkeley after the war. Initially, Harvard University provided some support for Tarski, and the City College of New York made Tarski a visiting professor. In his third year in the United States, the Guggenheim Foundation provided a fellowship that he took in residence at the Institute for Advanced Study.

Evans was well aware early on of Tarski's situation, as well as his eminence as a mathematician and philosopher, and wanted to take advantage of this opportunity in March 1942 to bring Tarski to Berkeley. Since it would obviously be inappropriate to appoint Tarski as an instructor at a salary of $2000 per year, Evans thought of using the title of lecturer. In addition, at the suggestion of Oswald Veblen, at Institute for Advanced Study, Evans approached the Rockefeller Foundation for a grant of $1000 to bring the salary up to a total of $3000. Rockefeller came through, and deal was arranged for Tarski to come to Berkeley as a lecturer for the duration of the war at a salary of $3000 with split funding between the university and Rockefeller. Sproul went along with the proposal [UA, CU 5, series 2, 1942, 400-Mathematics].

Tarski brought a rather different approach to logic than that represented by Benjamin Bernstein and Alfred Foster, who was then working closely with Bernstein. It would not be long before Tarski clashed with these two colleagues, and it happened in an algebra seminar in August 1943. Tarski was presenting material on the equivalence of sets of axioms for Abelian groups. Bernstein and Foster pointed out an error that they believed Tarski had made. Tarski maintained that there was no error. It is hard to tell from Bernstein's account in his papers of the disagreement [Ber, Box 1] whether there had been an error or just a misunderstanding in terminology, but the episode had a lasting negative effect on the relations between Tarski and his two colleagues.

It was not long—in fact, in February 1944—before Evans was recommending regularization of the appointment of Tarski as full professor. This recommendation did not succeed, but Evans renewed the request the following year, and Sproul agreed to an appointment as associate professor for Tarski effective July 1, 1945. Evans persevered, and the following year the long overdue promotion to full professor was approved. Tarski attracted many talented students to work under him and ultimately oversaw the creation at Berkeley of the premier logic group in the entire world. He reached retirement age in 1968, but he was recalled to teach for five more years. He remained active until his death in 1983, and in fact, the last of his 27 graduate students completed her degree in 1984. Tarski's efforts to assemble a group of scholars working in logic will be discussed in subsequent chapters. The logic group that Tarski and his successors have created at Berkeley remains the preeminent one. This group remained in the mathematics department and did not split off as a separate department as Neyman was to do with statistics. What was created was an interdepartmental "graduate group in logic and methodology of science." For more details on the life and work of Tarski, the reader should consult the recent full-length biography of him by the Fefermans [Fef].

Relations between Tarski and Neyman, both of whom had received their doctorates at the University of Warsaw in 1924, were never warm. Neyman had in fact opposed the original appointment of Tarski and had recommended to Evans that the appointment go to Antoni Zygmund instead, who then held an appointment at Mt. Holyoke College—a liberal arts college that did not have a graduate program in mathematics [Reid 1982, p. 181]. Zygmund had received his doctoral degree in Warsaw in 1923, and Neyman and Zygmund were clearly old friends who worked in related areas of mathematics.

Those members of the Berkeley faculty on active duty in the military and those working at Aberdeen were not the only ones involved with war work. In February 1942, Neyman received a contract from the National Defense Research Council to perform statistical analyses of bombing accuracy and assessment of bomb damage. This work, which engaged his whole laboratory, also necessitated frequent absences from Berkeley. In 1944 he was asked to go to Europe as part of a team to work on assessment of bomb damage. His students were also recruited directly. Mark Eudey was assigned to the 9th Air Force to do statistical work, and Joe Hodges and Erich Lehmann were assigned to the command staff of General Curtis Lemay in Guam as statistical analysts. Evans himself did consulting work for the government as well and was in Washington D.C. and absent from Berkeley for periods of time.

As the war was ending in 1945, Evans had an opportunity for an additional appointment at the entry level. He selected Abraham Seidenberg for this appointment. John McDonald had just retired on July 1, 1945, and as already noted, Bing Wong had been incapacitated by a stroke and had retired. These two represented the department's tradition in algebraic geometry, albeit of a very classical style. There was a new school and approach to algebraic geometry led by Oscar Zariski that would flourish in the coming decades. Seidenberg was a student of Zariski and so represented this new school.

Abraham Seidenberg was born June 2, 1916, in Washington, D.C., and received his baccalaureate degree from the University of Maryland in 1937. He then went to Johns Hopkins University for his doctoral work. His dissertation, completed under Zariski in 1943, was entitled "Valuation Ideals in Rings of Polynomials in Two Variables." He was Zariski's third and last student at Johns Hopkins before Zariski moved to Harvard. Zariski's first two students were Harry Muhly and Irving Cohen. Seidenberg had an additional connection with Zariski, since his wife, Ebe, and Zariski's wife were sisters.

Seidenberg first found an instructor position at Hamilton College and then one at Williams College. In September 1944, he held a position as a mathematician in a research group at the MIT Radiation Laboratory for a year before coming to Berkeley. His research covered a number of areas including commutative algebra and algebraic geometry. He was coauthor with Irving Cohen of a very influential paper on ideal theory. His papers on algebraic geometry include a very well known result on the normality of the general hyperplane section of a normal projective variety. He also made notable contributions to differential algebra, and another famous result is the Tarski–Seidenberg theorem on the existence of a decision procedure for algebra over the real field. In his later years his research interests expanded to include topics in the history of ancient mathematics, a field in which he made some notable contributions. Seidenberg was promoted to assistant professor in 1947, to associate professor in 1953, and to full professor in 1958. He supervised the doctoral work of 12 students during his career, retiring in 1986 after 41 years of service to the university. He died two years later.

In 1945 it had been just over 12 years since the 1933 decision to remake the department. At that time there had been 18 faculty members in the ranks from instructor to full professor and who were regarded as continuing faculty. Counting the retirement of Haskell, there had been 10 retirements during the period, and there had been 12 appointments, counting the appointment of Evans himself, so 11 over which Evans effectively had presided. Of the 12 appointees, only one, Ralph James, had been lured away by an outside offer. The others remained for the remainder of their careers at Berkeley. Thus, in these twelve-plus years, the department had grown by a net of one full-time equivalent, not counting temporary instructors and lecturers who had been hired, especially during the war years. The department had grown immensely in stature. Its strengths were in partial differential equations and analysis with Morrey, Lewy, and Evans, plus Morse and Wolf, and in statistics with Neyman and in logic with Tarski. The latter two areas, especially statistics, would grow by addition of faculty in the coming years. Lehmer and Robinson provided real strength in number theory, especially the computational aspects of it, but additional faculty would not be added in this area, and the field was not adequately represented in terms of numbers of faculty. Geometry and topology were not represented at all, and algebra was only weakly represented. These gaps would not be filled adequately to create a truly balanced department for another fifteen years.

Chapter 7

The Postwar Years: 1946–1949

With the end of the war and the return of veterans to civilian life, the GI Bill, which offered tuition assistance to veterans, caused enrollments in higher education across the country to skyrocket. Faculties everywhere had to expand, and the mathematics faculty at Berkeley was no exception. It expanded from 19 regular full-time equivalent (FTE) positions in the professorial ranks in 1945–1946 to 28 FTE in the professorial ranks in 1949–1950, which is when Evans stepped down as chair. There were ten new appointments during this period and one separation (Goldsworthy's death in 1949), resulting in a net growth of nine faculty positions. There were in addition a number of temporary instructors and lecturers on the payroll to meet the enrollment demands.

Of the ten appointments, five were in the general area of statistics (Erich Lehmann, Charles Stein, Michel Loève, Edward Barankin, and Joseph Hodges), and so by 1949 Neyman had built a substantial group in statistics. The other five were in various other areas (Edmund Pinney, Stephen Diliberto, John Kelley, Robert James, and Paul Garabedian). We shall discuss these five appointments first before turning to the development of the statistics group.

At this time, Evans was considering the proper role that applied mathematics should play in the department. He felt that it should play a larger role, and in American mathematics generally. There were several reasons that had him to this conclusion. One of them was the role that mathematicians had played during the war, specifically the involvement of several Berkeley mathematicians at the ballistics laboratory at Aberdeen, and the involvement of Neyman and members of his group in assessing the effects of bombing. Mathematicians had been involved in many other aspects of the war as well, including the MIT Radiation Laboratory, the Manhattan project, and in cryptography. Evans had some involvement himself and had also joined the Army Signal Corps in World War I to work as a mathematician. Evans's own work on mathematical economics represented an important application, and his decision to build up statistics at Berkeley by hiring Neyman and by supporting his program represented another effort to build outside of pure mathematics. But Evans had grander plans for building up applied mathematics, especially in areas of applications of differential equations. As we have remarked previously, Evans had a very broad view of mathematics and of what areas of mathematics and its applications should be included in a mathematics department.

One of his first postwar appointments, in 1946, was that of Edmund Pinney. Pinney was born on August 19, 1917, in Seattle, Washington. He attended Caltech, where he received his baccalaureate degree in 1939 and a PhD in mathematics in 1942. His dissertation, "Calculus of Variations in Abstract Spaces," was written under A. D. Michal, which made him a mathematical grandson of Evans. Pinney also worked with Harry Bateman. After graduation he worked for a year at the MIT Radiation Laboratory and then for two years was a research analyst at Consolidated Vultee Aircraft Corporation. Following a year as an instructor at Oregon State College, he came to Berkeley as an instructor. Promotion to assistant professor followed in 1948, to associate professor in 1952, and to full professor in 1959.

An extended quotation from his memorial article provides a good overview of his research work:

> Pinney was an applied mathematician of great versatility and wide research interests. Much of his early work was focused on electro-magnetics (radar and wave-guides) and was inspired by problems of national security interest. Other work involved analysis of forced aerodynamic oscillations (such as those that destroyed the Tacoma Narrows Bridge in Washington State). In the 1950s Pinney was the Director of a Berkeley project sponsored by the Office of Naval Research. In this connection he investigated the propagation of large-amplitude waves through soils and other geophysical media (this was prompted in part by the need for a better understanding of the effects of the Bikini bomb test). In the 1960s he undertook studies of theories of ionized plasmas, which have application to radar detection of rocket plumes.... Much of his research in later years was focused on the difficult "angle problem" for Rayleigh waves. This concerns the propagation of reflection of seismic waves at corners such as the continental shelf. He continued this work after retirement and achieved partial results using asymptotic methods before finally turning to numerical techniques.

Pinney supervised the doctoral work of six students during his career and retired from active duty in 1988 after 42 years of service. He died in 2000.

In the following year, 1947, Evans reached out in a slightly different direction to hire Stephen Diliberto. Stephen P. L. Diliberto was born on November 13, 1920, in Orange, New Jersey, and graduated from Princeton University in 1943 with a bachelor's degree in mathematics. He was interested early on in aeronautical engineering, but ultimately decided on mathematics. He entered graduate school in mathematics at Princeton and completed his degree in 1947 with a dissertation entitled "Reduction Theorems for Systems of Differential Equations." Although

Solomon Lefschetz was his advisor, Diliberto found his dissertation topic on his own and worked rather independently, without much direct supervision from Lefschetz. The general topic of his work was differential equations in the large and stability issues concerning periodic solutions, work that was very much in the direction of that of Poincaré and Lyapunov. This was an area of research that was not represented at Berkeley, and Diliberto's interest and background in aeronautical engineering and in applications was an added attraction for Evans.

Evans also had another consideration in mind. He realized that the department was missing representation in topology and modern differential geometry, and was hoping that Diliberto might contribute perhaps indirectly to building up in this area given the evident connection between differential equations and these topics. Lefschetz and others (Marston Morse and Claude Chevalley) recommended Diliberto to Evans, and so Diliberto was hired as an instructor in 1947. He was promoted to assistant professor in 1949 and to associate professor in 1955. An offer from Lefschetz for Diliberto to become a research professor at a very attractive salary in the Research Institute of Advanced Studies (RIAS), which Lefschetz had established, led to Diliberto's promotion to full professor in 1959. His research in ordinary differential equations developed along the lines initially laid out in his early work on stability, and he was a popular dissertation advisor, supervising the work of 24 doctoral students during his career. Diliberto was an active citizen of the campus, serving as assistant dean of the College of Letters and Sciences under Deans Davis and Constance from 1953 to 1956. He also served on several administrative and Academic Senate committees and was a consultant to the China Lake Naval Station. He retired in 1991 after 44 years of service.

Evans pursued his goal of building up geometry and topology more directly with the appointment of John L. Kelley in 1947. During the war, Kelley had served at the Army Ballistics Research Laboratory at Aberdeen. Professors Morrey, Lewy, and Morse, from Berkeley, were there as well and had used the opportunity to scout future faculty members. Kelley and Morrey had briefly been roommates at Aberdeen before their respective families arrived to set up housekeeping. Kelley and Morse became particularly close friends, and Morse, when he returned from Aberdeen after the war, had been advocating to Evans the appointment of Kelley at Berkeley.

John Leroy Kelley was born on December 6, 1916, in rural Kansas. His father was an itinerant minister and his mother a schoolteacher. The family was poor and migrated to California in 1930, where the California educational system provided him opportunity. Kelley graduated from high school at age 14 in 1931 and then enrolled at Los Angeles Junior College, transferring to UCLA, where he received an AB in mathematics in 1936 and an MA in 1937. One of his professors at UCLA, W. M. Whyburn, recognized Kelley's talent, and since UCLA had no doctoral program at the time, he arranged for Kelley to be accepted with a teaching assistantship into the doctoral program at the University of Virginia, where his brother Gordon T. Whyburn was a faculty member.

Kelley flourished at Virginia under the influence of Whyburn, G. A. Hedlund, and E. J. McShane, completing a dissertation in 1940 in point-set topology under Whyburn entitled "A Study of Hyperspaces." He also had several publications in first-rate journals (*American Journal of Mathematics, Duke Mathematical Journal,* and the *Proceedings of the National Academy of Sciences*) to his credit while still a graduate student. Next came appointment as assistant professor at Notre Dame, but when the United States entered World War II, one of his mentors, E. J. McShane, became head of the theory group at the Ballistic Research Laboratory at Aberdeen and recruited Kelley to join the group. During his time there, Kelley made significant contributions to understanding yaw in projectiles and later coauthored a book on exterior ballistics with McShane and F. J. Reno.

After the war, Kelley was appointed assistant professor at the University of Chicago beginning in the fall of 1945 and was given six months of leave to the Institute for Advanced Study. During this time, joint work with Everett Pitcher led to a significant publication in the *Annals of Mathematics* of a paper on algebraic topology. Joint work with Richard Arens on a characterization of the space of continuous functions on a compact Hausdorff space appeared in the *Transactions of the American Mathematical Society* and represented a new direction for Kelly's research. During this year he also became acquainted with the work of Gelfand, Shilov, Raikov, and Stone in which Banach algebras and Banach spaces were the key players in analyzing topological and algebraic problems. Also, building on his work at Aberdeen, Kelley became a consultant for the Atomic Energy Commission (AEC) and was granted an AEC security clearance.

Mutual interest between Kelley and the Berkeley mathematics department warmed in 1946, and in October, Evans recommended to the administration Kelley's appointment as assistant professor effective July 1, 1947, at a salary of $4500, a figure that was at the top of the range for this rank and that matched Kelley's salary at Chicago. Owing to administrative bumbling or oversight, the campus review of the recommendation was delayed, and the offer was not finally proffered until March 1947. Evans wrote to Kelley specifically about the offer that "The idea we had in inviting you to California was that you would be responsible for the development of our graduate work in topology in broad directions." Then began a period of negotiation. In a decision prior to the Berkeley offer, Chicago had raised Kelley's salary for the following year by $500. Also, the "Stone Age" had begun at Chicago, during which Marshall Stone would reshape that department. Kelley notes in his correspondence with Evans that Mac Lane and Zygmund were coming to Chicago the following fall (1947) and that André Weil would be a visiting professor. Hence, the attractiveness of Chicago was considerably increased, and on April 16, Kelley decided to decline the Berkeley offer. However, within ten days Evans succeeded in getting the Berkeley offer raised to an associate professorship (with tenure, such as it was at UC) at a salary of $4800. With great enthusiasm, Kelley immediately accepted this revised offer. Kelley had at least one other offer at the time: an associate professorship at Ohio State University.

Kelley went on to play a major role in the department in future years to which we will turn presently. He was also one of the faculty members fired by the regents in August 1950 for failure to sign the special regental loyalty oath. After a successful legal challenge to the regental action, he was reinstated in 1953 and at the same time promoted to full professor. Clark Kerr, the first UCB chancellor, recruited Kelley to serve as one of his academic assistants in the chancellor's office in 1956–1957. Kelley also served as vice chair of the department for 1954–1957, and assumed the position of chair for 1957–1960. Under his leadership, the department made major steps forward both in appointments and curriculum and assumed a new complexion that it has had ever since. He served a second term as department chair from 1975 to 1978.

Although Kelley was appointed to build up topology at Berkeley, his research interests shifted away from early work in point-set topology and algebraic topology toward the study of Banach algebras, especially Banach algebras of functions on topological spaces. Berkeley was to remain isolated from major developments in algebraic topology and modern differential geometry for some time, in fact, until Kelley made a series of senior appointments during his first term as chair in 1958 and 1959.

On Kelley's teaching, one should mention his famous book *General Topology*, which was a staple for many generations of graduate students and established a model for graduate texts in mathematics. As to other aspects of his teaching, one cannot do better than to quote from his memorial article [IM]:

> Kelley was an exciting teacher; popular with students at all levels. He supervised eight doctoral dissertations and aided many other students with informal advice. His door was always open to students. For many years his Tuesday afternoon Functional Analysis Seminar, invariably followed by beer at LaVal's, was a focal point of the department's modern analysis group.
>
> Kelley's interest in teaching extended beyond the boundaries of the University. As a member of a national school mathematics study group he joined workshops of teachers and professors who gathered to write non-traditional math textbooks and infuse the school math curriculum with ideas that came to be called "New Math." In 1960 he took an extended leave to serve as the National Teacher on NBC's Continental Classroom TV program. His tweed jackets, with pipe ready to be puffed in the close-

ups, helped instill confidence among viewers who might have been taken aback by the new ways of teaching he promoted. Back in Berkeley he developed a new course in the Mathematics Department for prospective elementary teachers, and co-authored a text for such courses given elsewhere. In 1964 he introduced a new Math-for-Teachers major so that Berkeley undergraduates could prepare to teach high school math with a strong understanding of the subject; and he devised an internship program to follow their B.A. degree, enabling them to qualify for a teaching credential with a minimum of education courses that had turned off many math students. During 1964–65 he took leave to help set up the mathematics department at a new Indian Institute of Technology in Kanpur, India.

Kelley retired in 1985 and continued his research activities in collaboration with his old friend T.P. Srinivasan. He died in 1999.

In 1947, Evans appointed Robert Clarke James as an instructor in the department. James had received his doctoral degree under A. D. Michal at Caltech in 1946, so he was another mathematical grandson of Evans. James's doctoral dissertation was entitled "Orthogonality in Normed Linear Spaces." He had started publishing articles in highly regarded journals in 1943, several years before receiving his degree, and he spent some time at Harvard before coming to Berkeley. With an active research program concentrating on the geometry of Banach spaces, he won promotion to assistant professor in 1949. However, his stay at Berkeley was brief, and he left the university in 1951, settling ultimately at the Claremont Graduate University.

In 1949, Evans appointed an extraordinarily promising young mathematician, Paul Roesel Garabedian, as an assistant professor. Garabedian was born in Cincinnati, Ohio, on August 2, 1927, and graduated from Brown University with a BA in 1946 at the age of 18. He went on to receive his doctoral degree at Harvard in 1948 under Lars Ahlfors with a dissertation entitled "Schwartz's Lemma and the Szegő Kernel Function." Harvard elected him a junior fellow, but he took a year's leave from it to accept a National Research Council fellowship that he took at Stanford, working with Donald Spencer. He then resigned as junior fellow to accept the Berkeley appointment, but his stay at Berkeley was brief, and he left after only one year, when the regents imposed their loyalty oath in 1950, to return to Stanford. He subsequently left Stanford for the Courant Institute at New York University, and his name is associated with New York University, where he went on to a distinguished career working in partial differential equations.

Thus, of the five non-statistics appointments, only three ended up remaining at Berkeley: Pinney, Diliberto, and Kelley. Two were generally in applied mathematics, while Kelley, who started out in topology, ended up in functional analysis, and the department still had large holes in many areas of pure and applied mathematics. We turn now to the development of statistics during the postwar period.

As already noted, Neyman had been away from Berkeley on war-related work for extended periods of time, and after the war he looked forward to a return to teaching and research. He also longed to fulfill his dream since 1938 of building a major statistics center at Berkeley, which in his vision would include a department of statistics separate from mathematics. He had built a thriving statistical laboratory with many assistants and students, but he was the only regular faculty member. It was only after the war that statistics evolved into a distinct discipline, in good part because of the role that statistics played in war-related research. Creation of separate statistics departments was a constant topic of discussion nationally.

Neyman learned in the spring of 1946 that Harold Hotelling, a leading statistician and the mainstay of Columbia's program in statistics, was leaving Columbia for the University of North Carolina (UNC) that fall. Hotelling had for years urged and agitated, with no success, for an independent department of statistics at Columbia. Now UNC had recruited him to lead their newly created independent department of statistics. Columbia's response to this grievous loss was to create their own department of statistics with Abraham Wald in charge. Wald was also given a high-level visiting professorship for one semester that he could fill. It was quite natural for him to offer this position to Neyman for the fall semester of 1946.

In January 1946, Neyman had been again called away from Berkeley to join an international commission to oversee elections in Greece. On the way back in March, he stopped in Paris to give lectures and meet with colleagues. During this brief visit he was particularly impressed with the young probabalist Michel Loève, who was regarded as the "intellectual son" of the great French mathematician Paul Lévy.

When Neyman returned to Berkeley, and requested a leave of absence in order to take the visiting position at Columbia that Wald had offered, Sproul was reluctant to grant the leave, noting that Neyman had not spent much time in Berkeley in the last few years. This negotiation led to a discussion of the future of statistics at Berkeley, in which Neyman argued that Berkeley should either give up the idea of building a center or greatly expand the scope of what existed. No doubt Neyman had the events at Columbia and UNC in mind. Sproul then urged Evans and Neyman to expand the teaching and research program and to consider a major appointment [Reid 1982, p. 211]. Sproul then approved Neyman's request for leave the coming fall.

One of Neyman's star students, Erich Lehmann, was just finishing his doctorate, and he was appointed as instructor in the mathematics department starting in the fall of 1946. Neyman left him in charge of the statistical laboratory for the fall while he was on leave. This was the first faculty appointment in statistics since Neyman's appointment in 1938.

Both Neyman and Evans favored the idea of recruiting George Pólya to Berkeley. Pólya had come up from Stanford frequently to fill in for Neyman in his courses and seminars during the war and had done some research in statistics (among many other fields). Efforts to move him to Berkeley did not work out and succeeded only in getting Pólya a long overdue promotion at Stanford.

Neyman and Evans also took steps to attract Michel Loève to Berkeley. On July 26, 1946, Loève was offered a position as lecturer in mathematics and research associate in the statistical laboratory effective in the fall. Loève, however, had already accepted a position at the University of London as a lecturer, but he asked whether the offer could be deferred. Berkeley kept on pursuing him with the possibility of an offer the following year. In December 1946, Neyman wrote to Evans from his visiting position at Columbia, citing the importance and recognition of Loève's work and urging that Loève be offered a full professorship.

But other wheels were turning. Not unexpectedly, in the fall of 1946, Wald recommended Neyman for a permanent appointment at Columbia. The formal offer came through from the Columbia administration in January 1947. It was a very attractive offer in terms of salary and teaching responsibilities (four hours per week versus nine hours per week at Berkeley). Again, not unexpectedly, Neyman used this offer to try to better his position at Berkeley, and in particular, he renewed his request for an independent statistics department as the only way to raise Berkeley to the level of Columbia. Evans remained steadfastly opposed to an independent department, but over the next two months Neyman and Evans together with Sproul and Provost Deutsch worked out an agreement.

Henceforth, under this agreement Neyman would submit the budget for the statistical laboratory directly to the administration and not through Evans. Neyman would have authority over research appointments in the laboratory, and teaching appointments in the department would be negotiated jointly. In particular, there would be two new appointments: one at the junior level and one at the senior level. Neyman already had the candidates picked out, namely Loève for the senior appointment and Charles Stein, a brilliant student he had met at Columbia who was just finishing his doctoral degree under Wald, for the junior one. And so, with these arrangements in place, Neyman declined the Columbia offer in March.

Stein was offered a position as lecturer starting in the fall of 1947. The appointment was converted to an assistant professorship the following year. Loève was offered a full professorship for the fall of 1948, which he accepted. Thus, by 1948, the professorial faculty in statistics had grown from one to four: Neyman, Lehmann, Loève, and Stein, with more to come.

Erich Leo Lehmann was born in Strasbourg, France, on November 20, 1917, while his father was stationed there as an officer in the German army. He grew up in

Frankfurt, Germany, and the family left Germany in April 1933 almost immediately after the Nazis seized power, going first to Sweden and then to Zurich. Lehmann completed his secondary education in Zurich and then enrolled in courses at the University of Zurich. On the advice of Edmund Landau, a family friend, Lehmann went to Cambridge University to study under Hardy and Littlewood. This did not work out, and Lehmann then emigrated to the United States in 1940 and came to Berkeley. He hoped to enroll as an undergraduate student, but Evans felt that Lehmann's experience at Cambridge was sufficient to justify admission as a probationary graduate student in mathematics even though he had no undergraduate degree. Lehmann prospered as a student, and Evans convinced him to study statistics. Lehmann received an MA degree in 1943, and after interruption by war work with the US Air Force, he earned a PhD in 1946. His dissertation, written under Neyman, was entitled "Optimal Tests of a Certain Class of Hypotheses Specifying the Value of a Correlation Coefficient."

George Dantzig, whose graduate studies were also interrupted by the war, finished his doctoral degree under Neyman the same year, and Lehmann and Dantzig were Neyman's second and third students. After one year as instructor, Lehmann was quickly promoted to assistant professor in 1947, to associate professor in 1951, and to full professor in 1954. He went on to have a distinguished career at Berkeley, training 40 doctoral students, including Peter Bickel and Kjell Doksum, who subsequently became his colleagues. According to the "Mathematics Genealogy Project," Lehmann has, as of this writing, 384 mathematical descendants, which constitute almost half of Neyman's 805 mathematical descendants. Lehmann is a world-famous statistician whose work on a wide variety of topics has placed him in the forefront of modern statistics and earned him election to the National Academy of Sciences.

Michel Loève was born to a Jewish family in Jaffa, Palestine, on January 22, 1907. Following education in French schools in Alexandria, where he acquired Egyptian citizenship, he then went to the University of Paris. He received a baccalaureate degree in 1931 and then a Docteur des Sciences (Mathématique) in 1941 under Paul Levy. He was imprisoned during the German occupation, and after the war, he held a research position in the Institut Henri Poincaré. Because he was not a French citizen, his professional opportunities in France were limited, and so offers from institutions in England and the United States were attractive to him. He accepted a position at the University of London, but after two years he came to Berkeley in 1948 as full professor of mathematics. Loève had spent the spring semester 1948 at Columbia as a visitor before coming to Berkeley. In 1951, Columbia made an effort to recruit him away from Berkeley with a very attractive offer, which came in the middle of the loyalty oath controversy. Happily for Berkeley, Loève declined this offer. Loève supervised the doctoral work of five students, including Leo Breiman, who subsequently joined the faculty at Berkeley. When an independent statistics department was created in 1955, he split his appointment between the two departments until his retirement in 1974. He died suddenly in 1979.

An extract from Loève's memorial article refers to his scientific contributions:

> Mathematicians everywhere were inspirited by Loève, who worked with them to help create the modern theory of probability. His influence extended far beyond his numerous research contributions because of the dominant position, both as a reference and a graduate text, achieved by his book *Probability Theory*.

Another extract from the memorial article describes his legacy:

> The French school of probability from which Loève emerged was world famous. Jerzey Neyman, Berkeley's premier statistician, brought Loève here with the hope that he would found an equally renowned cluster of scholars and course in that discipline at Berkeley. The hope was realized, and Berkeley became a mecca for students and scholars of probability from all parts of the world.

Charles Stein was born on March 22, 1920, in Brooklyn, New York, and received a BS in mathematics from the University of Chicago in 1940. After some brief graduate work in mathematics at the University of Chicago, he served in the US Army Air Force during World War II. He was a member of a group of mathematical scientists that included Kenneth Arrow, Gilbert Hunt, George Forsyth, and Murray Geisler stationed at the Pentagon working in meteorology. As a result of a conversation with Arrow and scanning a paper of Wald, Stein obtained a result on the two-sample test for Student's hypothesis, which was published in the *Annals of Mathematical Statistics* in 1945. After being discharged from the army in February 1946, he enrolled as a graduate student at Columbia and obtained his degree in 18 months in 1947. His dissertation, completed under Abraham Wald, was entitled "A Two Power Test for a Linear Hypothesis Having Power Independence of the Variance." As already noted, Neyman met Stein when he was visiting Columbia in the fall of 1946, was impressed by him, and immediately recruited him to Berkeley as an instructor in 1957. Neyman secured his promotion to assistant professor the following year.

Stein had been awarded a National Research Council Fellowship for 1949–1950, which he took at the Institut Henri Poincaré. This was the year of the loyalty oath; see Chapter 9. Stein was unwilling to sign the oath and simply resigned before he would have been fired by the regents. He was appointed as an associate professor of statistics at the University of Chicago in 1951 and then as professor of statistics at Stanford University in 1953. After the loyalty oath controversy faded from prominence, Neyman made efforts to try to lure Stein back to Berkeley, but he had no success. Stein's departure was a grievous blow to the mathematics department and to the future statistics department, for Stein is regarded by many as the leading scholar in theoretical statistics of his generation.

Two further appointments in statistics followed. In 1947, Edward Barankin, who as a graduate student had joined the statistics group during the war years, was appointed as an instructor in the mathematics department. Barankin was born in Philadelphia in 1920 and received his baccalaureate degree from Princeton University in 1941. He came to Berkeley as a graduate student in mathematics, completing his degree with a dissertation written under the direction of Alfred Foster entitled "The Characteristic Values of Linear Transformations." The results concerned bounds on the eigenvalues of a matrix in terms of the matrix coefficients. He spent the year 1946–1947 as Hermann Weyl's assistant at the Institute for Advanced Study, and then returned to Berkeley as a faculty member with the title of instructor. His work spanned many areas, including game theory, linear and quadratic programming, and—within statistics proper—the theory of sufficient statistics. Promotion to assistant professor came in 1949, to associate professor in 1952, and to Professor (of Statistics) in 1959. He supervised the doctoral work of nine students during his career. He died on May 1, 1985, after an illness of several months.

His memorial article provides a summary of his scholarly work:

> His early work on sufficient statistics was highly regarded at the time and is still cited. About 1950 he started developing a new and rather complicated theory of stochastic processes and behavior, and although he continued to do excellent work in other areas, including sufficient statistics, and programming in operations research, his dominant interest for the rest of his life was his process theory. In his theory, as in the theories of Keynes, Carnap, and Jeffreys, when the relationships among events are adequately described, the probabilities of events can be calculated. Most of his United States colleagues never fully understood his approach to stochastic processes, but his work was highly regarded in Japan and was published in several Japanese statistical journals.

In 1949, another faculty member in statistics was added to the faculty: Joe Hodges. Joseph Lawson Hodges, Jr., was born in Shreveport, Louisiana, on April 10, 1922, and entered UC Berkeley as a freshman at age 16 in 1938. After receiving his baccalaureate degree in mathematics in 1942, he continued on as a graduate student in mathematics. Although initially attracted to the recently arrived Professor Tarski, he soon joined the Neyman group that was working on statistical problems concerning bombing patterns. He was then part of a group, which also included Erich Lehmann, that was sent to Guam to do operations analysis for General LeMay's 20th Air Force. He continued this work in Washington, D.C., for a brief period after the war before returning to his graduate studies in Berkeley.

After receiving his degree under Neyman in 1949, Hodges was immediately appointed as an assistant professor, without the usual preliminary appointment as instructor. His dissertation consisted of two parts: "(I) Initial Sample Size in the Stein Procedure (II) Stringency in Acceptance Sampling (Studies in Optimal Statistical Procedure)." He was promoted to associate professor in 1953 and to full professor in 1958. He and Erich Lehmann were lifelong friends and frequent collaborators. Both were on leave during 1951–1952, and both entertained offers from other universities, which, happily for Berkeley, they declined.

To quote from his memorial article:

> In statistics, Joe is best known for his work in non-parametric inference. In a technical report (with Evelyn Fix), written in 1951, but published only in 1989, he pioneered non-parametric density estimation. He is the inventor of the Hodges bivariate-sign test and (with Lehmann) of the Hodges–Lehmann estimator. One of his most striking discoveries was the phenomenon of super-efficiency, which exploded long-held beliefs about maximum likelihood estimation and had a profound effect on asymptotic theory. His bibliography lists a steady stream of publications, at least one, but usually more per year until the 1970s when administrative duties began to consume his energy.

During his career he supervised the doctoral work of five students. Hodges retired from active duty in 1991 but remained a notable presence on campus. On March 1, 2000, he died unexpectedly of a heart attack after 62 years of affiliation and service with the campus as a student and faculty member.

Evans stepped down as chair in 1949 after 15 years of service and turned the reins over to Charles B. (Chuck) Morrey. Evans had wanted to defer to a younger man, and Morrey, having served for several years as vice chair, was the natural choice. This act also initiated a tradition of a rotating chairmanship in the department, and in fact, when Clark Kerr became Berkeley chancellor in 1952, he instituted a general practice of rotation for all departments on campus.

In his fifteen years as chair, Evans had reshaped the department, having made or overseen 21 appointments to the professorial ranks. These were Charles Morrey, Alfred Foster, Ralph James, Hans Lewy, Jerzy Neyman, Raphael Robinson, Anthony Morse, Derrick Lehmer, Alfred Tarski, František Wolf, Abraham Seidenberg, Erich Lehmann, Edmund Pinney, Stephen Diliberto, Robert James, John Kelley, Charles Stein, Michel Loève, Edward Barankin, Joseph Hodges, and Paul Garabedian.

Only one of these, Ralph James, had been recruited away, so 20 remained and with Evans himself they constituted 21 of the 28 faculty in professorial ranks in the department as of July 1, 1949. The other seven dated from the pre-Evans era and were all to retire by 1954. The contributions of these seven were confined to teaching and service. Among the new faculty, Stein and Garabedian would resign within the year because of the loyalty oath controversy, and Robert James would depart in 1951 by resignation.

Among those new faculty who were not in statistics (15 of the 21), only one, Robinson, had received his doctorate at Berkeley, thus reversing the ingrown nature of the pre-Evans department. Of the six appointments in statistics, three were Berkeley doctorates: Lehmann, Barankin, and Hodges. Neyman's argument in favor of this practice was that Berkeley was just about the only university training capable people, and he was generally correct, since Berkeley and Columbia were now recognized as the top two departments in statistics in the country. When statistics finally fissioned off from mathematics as a separate department in 1955 with ten professorial faculty members, six had their doctorates from Berkeley.

The department had clearly reached preeminence in statistics and had great strength in partial differential equations with Morrey, Lewy, and Evans, although Evans's work by this time was concentrated in mathematical economics. Functional analysis was represented by Wolf, Kelley, and Robert James. Lehmer, together with Robinson, provided strength in number theory, especially in its computational aspects. Tarski was world-renowned in logic, and his presence immediately made Berkeley a center for the subject and was a base on which a preeminent group in logic would grow in the future. Foster straddled logic and algebra, but was not close to Tarski intellectually, while Seidenberg was squarely in commutative algebra and modern algebraic geometry. Pinney and Diliberto worked in differential equations and applied mathematics. Geometry and topology were not represented at all, and although Kelley was hired to provide coverage to this area, his interests were shifting to other areas. Thus, geometry and topology were major weaknesses, and both algebra and applied mathematics needed significant additional strength. Building additional strength in applied mathematics has been a constantly recurring issue from at least this time up to the present day.

The standard teaching responsibilities for faculty members in the department were three three-hour courses each semester. It was only later that this level of teaching responsibilities would be lowered in response to competitive pressures. The undergraduate and graduate curricula did not look qualitatively different from what they are today, but there were significant differences. A number of what we would call remedial courses, including intermediate algebra (today's second-year high-school algebra) were taught for credit. These courses are no longer offered for credit, although Mathematics 32, which covers the fourth year of high school mathematics, is offered for credit. Four semesters of calculus went only as far three semesters go today. All lower-division courses were taught in small sections of perhaps 35 students by teaching assistants, lecturers, instructors, and regular faculty. The lecture/discussion section format was to come later.

The upper-division curriculum did not have the richness and variety that it does today, and the graduate curriculum was likewise rather thin. There were the three standard graduate courses in real variables, complex variables, and algebra. The year-long course in topology (numbered 215) stopped far short of what would be taught today. There was a year-long course in differential geometry (numbered 240) that covered a number of the current topics, and a year-long course in algebraic geometry that mostly covered algebraic curves: the Riemann–Roch theorem, etc. In addition, there were graduate courses in statistics and a number of graduate seminars. Both the undergraduate and graduate statistics courses were listed separately rather than being listed together with the mathematics courses. The department sponsored two undergraduate majors, one in mathematics and one in mathematical statistics, and there were likewise separate graduate programs.

It has already been noted that the first PhD granted by the department was in 1901 and that there was a gap until 1909 before the second one was awarded. In the 27-year period from 1909 to 1935, inclusive, the department awarded 46 doctorates, for an average of rather less than two per year. It was remarked as well that 43 of these doctoral students worked under three faculty members: D. N. Lehmer super-vised 19, Mellen Haskell 13, and John McDonald 11, including Raphael Robinson, who received his degree in 1935. By 1936, the effect of Evans and the newly hired faculty was felt in the graduate program. In the 14 years from 1936 to 1949, the department awarded 61 doctorates, or about 4.5 per year on average. This included the four war years 1942–1945, when only six degrees were awarded. The degree production expanded further in the following decade, and in the 11 years from 1950 to 1960, inclusive, 114 degrees were awarded in both mathematics and statistics for an average of a little over 10 per year. In 1949–1950, the department enrolled 108 doctoral candidates and 73 other graduate students.

The department occupied about 6000 square feet of space in Wheeler Hall, which had been the location of the department since the building had been com-pleted in 1917. In addition to the 28 faculty members in professorial ranks, there were 12 FTE in teaching positions (instructors, lecturers, and associates) and 38 FTE teaching assistants, which represented 76 individuals, since graduate-student teaching appointments were half-time. Total undergraduate enrollment on the cam-pus still reflected the postwar enrollment surge and was over 19,000, within about 3,000 of where it is today. Space was very tight, and instructors were packed often four to an office. The location of the department was spatially isolated from the related disciplines of physics, chemistry, and engineering with which mathemati-cians and statisticians would naturally interact.

By this time, plans were underway for the construction of the new "classroom building" on campus, or what was subsequently named Dwinelle Hall. The mathematics department along with the statistical laboratory was to move into Dwinelle Hall on its completion. The move took place in mid 1952, and provided somewhat improved quarters for the department, but at an even further remove from the physical sciences and engineering precinct located at the eastern end of the campus, since Dwinelle Hall was located west of Wheeler Hall.

The department had also petitioned the main library to create a mathematics branch library to be located in Dwinelle Hall. Neyman insisted that it be called the Mathematics and Statistics Branch, a suggestion that carried the day. The mathematics department and the statistical laboratory had small collections in reading rooms in their quarters, but altogether the collections did not exceed 1500 volumes. The program planning guide for Dwinelle Hall included an area 38 feet by 22 feet for the library with a storage capacity of 6800 volumes. The main library agreed to the creation of the new branch library and to a phased transfer of mathematics and statistics holdings in the main library to the branch library. The library further agreed to provide staffing at the level of one senior library assistant, and with a vague promise to provide a professional librarian at an appropriate time in the future. The Mathematics and Statistics Branch opened for business in October 1952. The department made repeated and strenuous appeals for special one-time funding to buy books for the branch library. A year later, in October 1953, a report listed the holdings in the branch at 1973 volumes and 50 periodicals. Over time, the holdings would grow considerably.

Recall that in 1935 Evans had proposed the creation of a mathematics journal to be published by the University of California Press and to be called the *Pacific Journal of Mathematics*. In making this proposal, he had in mind the recently created *Duke Mathematical Journal*. But his proposal was turned down for lack of funds during the depression years. In the postwar years, the proposal was brought forward again under sponsorship by UC Berkeley and UCLA as well as other California institutions, and the *Pacific Journal of Mathematics* was ultimately launched successfully as a quarterly publication starting in March 1951.

It is difficult to estimate the "ranking" of the department in comparison to other mathematics departments in the United States at the moment in time when Evans stepped down as chair. In the one area of statistics, the ranking would have been either first or else second after Columbia, although Stanford would soon emerge as a major competitor. Overall, the department was easily in the top ten, but it would probably be hard to argue for a ranking higher than about five. Berkeley was certainly ranked well below Harvard, Princeton, and Chicago, the three departments that had for several decades been universally recognized as the top three in the country, and would continue to be so recognized until about 1960.

Chapter 8

Postwar Initiatives in Applied Mathematics

At the beginning of the previous chapter, we noted that Evans clearly saw the need to build up applied mathematics at Berkeley, and, in doing so, he had to move counter to the traditional penchant for pure mathematics in the United States. He made a few appointments of individuals working in applied areas, but in this chapter we discuss several efforts that Evans and then Morrey made in the years following the war to augment applied mathematics through institutional enhancement and infrastructure building. Unfortunately, none of these efforts or projects was successful, and fuller development of applied mathematics at Berkeley had to wait a number of years. The three projects were first an attempt to create a program in actuarial science; second, an attempt to attract a major computational center to Berkeley; and finally, an attempt by Evans's immediate successor, Charles Morrey, to bring Courant and his entourage to Berkeley in the fall of 1949. One reason that these initiatives came to naught may have been an institutional reluctance to embrace research institutes, especially when programs were proposed to be built on "soft" (i.e., grant and contract) money. Berkeley had had a long tradition of insisting that programs with an implied institutional commitment be funded with "hard" (i.e., state) money, and the campus's general unwillingness to relax this basic principle put it at a disadvantage in competing in the postwar environment for various projects. Ernest O. Lawrence's Radiation Laboratory was to some extent an exception to this principle, but it was a separate entity and Lawrence was a man of unusual influence and persuasiveness.

In the years after the war, Evans had conceived grander ideas concerning the development of applied mathematics far beyond what had already been achieved in statistics since Neyman's arrival in 1938. Reid [Reid 1982, 1976, p. 213] writes that Evans "conceived of a 'Mathematics Center' [at Berkeley] which would ultimately include an Institute of Statistics, a Bureau of Computation, an Institute of Actuarial Science, and an Institute of Applied Mathematics and Mechanics." It would seem that Evans had in mind that these institutes would all reside within the mathematics department, in accord with his broad conception of mathematics and what a mathematics department should comprehend. This goal of retaining such entities within the mathematics department was very optimistic, especially given the experience with Neyman and Neyman's determined effort to have statistics become an independent department, in which he was to be ultimately successful. There is no reason to believe that other leading figures who might be attracted to Berkeley to head up such other institutes would be any less eager to seek independence.

There were indeed several proposals for the creation of institutes on the Berkeley campus in the postwar years, but the university lacked any clear policy or guidelines. This situation led Alva Davis, dean of the College of Letters and Sciences at Berkeley, with clear encouragement from Sproul, to draft a paper, "The Place of the Institute in University Organization," which he presented to the regents at their meeting on January 21, 1949 [UA, CU 5, series 4, Box 11, folder 5]. In it he outlines for the first time formally what has been university practice on organized research units (ORUs). The report recognized that institutes would play an important role in the university, and he attempted to delineate this role. In particular, institutes were intended as a vehicle to bring interdisciplinary groups of faculty together for research on a specified topic. Institutes would not have teaching responsibility, and would not have any FTE faculty positions. The members and the director of an institute would hold regular faculty positions in academic departments, but might have released time supported by institute funds. Unfortunately, Evans's conception of institutes did not fit this model.

In 1947, Evans and Neyman took concrete steps toward the idea of developing actuarial sciences at Berkeley. They could not identify anyone in the United States in this field who had the stature suitable for a faculty appointment at Berkeley in mathematics. But they were drawn to a natural target abroad, namely Harald Cramer, who was professor of actuarial mathematics and mathematical statistics at the University of Stockholm. Cramer visited Berkeley to give a course in the 1947 summer sessions, and Neyman approached him about the possibility of coming to Berkeley permanently. Evans then subsequently raised the issue more formally with President Sproul and was able to write to Cramer in November 1948 to say that he had authorization from the president to approach Cramer about coming to Berkeley to head a group in actuarial science. Neyman also wrote to Cramer encouraging him. While the administrative plans were not firm, it appears that what was planned was an autonomous unit or institute devoted to actuarial science within the mathematics department similar to the statistical laboratory.

Cramer hesitated for some time, and ultimately events in Sweden overtook matters. Cramer was offered and accepted the post of rector of the University of Stockholm, and he then declined the Berkeley invitation in fall 1949. He did, however, visit Berkeley in the summer of 1950, when he attended the Second Berkeley

Symposium on Mathematical Statistics and Probability, and met with a group of faculty who were thinking about the future of actuarial science at Berkeley.

In a related direction, the campus had sponsored for decades a small program in insurance, first located in the economics department, but which subsequently moved to the School of Business Administration. The program consisted of a single faculty member, A. H. Mowbray. Albert Henry Mowbray graduated from UC Berkeley in 1904 and worked in the insurance industry and for several governmental agencies concerned with insurance for a number of years before joining the Berkeley faculty in 1923. He was an eminent, nationally recognized expert on insurance matters who served on many governmental advisory boards on insurance issues. However, he had no formal background in mathematics or statistics. For many years he oversaw the training of students at Berkeley who would go into the insurance industry.

Mowbray was scheduled to retire in 1949, and as early as 1946, the School of Business Administration was already thinking about his successor. Stefan Peters, who was working as a research actuarial analyst for the West Coast Life Insurance Company in San Francisco and who was teaching an insurance course in University Extension, came to their attention. Peters was born in Posen, Germany (now Poznań, Poland), on June 27, 1909. After graduating from Gymnasium in 1927 he seems to have been a peripatetic student, studying mathematics at the universities of Heidelberg, Berlin, Freiburg, Erlangen, and Göttingen. He received his PhD in Erlangen in 1930 with a dissertation under Wolfgang Krull on infinite Abelian groups. After teaching briefly at Göttingen and Erlangen, he departed or was forced to depart academic life in March 1933 and went into actuarial work, first in Germany and then in Italy. In 1938 he emigrated to the United States and was employed in actuarial work in New York until he joined the US Army in 1943. He became a naturalized citizen in 1944 and after discharge from the army moved to Berkeley and resumed actuarial work [UA, Peters file].

After Peters had been identified as a possible successor to Mowbray, Evans met him and was sufficiently impressed by him that he hired Peters as a temporary part-time lecturer in mathematics beginning in the spring of 1948 to teach new courses on life contingencies and population statistics, the start of what might be an actuarial curriculum. That fall, as the date of Mowbray's retirement neared, the School of Business Administration began a formal search for his replacement. With Mowbray's endorsement, the business school selected Peters as their candidate for the position and proposed his appointment as an associate professor of insurance to be effective July 1, 1949. When Mowbray died suddenly of a heart attack over the Christmas 1948 break, the proposed Peters appointment was advanced by six months, being approved to begin on January 1, 1949, and structured at two-thirds time as associate professor of insurance and one-third as lecturer in mathematics [ibid.].

The oath controversy (see Chapter 9) intervened, and since Peters was one of the nonsigners, he was fired by the regents in August 1950. After the successful lawsuit against the regents on the matter of the oath, to which Peters was a party, he was offered reinstatement to his faculty position in November 1952. However, he declined this offer and continued his career in actuarial work in the insurance industry in New York, where he had started out in 1938 when he arrived in the United States, and to which he had returned after being dismissed by the regents. The courses in life contingencies and population statistics continued to be taught and became part of the statistics department curriculum when the two departments separated in 1955. Cramer finally came to Berkeley as a visiting professor for the fall semester of 1953, a visit that was made possible by a donation from Walter Heller, one of the long-time patrons of the university [UA, CU 5, series 3, Box 37, folder 44]. The School of Business Administration hired a tenured faculty member to replace Peters, but no program or institute in actuarial science was ever developed in the mathematics or statistics department.

The second element of Evans's dream, a bureau of computation, has an interesting story involving one of the first competitions between Berkeley and UCLA. After the war, it had become increasingly clear that large-scale computation would have a profound impact on mathematics and science generally. The first fully electronic digital computer to be built was the ENIAC (Electronic Numerical Integrator and Computer), built between 1943 and 1945 at the University of Pennsylvania School of Engineering under contract with the Army Ordnance Bureau and the Ballistics Research Laboratory (BRL) at Aberdeen. This machine was to be used for ballistic calculations, and the preparation of firing and bombing tables. After the machine was constructed and tested, it was moved from Philadelphia to Aberdeen. As already noted, several Berkeley mathematicians—Morse, Lewy, and Morrey—and a future Berkeley mathematician, John Kelley, had served at Aberdeen during the war. In addition, Leland Cunningham, an expert in celestial mechanics on the faculty in the astronomy department at Berkeley, had been at Aberdeen leading a computational group.

In the spring of 1945, Col. Leslie Simon, the director of BRL and the person who was responsible for getting the ENIAC project started, was looking for help in numerical analysis and the use of the new machine. Following a suggestion of Evans, he recruited Dick Lehmer to come to Aberdeen to help out with the ENIAC. As already noted, Lehmer, with his father, had been an early pioneer in constructing electromechanical computing devices, sieves, for examining problems in number theory. Lehmer accepted the invitation and spent the 1945–1946 academic assigned to Aberdeen. Writing to Evans during his time at Aberdeen, he reported that

> The ENIAC is running much better than at first and breakdowns are getting rare and short…. I am learning about electronic computation the hard way…. I have been lay-

ing a few plans [not on company time of course], rather grandiose perhaps, for two or three special units for dealing with number theory problems which would be better by a factor of 1000 than anything we now have. I hope that the new plans for department space will include a room for an electronics laboratory. [DM Lehmer file]

Even before the advent of electronic digital computers, many faculty members found that they had to confront large data sets in their work and were beginning to use IBM punch cards for recording and processing data. In 1943 and 1944 the need was being felt for the creation of a centralized data-processing or computing facility with IBM-type equipment that would be available to the faculty, perhaps on a recharge basis. What was envisaged was data processing and simple calculations rather than "number crunching" as it would be called today. Harry Wellman, a future acting president of UC, was tasked by Sproul in 1944 to chair a committee to look into setting up such a computing or data center. Neyman, whose war work involved processing large amounts of data, urged action, but in spite of the positive recommendations of the Wellman committee in November 1944, not much happened. A group at UCLA had also urged creation of such a center as early as 1943 [CU 5, series 4, Box 11, folder 4].

In April 1946, C. D. Shane, the director of Lick Observatory, wrote to Sproul urging creation of such a center, citing the benefits for astronomy. Also in April 1946, Samuel Herrick, a young faculty member in astronomy at UCLA, again urged creation of such a data and computing center. The dean of engineering at UCLA, L. M. K. Boelter, lent his enthusiastic support, and more importantly offered space. Boelter also proposed a direct approach to Thomas Watson, the head of IBM, to ask for a donation of equipment from IBM similar to the donation of equipment IBM had made to create a data and computing center at Columbia University. Sproul gave instructions to Boelter to back off from IBM, saying that he, Sproul, himself would make the approach, which in the end was unsuccessful [ibid.]. In general, it appears that UCLA was rather better organized and more aggressive in pursuing computing than was Berkeley. Dean Boelter was already playing a large role and would play an even larger role in the future. Boelter had been on the Berkeley engineering faculty since 1919, and in 1944, when he was associate dean, he was selected to organize the new College of Engineering at UCLA. Similar administrative leadership in developing computing at Berkeley seemed to be lacking.

Lehmer returned from his work with ENIAC in July 1946 with great enthusiasm for computing. In particular, he wanted to build on and expand his work with his father in the 1930s, when they had constructed electromechanical sieves for use in number-theoretic calculations by taking the next step of constructing an electronic sieve. He wrote to Vannevar Bush at the Carnegie Institution about his idea and received an encouraging response [DM, Lehmer file]. Another early Berkeley advocate for electronic computers was Paul Morton, in the Electrical Engineering Division of the College of Engineering. In 1948 he started construction of what he called the CALDIG (or California Digital Computer). In subsequent years there are notes indicating that the electrical engineering shops were working on CALDIG and on Lehmer's electronic sieve, but there is no indication that either project was completed.

But then in July 1946, an entirely new option opened up. On July 8, 1946, President Sproul received a letter from J. H. Curtiss, assistant to the director of the National Bureau of Standards (NBS), announcing the intention of the NBS to create a mathematical computation laboratory tied "with an educational institution with a strong mathematical department, good library facilities and a research program which gives rise to a natural interest in computational facilities." The letter also suggested a desire, based on geographical diversity, to locate this facility on the Pacific coast, and it went on to say that representatives would come calling to discuss UC's interest in hosting such a facility. The letter closes by noting, "I have already contacted D. H. Lehmer of your faculty on the matter of this computational laboratory." The funding for this facility was to be partly NBS money, but a substantial part of funding would be money from the military that was funneled through NBS. It was clear that the military would have a large say in the form and location of the facility. It was well recognized that the army's Department of Ordnance had an excellent facility (ENIAC), while the navy and the army air force had nothing. The other services were therefore eager to enter the electronic computing field. The NBS had the virtue of being a neutral party through which to route money. (See [UA, CU 5, series 4, Box 11, folder 4] for the documentation for this and the following five paragraphs.)

This invitation immediately set off a competition between Berkeley and UCLA in which Sproul sought to be meticulously neutral. Berkeley faculty and administrators met with the visitors and made a proposal. On August 21, Evans wrote to Curtiss supporting the proposal, and Lehmer on August 26 wrote to Wellman, citing the benefit to mathematics of having the center at Berkeley. But by mid-September, the wind was already blowing toward UCLA when Sproul received a letter from Col. Mesick, of the army ordnance department based in Pasadena, saying that while Caltech and USC had some interest, they were not as far along as UCLA. He went on to say, "By personal conference with Dean Boelter [of UCLA], I feel sure that he is in a position to take the initial steps toward the organization of a cooperative computing center as outlined in the minutes of the previous meeting." Mesick concluded by stating his personal opinion that the center should go to UCLA. No mention was made of either Berkeley or Stanford in this correspondence. Discussions and interviews with institutions and individuals continued through the fall. Professor C. D.

Shane was selected to be the lead person on the Berkeley proposal. The lead person at UCLA was Samuel Herrick, and UCLA provost Dykstra sent in their proposal in what is described as a carefully prepared and effective letter. Stanford was very much in the picture, for they apparently were willing to provide space for the facility on very generous terms. Curtiss himself visited California in late November and early December to talk to representatives of the institutions competing for the facility.

Then on January 17, Curtiss announced that the field has been narrowed to Stanford and UCLA. In a meeting on January 21, President Sproul and Provost Deutsch, of the Berkeley campus, discussed the situation. Sproul memorialized the conversation as follows:

> I told him [Deutsch] of Curtiss' report on the Computing Center, in which he stated that it is likely to go to Stanford because of the building facilities available there, but that if a perfectly free choice were made, it would go to UCLA, "because of the up and coming, progressive men on that faculty, so different than [the] confused and stodgy attitude of the Berkeley faculty." Dr. Deutsch was disposed to believe that what Mr. Curtiss, who was a graduate of Berkeley, complains of, is really a credit to us, in that it reflects the scholarly standards of our group and their abhorrence of promotional methods. Nevertheless, he agreed with me that we should do more to let our light shine before men.

It is clear from the tone of these memoranda that Sproul was distressed that Berkeley had been muscled aside in this competition.

In subsequent correspondence in February to Provost Dykstra and Professor Herrick, at UCLA, Curtiss made clear some of the politics of the selection process, namely, that if the Office of Naval Research and the Commerce Department (in which NBS resided) predominate in the decision, sentiment would favor Stanford, but that if the army, and especially the air force, participate, then the balance would swing to UCLA. Curtiss suggested that UCLA find ways of getting the army involved, and indeed, UCLA immediately formulated a six-point program of efforts through various agencies and individuals to influence the decision in favor of UCLA. It is also noted in these plans that that on July 1, 1947, Stanford would lose Dr. Schaeffer, who was looked on as a drawing card by Curtiss. Finally, UCLA put on the table a very generous offer in terms of space for the institute, in fact offering it temporary space and then permanent space in a planned new engineering building complex.

The fact that UCLA was offering up state-funded academic program space was an irritation at the Berkeley campus and of concern to President Sproul. In early April, Provost Deutsch met with a group of senior faculty members, including Shane (director of Lick), Evans, Lehmer, Cunningham, O'Brien (dean of engineering), and Howe (associate dean of engineering). These faculty members were unenthusiastic about allocating valuable permanent academic space to a government agency.

Although they also acknowledged that such computing equipment was essential, they felt that it would be best for the university to plan and acquire such equipment on its own and that the NBS lab might prejudice such development. Deutsch, perhaps reacting to his conversation with Sproul, wrote to Curtiss proposing that the lab be located at Berkeley and that while Berkeley could not provide space, it could provide land on which NBS could construct a facility. This offer was politely rejected, since it was clearly not competitive.

In the end, UCLA won the competition. Edward Condon, the director of NBS, came to California in late May and June, first meeting with Sproul and then with Dykstra to "close the deal." Sproul was of the view that Stanford was just as much out of the picture at the end as was the Berkeley campus, and was kept in the foreground of the discussions merely for trading purposes. The formal name of the facility was to be the Institute for Numerical Analysis (INA), and indeed the INA was only one part of a larger initiative of the NBS.

The bureau announced with some fanfare the creation on July 1, 1947, of the National Applied Mathematics Laboratories of NBS, which would consist of four operating units: (1) the Institute for Numerical Analysis at UCLA; (2) the Computation Laboratory, then in New York, but to be moved to NBS in Washington, D.C., in 1948; (3) the Statistical Engineering Laboratory at NBS; and (4) the Machine Development Laboratory, also at NBS. For a description of the program, see [Cur].

By the summer of 1948, the INA was up and running in temporary space, and the most important event was the arrival from NBS in Washington of Harry Huskey, who would assemble a team and build the electronic digital computer that would be the centerpiece of INA. We will meet Harry Huskey a little later, when he joined the Berkeley mathematics department. Suffice it to say that Huskey and his team completed the construction of the machine in under two years, and it became operational in July 1950 [Hus]. It was of relatively modest size physically (as contrasted to the behemoths that were typical of early computers), but it was the only digital electronic computer on the West Coast and was briefly the fastest machine around.

Huskey's SWAC (Standards Western Automatic Computer) was a success scientifically and was used for a variety of problems in science, mathematics, and engineering including especially problems in number theory. In 1951, INA succeeded in attracting D. H. Lehmer to become its director. This was in the middle of the oath controversy, and Lehmer no doubt welcomed the opportunity to take leave without salary from Berkeley. As director of INA he was an employee of the federal government. Lehmer returned to Berkeley after two years as director after the oath controversy had died down. Professor C. B. (Charles Brown, or just plain Tommy) Tompkins succeeded him and became director of numerical analysis research, the successor organization to the INA, which resulted when, in 1954, NDS could no longer launder money from defense agencies to fund INA. It became a UCLA research unit, and under Tompkins's leadership, UCLA developed a strong presence in computing. In 1953–1954, INA was a sizable operation with an annual operating budget of over $700,000 and 70 employees [CU 5, series 3, Box 23, folder 2]. Thus, Evans's idea of a bureau of computation had gone to UCLA, probably in a much grander form than he had imagined; and Dick Lehmer had almost gone with it.

It was widely recognized at the time that Berkeley was very much behind the curve in computation, and in fact, the Wellman committee recommendation from 1944 to establish a data-processing and computing facility had not been acted on for many years. On February 4, 1949, William Dennes, dean of the graduate division, wrote to Sproul urging the creation of such a computing center at Berkeley. On April 4, 1950, Sproul authorized Dennes to set up a committee to study the matter. The committee consisted of Shane, Jones, and Wellman, plus Lehmer and Morton as technical experts. They recommended setting up such a facility in the new electrical engineering building that would operate on a recharge basis. The facility would include the "electronic digital computer now being completed by Professor Morton." On October 4, 1950, Sproul authorized creation of the facility, and it went into operation on November 6. However, Berkeley would continue to lag in computing for many years to come.

Attempts to develop the third part of Evans's dream of an institute of mathematics and mechanics began after Evans stepped down as chair on July 1, 1949. Indeed, one of the first projects undertaken by Morrey as the new chair was the implementation of Evans's idea of an institute of applied mathematics and mechanics. When Richard Courant was passing through Berkeley in August 1949, Morrey, as new chair of the department, asked him whether he and some of his colleagues might be interested in moving to Berkeley. Courant had been impressed with what Evans had achieved in building up Berkeley with strong appointments, notably Morrey and Lewy, among many others. Morrey by this time had become a deeply respected leading figure in partial differential equations, and Hans Lewy, who also participated in the discussions with Courant, was also widely and deeply respected for his work in partial differential equations and was a student and colleague of Courant from their Göttingen days.

Courant's relationship with Berkeley went back many years, to the spring of 1932, when he visited Berkeley as a part of a lecture tour of the United States. Constance Reid, in her biography of Courant, reports [Reid 1976, p. 135] that "the American University which charmed him most was Berkeley," and she reports that in response to a question whether on this trip he ever considered remaining in the United States, he replied, "yes, fleetingly at Berkeley" [Reid 1976, p. 136]. This is an interesting observation, since Berkeley at the time was rather weak as a research department, but perhaps Courant saw that mathematics at Berkeley could make great strides under a strong leader, as it did ultimately under Evans. Courant, as mentioned earlier, was also offered a one-semester visiting appointment at Berkeley in 1934 after he had left Germany, but declined when New York University (NYU) made an offer that had greater prospects for permanence.

Courant appeared to take the initiative from Berkeley in 1949 very seriously. According to Reid's biography, Morrey had initially thought in terms of attracting Courant, Freidrichs, and Stoker to Berkeley, but the initiative soon expanded to include nine full- and part-time faculty, a group of students, and support staff [Reid 1976, p. 272]. Such an infusion of new faculty and programs at Berkeley would be the fulfillment of Evans's dream of an institute of applied mathematics and mechanics at Berkeley. Discussions and negotiations continued throughout the year between Berkeley and Courant, and various options of shared appointments were explored. According to Reid, Kurt Friedrichs believed that "Courant really wanted to be in both places. That was always what he wanted. ... When he came to a fork in the road he would quite literally want to go both ways. You could actually feel that!" [Reid 1976, p. 275].

Morrey brought President Sproul into the discussion about moving the Courant group to Berkeley in October. See [CU 5, 1950, 400-Mathematics] for documents related to the next four paragraphs. In January 1950, Morrey made a formal proposal to President Sproul outlining in detail his proposal for the creation of an institute of mathematics and mechanics as a separate budgetary unit headed by Courant. Courant had an ongoing grant from the Office of Naval Research (ONR) that provided major support for the group at NYU (about $140,000 per year), and it was assumed that his grant would be transferred to Berkeley if the group moved. Morrey proposed that altogether twelve mathematicians, essentially the staff at NYU, be brought to Berkeley—namely Richard Courant, Kurt Friedrichs, J. J. Stoker, Fritz John, Harold Grad, Joseph Keller, Harold Shapiro, Peter Lax, Clifford Gardner, Arthur Peters, Eugene Isaacson, and Bernard Friedman. It was proposed they would hold fractional appointments in the Department of Mathematics ranging between 50% and 67% for a total of just under seven FTE and departmental salaries of $43,800. The balance of their academic-year salaries and summer salaries would be paid by the ongoing ONR grant, although the university would take responsibility for the full academic-year salaries, thus "backstopping" the ONR support. Additional monies would be needed for the support budget and graduate-student support.

This proposal represented a major expansion of the mathematics department (12 new FTE on top of the 29 FTE in the department), although about five of these would be paid for by ONR on soft money, at least as long as the soft money continued to flow. The net ongoing additional cost to the University would be realistically on the order of $60,000, exclusive of the backstop of the ONR part of the salaries, which was about another $30,000. On March 8, Sproul sent Morrey's proposal to the Academic Senate's Committee on Educational Policy for comment. In response to questions from the senate, senior staff of the department had met to discuss the proposal and to determine how much the department could "contribute" toward the proposal. There were five retirements coming up in 1951–1955, and the department determined that it could forgo replacements for two of these, and that they could possibly forgo one of four instructorships that were available. Moreover, Morrey corresponded with Courant, relaying some other questions from the senate committee reviewing the proposal.

The senate committee recommended against the establishment of the institute as proposed on May 1, 1950. The committee's reasoning is an interesting commentary on a deep conflict that was developing concerning the mission of the university. On the positive side, they wrote that "the field of applied mathematics is eminently worthy of cultivation," and noted that "the field developed largely in Europe, and it was only recently that attempts were begun to develop it in this country." The committee opined, "If judging on the grounds of content alone, we would be inclined to approve the project."

But then on the negative side, the committee wrote at some length, "Our main concern, however, is with the educational implications of the proposal. This is not the only question of this sort which has been before us this year. We are coming to feel more and more strongly that the establishment of institutes should be viewed with great skepticism." They expanded on this point:

> If by the creation of Institutes the University puts more and more weight on research, it is in danger of neglecting its primary function of instruction of youth of the State. An Institute like this one will in all likelihood displace some other and more fundamental activity. It will not be an addition to our activities, but it will involve a subtraction elsewhere. We wonder if the people of the State will feel that the University is discharging its educational obligations if it permits the sort of shift which seems to be underway from teaching to research to continue.

The committee also complained that the institute was not interdisciplinary, since the proposed appointees were all mathematicians and could be accommodated within the Department of Mathematics. This discussion took place just a few weeks after Dean Davis's presentation to the regents on institutes, and it is possible that Morrey's proposal had been a factor in motivating Sproul and Davis to plan this presentation.

This report effectively ended the discussion at the Berkeley end, although one suspects that the idea was also essentially dead at the New York end by this time. It appears that a significant portion of the University of California was not ready for such an institute. The university was in the midst of a painful transition to readjust the balance between teaching and research. Ten years later, the reception of such a proposal might have been very different.

Of course, an obvious question that one can ask in retrospect on this whole episode is whether Courant was more interested in trying to improve his position at New York University than in bringing his group to Berkeley. Reid's assessment is that, overall, Courant was very serious about Berkeley. Morrey, who was in the best position to judge Courant's intentions, went to considerable lengths and effort to develop and to sell the idea to President Sproul and others and would presumably not have done so unless he believed that Courant was serious. But Courant also initiated discussions with the leadership of New York University and succeeded in improving the situation for himself and his group. Subsequently, he was able to win for NYU an Atomic Energy Commission contract for the installation of a large computer, a UNIVAC IV, and in 1953, he parlayed that coup into the realization of his goal of creating within New York University an Institute of Mathematical Sciences that contained within it a sizable computing facility. In the fall of 1953, the institute moved into new quarters along with its new computer.

The developing loyalty oath controversy, starting in the late spring of 1949 and culminating in the summer of 1950, no doubt played some role in the failure of this attempt to attract Courant and his group to Berkeley. The oath controversy, as we shall see in the following chapter, occupied the time and energy of faculty and administrators, diverting it from matters of recruitment, and also cast something of a stigma on Berkeley. Nevertheless, it is fascinating to speculate "what if" this initiative had been successful. What would have been the impact on Berkeley and on the development of American mathematics?

Reflecting on the Berkeley possibility years later, Courant remarked to Alexandre Chorin that it was too bad it that it did not work out, for applied mathematics would have been less "ghettoized" if he had come to Berkeley in 1950 [Chorin, personal communication]. Joe Keller made the same point in different words, as reported by Reid: "If we had come to Berkeley in 1950, we would immediately have become part of the leading mathematical establishment in the country and we would have had a ten or fifteen year head start on what we have been able to do here" [Reid 1976, p. 276]. Keller perhaps confuses the standing of the Berkeley mathematics department in 1950 with its standing in 1960 or 1965. On the other hand, if Courant and his group had come to Berkeley, it is hard to avoid the conclusion that, like Neyman, Courant would have sought to create an independent department, and applied mathematics would now be quite different at Berkeley, since one of the trademarks at Berkeley has been the seamless continuum of pure and applied mathematics in one department.

Chapter 9

The Oath Controversy

What became known as the California Oath Controversy began quietly in March 1949 when the Board of Regents, upon the recommendation of President Sproul, added to the standard oath of allegiance that had for some years been required of all public employees in California, including university employees, a disclaimer oath requiring university employees to attest to nonmembership in any organization that advocated the overthrow of the government. Shortly thereafter, the text was modified at the June 1949 regents' meeting to include specific mention of nonmembership in the Communist Party. All university employees would be required to sign the new oath as a condition of continued employment. This requirement aroused the ire and passion of many members of the faculty, and opposition to the imposition of this disclaimer oath by the regents without prior consultation grew among the faculty. Anger at the regents also grew among the faculty, while at the same time there was considerable dissension among the faculty about how exactly to respond to the oath requirement. Basic academic freedom and the principles of tenure were seen to be at-risk, and indeed tenure at UC existed only by the force of custom rather than formal policy—see the end of this chapter. Indeed, in this controversy, fundamental issues of university governance ultimately came to the fore. The oath controversy peaked in August 1950 with the dismissal of a group of faculty members who had declined to sign the oath, but the controversy continued on for several years afterward. David Gardner's book [Gardner 1967], which is based on his UC Berkeley doctoral dissertation, is an indispensable source for this controversy.

The oath that was required is as follows, where the text up to the semicolon is the standard (and noncontroversial) oath of allegiance previously required of all public employees. The remainder was the new disclaimer oath:

> I do solemnly swear (or affirm) that I will support the Constitution of the United States and the Constitution of the State of California, and that I will faithfully discharge the duties of my office according to the best of my ability; that I am not a member of the Communist Party or under any oath, or a party to any agreement, or under any commitment that is in conflict with my obligation under this oath.

The controversy began over the new disclaimer oath itself, but then soon shifted to the issue of the legitimacy of the regental policy that had been in place since 1940 that banned employment at the university of members of the Communist Party. Regents saw the new oath as a way of "implementing" the 1940 policy on nonemployment of members of the Communist Party, and so the issues were connected. And finally, the issue became a power struggle between the faculty and the regents over who had the authority to judge the qualifications of an individual to be a member of the faculty. The oath controversy, as it evolved and grew over the following time, ended up dominating faculty life on and off campus. Professor George Stewart, in *The Year of the Oath* [Ste, p. 9] wrote,

> In that year [1949–1950] we went to oath meetings, and talked oath, and thought oath. We woke up and there was the oath with us in the delusive bright cheeriness of the morning. "Oath" read the headlines in the newspaper, and it put a bitter taste into the morning coffee. We discussed the oath during lunch at the Faculty Club. And what else was there for subject matter at the dinner table. Then we went to bed, and oath hovered over us in the darkness, settling down as a nightmare of wakefulness.

The short history is that after much faculty turmoil, many strained and inconclusive negotiations between faculty committees and the regents seeking a compromise, and attempts at mediation by other groups, all but a very small number of faculty signed the oath, some happily, but many with grave misgivings and disgust, linked with fear of unemployment. The controversy reached a climax in August 1950, when the regents, in a sharply divided vote, ended up firing the small group of 24, mostly tenured, faculty members who had declined to sign the oath (the nonsigners). This regental action was in most cases contrary to the recommendations from the Academic Senate's Committee on Privilege and Tenure. In his discussion of the loyalty oath controversy in his memoirs [Kerr 2001, pp. 27–47], Clark Kerr asserts that "the loyalty oath caused the single greatest confrontation between a faculty and its board of trustees in American history."

It was never alleged that any of the nonsigners were members of the Communist Party or were disloyal or were professionally unfit. Their sole offense was refusal to sign the disclaimer oath. The nonsigners then sued the regents, and ultimately, in late 1952, the California Supreme Court held the disclaimer oath unconstitutional and ordered the nonsigners reinstated, provided they signed a new oath (the Levering oath) that the legislature had subsequently required of all California public employees. Most of these faculty members resumed regular duties in 1953, although a few resigned because they had become established in other positions and did not wish to return to UC.

From the perspective today of more than fifty years later, many aspects of the loyalty oath controversy have a degree of strangeness and unreality. But, these events should be placed in the context of their time: the postwar years, the beginning of the Cold War, and the onset of a virulent species of anticommunism as a political phenomenon. This was a time of turbulence in international affairs marked by concern about expansion of Communist rule and power, and concern about the effectiveness of the basic American foreign policy of containment. At the same time, there was growing concern about the possibility of domestic Communist subversion and espionage, and its impact on national security. The nature and level of the concern varied with international developments and with the flow of domestic politics. All of these contemporaneous contextual factors played a role in shaping the oath controversy, and our understanding of the oath controversy should be informed by a brief description of them.

The Soviet Union had recently consolidated its hegemony over Eastern Europe and was attempting to dislodge the Western powers from Berlin by blockading ground access to the city. This blockade was lifted only in the spring of 1949. Meanwhile, it was becoming clear that the Chinese Communist Party would defeat the Nationalists and seize power in China. The final collapse came in late 1949. The Soviet Union successfully tested an atomic bomb in August 1949, long before it was estimated that it would be able to achieve this goal. This event also raised the question of whether the USSR had gained atomic secrets via espionage, and it soon became clear that they had. The North Koreans, backed by China and the Soviet Union, invaded South Korea in June 1950, and during the initial months of the conflict, the war was going badly for South Korea and the United States. This initial phase of the Korean Conflict coincided exactly with the time that the loyalty oath controversy reached its peak of intensity, and the events in Korea no doubt influenced the outcome of the oath controversy.

One response to the perceived internal Communist threat of espionage or subversion was through administrative and judicial proceedings. The Truman administration instituted loyalty board hearings to screen federal employees and remove those who were members of the Communist Party of the United States (CPUSA) or were otherwise regarded as disloyal or unfit for service. Then, the federal government decided to bring criminal charges under the Smith Act against twelve CPUSA leaders for conspiracy to teach and organize the overthrow of the government by force and violence. The trial, which began in early 1949, ended in the convictions of the defendants in October of that year. The convictions were upheld unanimously by a federal appeals court in August 1950, and it was virtually a foregone conclusion that the US Supreme Court would agree, which it did in June 1951. Ultimately, more than a hundred leaders of CPUSA were convicted and imprisoned under the Smith Act.

In January 1950, Alger Hiss, a former official in the State Department, was convicted of perjury for denying membership in CPUSA and denying acts of espionage in his testimony before a grand jury (the charge was perjury because the statute of limitations had run out on the espionage charges). In February 1950, Klaus Fuchs, a British physicist of German birth who had been employed in the Manhattan project, confessed to passing atomic secrets to the Soviet Union, and received a prison sentence in Great Britain. His confession implicated other CPUSA members in the United States and led to several arrests in the spring and summer of 1950, culminating in the arrests in July 1950 of Julius Rosenberg and later his wife, Ethel. The Rosenbergs were subsequently convicted of involvement in espionage during the war and were sentenced to death. Both the Hiss case and Rosenberg cases stirred strong and lasting passions on both sides that persist even today. These judicial events were taking place at the same time as the oath controversy and no doubt influenced its outcome.

While these judicial proceedings, which were aimed at the CPUSA leadership or those members who had been involved in espionage, went forward, anticommunist feeling in the political realm reached hysterical levels, and many who were not members of CPUSA but who were perhaps former members, self-described progressives, or so-called popular-front liberals were swept up into the maelstrom. The controversy was used by many for partisan political gain. Both the Un-American Activities Committee of the US Congress and the Un-American Activities Committees of the California legislature (the Tenney Committee and the Burns Committee) were actively pursuing what they saw as subversion and espionage. Many former members of CPUSA as well as many popular-front liberals and "fellow travelers" were called to testify, their loyalty and integrity questioned and in many cases their reputations unfairly besmirched or careers ruined. A number were fired from their jobs, and some went to prison on charges of contempt for refusing to answer questions.

In February 1950, Senator Joseph McCarthy made his famous speech in which he asserted that he had a list of names of a number of State Department employees who were members of the Communist Party. His speech set off a vitriolic controversy that led to many congressional hearings and to ugly forms of political warfare and unjustified personal attacks on many innocent individuals. In his memoirs, Clark Kerr aptly summed up the context of the loyalty oath controversy: "The national context was an evil one, and the loyalty oath conflict internally became the greatest contaminant ever to enter the body politic of the University. The years 1949–50 witnessed the worst ravages of anti-Communist hysteria nationally and the height of the loyalty oath controversy inside the University" [Kerr 2003, p. 28].

A key underlying issue in the California Loyalty Oath Controversy and in the national debate was a disagreement over the nature of CPUSA and the fitness of its members for university employment. In its starkest form, one side saw CPUSA as an indigenous progressive party of the left whose positions and urgings should compete with those of other political parties under the traditional protections of the

First Amendment. The other side saw CPUSA as a criminal conspiracy, effectively under the control of a foreign government and devoted to the violent overthrow of the US government—a view reflected in the Smith Act cited above. In addition, it was claimed that CPUSA maintained a rigid party discipline of thought and speech, so that CPUSA members were thought to lack intellectual independence. Many individuals held intermediate or more nuanced views between these two extremes, but it is this latter view that dominated public opinion in this period.

Likewise, it was a widely, but not universally, accepted view that membership in CPUSA, in and of itself, was a bar to employment as an instructor at UC and more generally to any employment by UC. Indeed, as noted, the regents in 1940 had adopted a policy barring members of CPUSA from university employment. (As an interesting sidelight, the regents adopted this policy in the aftermath of their action to dismiss a teaching assistant in the Berkeley mathematics department, Kenneth May, who was a member of CPUSA. The dismissal had been recommended by the department chair, the senate Committee on Privilege and Tenure, and President Sproul [Gardner 1967, p. 276, n. 25].) The faculty never objected to or challenged this policy on nonemployment of members of the Communist Party.

It is clear in retrospect that Sproul had seriously misjudged the likely faculty response to the disclaimer oath that he proposed in March 1949. Perhaps one reason is that there was no protest from the faculty over the policy on nonemployment of members of the Communist Party, and so perhaps he did not expect a serious protest over what could be regarded merely as way of implementing this policy. Sproul did consult established Academic Senate advisory committees in the period between the regents' meetings of March and June 1949, when the oath came before the regents again for another vote. In response to a question from a regent about any flareback from the faculty on the oath, Sproul responded, "None from the Advisory Committees which you have authorized me to negotiate with on the problem. You can assume that there will be no likelihood of considerable flareback from the Senate" [as quoted in Gardner 1967, p. 46].

As an example within the university community of the sentiments regarding the criminal nature of CPUSA, Regent John Francis Neylan, who was the leader of the pro-oath group of regents, responded with brutal bluntness in a meeting with a faculty committee in September 1949 to a question about whether the ban on employing Communists would set a bad precedent: He said, "Communism is not a political party, but a criminal conspiracy and…the disqualification of a person in the communist movement would [not] set a bad precedent any more than the dismissal of members of Murder, Inc." [as quoted in Gardner 1967, p. 67].

Another expression of this view was contained in a speech by President Sproul on November 1, 1949, to the American Bankers Association, when he said,

What place in a university can appropriately be given to such purblind fanatics as those who use a false and brutal hope to persuade the young and gullible to sign away their American birthright? ... I am not thinking, of course, of men who merely hold unorthodox political or economic views, but of members of that close-knit, rigidly controlled conspiracy mislabelled Communist Party, every member of which is thoroughly indoctrinated, and cannot possibly be ignorant of the obligations he has undertaken, and the discipline to which he has committed himself. No man can be a member of this subversive organization without taking on the coloration of its leaders and sharing in their guilt. [as quoted in Gardner 1967, pp. 76–77]

Edward Tolman, a distinguished senior faculty member in psychology who became the leader of the nonsigners, wrote to the regents on July 18, 1950, concerning his reasons for not signing the disclaimer oath at the height of the controversy. In this letter, he endorsed the regents' policy on nonemployment of member of the Communist Party:

One must clear away matters on which there is no disagreement. ... Membership in the Communist Party or any other organization which advocates the overthrow of the government by force or violence disqualifies anyone from the privilege of teaching at the University of California. All of us recognize that loyalty to any doctrine of totalitarianism shackles the free pursuit of truth. [UA, CU 5, series 4, Box 39, folder 16]

This view was widely shared among the faculty. Indeed, in March 1950, during the controversy, the Academic Senate conducted a mail ballot of its members on the issue of whether UC should bar members of CPUSA from employment. The vote was 79% in favor of the policy banning the employment of CPUSA members.

Clark Kerr, among many other prominent faculty members, supported the policy. In his memoirs, he offers his thoughts on the issue:

I was both alienated by Communist attacks on me and by their blind support of a horrendous regime. I thought I was seeing a failed political ideology in its final death throes, not a threat to democracy or western capitalism. I never understood those who thought that the United States was threatened internally by a Communist conspiracy. Either they doubted the strength of our democracy, or they were using the Communist threat as away of attacking liberals. Yet as a Berkeley faculty member, I voted against employing Communists as members of the University faculty and applied the anti-Communist policies of the Board of Regents as chancellor and president. But I also insisted on proof, not just allegations of party membership, and fought the Burns Committee, at great personal cost. [Kerr 2003, p. 69]

The American Association of University Professors (AAUP) held a sharply contrary view on employment of members of the Communist Party to the effect that "professional unfitness could not be inferred from association nor determined by imputing to an individual the characteristics known generally to identify any category or organization" [as quoted in Gardner 1967, p. 13], and rejected the notion that Communists, in the absence of demonstrable professional unfitness, should be barred from teaching positions. Some faculty members shared this view, but as is evident from Tolman's letter cited above and from other evidence, this was a minority view at the time.

The context of the senate vote mentioned above was that a committee of the faculty charged with negotiating with the regents (the so-called Committee of Seven, of which Griffith Evans was a member) came to believe that a deal with the regents to end the oath controversy might be possible. This was predicated on the belief that what was really important to the regents was to have the faculty accept and endorse the regental policy on nonemployment of members of CPUSA and that the regents were perhaps less concerned with the oath itself, since it was simply one among possibly other ways of implementing the underlying policy. The deal that many imagined would be that if the Academic Senate would vote (by an overwhelmingly margin) to endorse the policy on nonemployment, then the regents would pull back on their insistence on the disclaimer oath. As noted, the senate did vote by a wide margin in support of the policy, and the results were reported to the regents. But, when the regents met on March 31, 1950, and considered a motion to rescind the oath, the vote was a 10–10 tie and the motion to rescind therefore failed. The faculty thereafter coined a term for this action: the "Great Double-Cross" [Ste, p. 39].

The administration had essentially stepped to the side in this controversy, although Sproul, who had initially proposed adding the disclaimer oath, ended up supporting the faculty position as a member of the Board of Regents, as did Governor Warren, also as a regent. The regents, as the vote on the motion to rescind shows, were themselves divided between a pro-oath faction, led by Regent John Neylan, and the group opposed to the oath that had formed around Governor Warren. Interestingly, Neylan was absent from the March 1949 meeting when the regents first voted to institute the disclaimer oath, and indicated in June when the matter came back to the regents that he opposed such an oath [Gardner 1967, pp. 40, 42]. Hence, both Sproul and Neylan had changed sides from their initial positions in early 1949.

At their next meeting, on April 21, the regents, in response to an effort by the Alumni Association to mediate the conflict, dropped the disclaimer oath as coupled with the oath of allegiance, and substituted for it an essentially identical statement that was to be included in the appointment letter that faculty were required to sign and return to the administration. This statement read as follows:

> Having taken the constitutional oath of office required of public officials, I hereby formally acknowledge my acceptance of the position and salary named, and also state that I am not a member of the Communist party or any other organization which advocates the overthrow of the Government by force or violence, and that I have no commitments in conflict with my responsibilities with respect to impartial scholar-

ship and free pursuit of truth. I understand that the foregoing statement is a condition of my employment and a consideration of payment of my salary.

Thus, in a technical sense, the disclaimer oath was rescinded, but it reappeared, just as real, in a different form. This form was perhaps felt by some to be a bit less objectionable. At this April 21 meeting, the regents also formally agreed that if a faculty member did not sign this statement, he or she was to be entitled to a full hearing before the Academic Senate Committee on Privilege and Tenure on the reasons for failure to do so. The regents agreed not to take action on dismissals until the recommendations of this committee had been submitted to the regents through the president. The resolution implementing these changes passed with only one dissenting vote.

The faculty was in considerable turmoil following this April 21 "Alumni Compromise," and a vigorous debate ensued about whether to "accept" it. Many moderate and conservative faculty members urged acceptance. However, positions generally hardened on both sides over the next months with a small group of faculty determined even more to resist and a group of regents determined to hold to their position. There was also discussion about what criteria the Committee on Privilege and Tenure should use in reaching their recommendations concerning dismissal of nonsigners, but no clear resolution of the matter.

The small group of faculty members had in the end declined to sign the modified disclaimer statement (oath), and were then subject to dismissal. The intellectual and moral leader of the nonsigners was Professor Edward Tolman, a distinguished psychologist who had served on the faculty for 32 years, had been selected as faculty research lecturer, and had served as president of the American Psychological Association. It was at this point that the controversy, which had already shifted from the oath itself to the regents' policy on nonemployment of Communists, shifted to the issue of who had the authority to determine the fitness of a person to be member of the faculty: the faculty senate or the regents. The Committee on Privilege and Tenure (of which Griffith Evans was a member), as stipulated by the regents, held a hearing for each of the 36 nonsigners and recommended against dismissal for 31 and for dismissal of the other 5. (In the end, but after the fact, the committee recommended against dismissal for all of them.)

In the course of the hearings, the committee inquired into the attitude of the nonsigners toward the Communist Party and possible membership in it. If the faculty member was forthcoming to the committee about his or her attitude toward the Communist Party and presented testimony or other evidence that they were not a member of the party or of other subversive organization and were under no commitment that would interfere with their loyalty to the United States, and if they presented compelling reasons for not signing the oath, the committee would recommend continuation of employment. Some nonsigners declined to discuss these issues, asserting a "presumption of innocence" and saying that until some evidence of membership in CPUSA was presented, employees should not be required to clear themselves of the imputation. Those who took this stance were generally not recommended for continuation by the committee. It should be said that the committee found no evidence of disloyalty in any of the nonsigners and found them all to be competent scholars and objective teachers.

President Sproul endorsed the recommendations that 31 nonsigners who had been cleared be retained and that the five nonsigners not cleared by the committee be terminated. At the July 21 regents' meeting, the recommendations for the dismissal of the five were accepted unanimously, while the recommendations for retention of the 31 were also accepted, but only by a vote of 10–9. The regents voting against the motion for retention felt that the criteria the committee had used diverged substantially from those that the pro-oath regents had imagined would be employed when they passed their resolution. Regent Neylan, cognizant of the fact the some members of his pro-oath faction had not been able to attend the meeting, announced that he was changing his vote and that he would ask for reconsideration at the next regents' meeting. This action effectively put everything on hold for a month. The final confrontation came at the August 25 meeting, when the regents rejected the recommendations of the faculty committee and President Sproul for retention by the narrow margin of 12–10 and acted to dismiss all the nonsigners.

The regents then gave the nonsigners ten days and one last chance to sign or to tender their resignations and receive a year's severance pay. The list of 36 was thereby reduced to 24 (including 21 who had been recommended for retention and 3 who had not), who were then dismissed from the faculty. Four of the 24 nonsigners dismissed were members of the Berkeley mathematics department: John Kelley, Hans Lewy, Stefan Peters, and Pauline Sperry. The department was thus definitely overrepresented among the nonsigners. Lewy, Peters, and Sperry had been recommended by the committee for continued employment, based on their providing evidence to the committee that they were not members of CPUSA and describing their principled reasons for not signing. Sperry's reasons for not signing were rooted in her Quaker religious beliefs, while Lewy and Peters were moved to resist by their experiences in Nazi Germany nearly two decades before.

In his statement before the committee Lewy began by writing,

> ...I am not a communist and have sworn so to the Federal Government on occasion [no doubt this was required for his work at Aberdeen during the war], and that I detest the stifling of free thought and regimentation which attend the establishment of dictatorships in the world. Why then not say so to my employers on my employment contract?

He wrote then that there were four reasons for his refusal to sign. The first three concerned procedure and the regents' intentions and authority in the matter.

But then he continued thus:

> There is a fourth and deeper reason. I have lived half of my thinking life in Europe, half in America. In Europe any person travelling, citizen or foreigner is required to keep at all times the police informed about his whereabouts, by presenting himself in person at the police station of every new town where he stays more than 24 or 48 hours, according to country—whether he stays at a hotel or with relatives or friends. Most Americans heartily dislike this system and would resent its introduction here. Yet it entails only a few minutes discomfort—less time than would be involved in signing an oath every morning before or after breakfast. Now, why should anyone whose motives for travelling are pure, object to signifying them to the constituted authorities? Especially when to do so makes the apprehension of crooks so much easier since they would take pains to show up at the police station of every city they choose to go to? The answer is evident. Most Americans prefer, to their great good fortune, a system based on general good faith rather than on general distrust. To the immigrant from Europe, this belief in good faith has been the inspiration of his loyalty to his adopted country; it is worthy of a sacrifice to help preserve it intact. Belief in good faith has, to its immense advantage been the spirit of the University of California in the past. It has inspired the loyalty of its faculty to the University. Now the shadow of suspicion has been cast over the campus. The University has lived its first year of the new atmosphere of suspicion. The harm done to the University is written all over it. Where is the alleged good it would do? [Lewy, Box 2, folder 51]

Kelley was one of the three who were not recommended for continued employment. His case was complex and deserves further elaboration. Unless otherwise noted, documentation for the narrative in the following seven paragraphs is contained in University Archives Bancroft Library [UA, CU 5, series 4, Box 39, folder 7]. During the war, Kelley had worked at Aberdeen Proving Ground on ballistics research, and then from 1945 to 1949, he been a consultant to the Manhattan Project and its successor, the Atomic Energy Commission (AEC), on ballistics. He performed classified research for the AEC under a security clearance. He received commendations for his work at Aberdeen and for his work for the AEC that were couched in the highest laudatory terms. While at Aberdeen, he and E. J. McShane had collaborated with a permanent employee of Aberdeen, Franklin V. Reno, on an army ordnance book on exterior ballistics. Mr. Reno later came under suspicion, since apparently he had been at one time a CPUSA member.

Kelley was interviewed by FBI agents in December 1948 about his association with Reno, and in the course of the interview, Kelley gave permission to the agents to examine his papers. The agents found what they believed were classified documents in his possession, which they believed were not being safeguarded according to AEC regulations. Kelley stated that he believed the documents were properly in his possession, and if they were now classified, then they had been subject to a reclassification without his knowledge. Subsequently, he received a letter from the AEC in which it was stated, "In view of your apparent indifference toward security regulations, it was felt that a renewal of your contract was not justified. No other considerations entered into this consideration." Kelley then applied to the AEC for a hearing in order to clear his name on this matter.

Kelley had also been subpoenaed to appear before a grand jury in New York in January 1949 that was investigating his coauthor F. V. Reno. Kelley testified under oath that he was unaware of Reno's membership in CPUSA. Reno was never indicted on any charges, but Kelley felt that both he and McShane had come under suspicion because of their scientific collaboration with Reno, a sign of the malicious tenor of the times. Kelley also testified under oath before the grand jury that he was not a Communist and was not affiliated with any subversive organizations, and in addition said that he had repeatedly so stated officially, as required in connection with his various governmental duties.

Morrey, his department chair, wrote to the Committee on Privilege and Tenure that on the basis of his long and intimate association with Kelley starting in 1942 at Aberdeen, he was prepared to swear under oath that Kelley was not a member of the Communist Party. Morrey also reported that Kelley had commented humorously that the Department of Justice could probably be paved with his sworn statements that he was not a Communist.

In a letter of Dean Alva Davis, of the College of Letters and Sciences, to Sproul of May 15, 1952 (nearly two years after the committee hearing), Davis stated that he was told that Kelley had appeared before the committee ready to present a statement that he was not then, nor had ever been, a member of the Communist Party, but that due to irritation caused by the nature of the questioning by one of the committee members, he withdrew his statement and refused to answer any more questions. William Duren, the chair of the mathematics department at Tulane, who had arranged a visiting faculty position for Kelley in his department after he was fired from UC, offered a statement indirectly confirming the possibility of such a scenario in a letter to Morrey in March 1953. He wrote, "I am positive from first hand knowledge of his [Kelley's] reaction to friendly advances from real or suspected Marxists that he has no common cause whatever with them though he would still refuse to try to prove this to you in any way except that required by law. ... [H]e lacks normal common sense in self protection" [DM Kelley file].

Kelley's refusal to make a statement to the committee about his nonmembership in the Communist Party bothered the committee, and they described his refusal as "a great mistake." The committee was also apparently disturbed by the fact that Kelley's security clearance had apparently been discontinued. Although every member of the committee agreed that they believed that Kelley was a loyal American and not a Communist, and furthermore had the highest regard for his scholarship and teaching, they recommended that his employment be terminated. President Sproul concurred, and Kelley's employment was indeed terminated by the regents at their July 21 meeting.

On July 31, 1950, the deputy general manager of the AEC, correcting earlier correspondence, stated unambiguously in a letter to Kelley that the AEC had never denied his security clearance. This letter was subsequently forwarded to the Committee on Privilege and Tenure, and on August 31 the chair of that committee wrote to Sproul, forwarding a copy of the letter, and stating that had the committee had this letter at the time of the hearing, it would have recommended in favor of continuance of Kelley's appointment, and that it so recommended now. This, of course, was moot as far as the regents were concerned, for they had already acted to fire all the nonsigners, irrespective of the senate's and Sproul's recommendations. But it was a significant action by the senate.

In his history of the controversy, David Gardner wrote,

> The irony was that not one of those dismissed was accused by any Regent of being a Communist or in sympathy with any other organization allegedly subversive. Furthermore each had been found by the Committee on Privilege and Tenure to be a competent scholar, an objective teacher, and untainted by disloyalty to the country. How the Regents of the University of California came to sever from the institution's service men and women against whom no charge of professional unfitness or personal disloyalty had been laid is an extraordinary study in futility. [Gardner 1967, p. 3]

Gardner further observes of the final confrontation in August,

> The authority of the Board of Regents and the Academic Senate in the appointment, promotion, and dismissal of members of the faculty would be tested, and those on the faculty and on the Board who held unyielding to their principles were in the final confrontation nearly isolated, resented by many of their colleagues, and condemned for an intransigence that their associates regarded only as harmful to the general welfare of the University. [Gardner 1967, p. 139]

In addition to the nonsigners who were dismissed, a number of faculty members resigned out of disgust and frustration as a result of the controversy. In addition, there were many documented cases of offers to come to Berkeley being declined owing to the fallout from the controversy. By 1949, Berkeley had risen to be ranked within a small group of the most distinguished private universities, but it was generally believed in Eastern circles that the loyalty oath controversy would finish Berkeley as a major research university. However, Berkeley proved to be more resilient than these dire predictions suggested and recovered from the grievous wounds of the controversy, regaining and enhancing its academic stature and then going on to earn the ranking of the "best balanced distinguished university in the country" in the 1964 American Council on Education rankings.

As noted, the Berkeley mathematics department lost John Kelley, Hans Lewy, Stefan Peters, and Pauline Sperry through dismissal. All four were ultimately offered reinstatement as described below. Kelley and Lewy resumed duties in 1953, while Sperry elected to retire after her reinstatement, and Peters never returned. In addition, two assistant professors, Charles Stein and Paul Garabedian, simply resigned from the university in 1950 before it became necessary for them to sign the loyalty oath. They were both permanent losses. When the department attempted to hire Henry Scheffe a year later, Scheffe turned the offer down because of the loyalty oath. Fortunately, Neyman did not give up, and Scheffe was successfully recruited in 1953 after the oath crisis had passed. Leon Henkin similarly declined a job offer, but accepted later after the controversy had been settled.

Garabedian took a position at Stanford and remained there for about a decade before departing for the Courant Institute at NYU. Stein went first to Chicago in a temporary position, but shortly thereafter was recruited by Stanford, where he has remained ever since and is now professor emeritus. Lewy spent some time at Stanford as a visitor, and it is not known whether Stanford made an attempt to recruit him. Thus, the main beneficiary in mathematics of the oath controversy was Stanford, and it is ironic that the person most involved in recruiting Stein and Garabedian, both of whom were extraordinarily grievous losses for Berkeley, was Albert Bowker, chair of the statistics department as Stanford, who twenty years later would become the Berkeley chancellor [Albert Bowker, personal communication].

Immediately after the regents' decision to dismiss the nonsigners, faculty support groups were formed, and many donated a percentage of their salaries to a fund that would provide financial support to their dismissed colleagues. Professor Raphael Robinson, of the mathematics department, served as treasurer of this group in the north [DM, Raphael Robinson file].

The loyalty oath controversy then moved in two different directions. First, it spread outward from the university more squarely into the political arena of the state, although in some sense, it was already there. Governor Warren called a special session of the legislature for September 21, 1950, asking it to enact a loyalty oath for all public employees. Within five days, the legislature enacted what became known as the Levering oath, which the governor then signed into law on October 3. It seems puzzling at first that the governor, who had voted in support of the faculty's opposition to the oath as a regent, was now actively supporting the Levering oath. However, the idea of a loyalty oath was immensely popular with the public, and it should be observed that Warren was in the midst of a campaign for a third term as governor.

The regents' disclaimer oath was in effect aimed particularly at UC faculty and was perceived as such, although formally it covered all university employees. The Levering oath covered all public employees—state, county, and municipal as well as all civil-defense workers. Both the faculty and those regents, including the governor, who supported them in the oath controversy, had argued against the regents' loyalty oath on the grounds that it singled out for special treatment university employees. Although there was thus a distinction between the two oaths that the governor could invoke, politics was probably more important.

The Levering oath was worded as follows:

I, _____, do solemnly swear (or affirm) that I will support and defend the Constitution of the United States and the Constitution of the State of California against all enemies, foreign and domestic; that I will bear true faith and allegiance to the Constitution of the United States and the Constitution of the State of California; that I take this obligation freely, without any mental reservation or purpose of evasion; and that I will well and faithfully discharge the duties upon which I am about to enter.

And I do swear (or affirm) that I do not advocate, nor am I a member of any party or organization, political or otherwise that now advocates the overthrow of the Government of the United States or the State of California by force or violence or other unlawful means; and within the five years immediately preceding the taking of this oath (or affirmation) I have not been a member of any party or organization, political or otherwise that advocated the overthrow of the Government of the United States or the State of California by force or violence or other unlawful means except as follows: _____ (if no affiliations, write in the words "no exception") and that during such time as I am a member or employee of the _____ (name of public agency) I will not advocate or become a member of any party or organization, political or otherwise, that advocates the overthrow of the Government of the United States or of the State of California by force or violence or other unlawful means.

After the Levering oath had been put in place, the regents confronted the issue of whether it should apply to university employees. An argument based on the constitutional independence of the university (Article IX, Section 9 of the California Constitution) could be made that it would not apply to the university. However, after considerable discussion of this constitutional issue, the regents determined that university employees were subject to the Levering oath. But at the same time, the regents did not withdraw their own loyalty oath. Hence faculty and staff now had to sign a second oath, which arguably was more intrusive and a greater affront to civil liberties than the original regental oath. Its only advantage was that it did not single out university employees, and this was an important point for many.

The second direction in which the oath issue moved was into the courts, and the narrative here is based on [Gardner 1967, pp. 223–244]. A group of 20 of the dismissed nonsigners brought suit against the regents on August 31 in the State

District Court of Appeals. The lead plaintiff was Edward Tolman, and the case was known as Tolman vs. Underhill, since Robert Underhill was the treasurer and secretary of the regents. The main arguments made by the petitioners were (1) that the California Constitution (Article XX, Section 3) specified an oath of allegiance for public officials and precluded any other oath such as the regents' disclaimer oath; (2) that the regents' oath violated the constitutional prescription that the university be kept independent of all political or sectarian influence (Article IX, Section 9); and (3) that the regents' action was procedurally flawed and furthermore violated understandings regarding tenure that dismissal would be only for moral turpitude or incompetence, neither of which had been demonstrated. Oral arguments were heard by the court on December 22, and the tenor of the hearing, especially the questions from the bench, gave some encouragement to the petitioners. On April 6, 1951, the court handed down a unanimous opinion in support of the petitioners based on grounds (1) and (2) above and ordered the regents to reinstate the nonsigners.

After a contentious debate, the regents decided not to appeal the Appeals Court verdict to the California Supreme Court, but some individual regents announced their intention to make an appeal as individuals. In the event, the Supreme Court, on its own motion, took the case on May 31 along with several other oath cases, including constitutional challenges to the Levering oath. The Supreme Court also suspended the decision of the Appeals Court and the reinstatement order, thus leaving the nonsigners high and dry for the moment. The Supreme Court heard oral arguments nearly immediately, on June 20, 1951, but the decision did not come down for another 16 months. It is worth noting that the Appeals Court finding on point (1) above argued by the petitioners would also call into question the constitutionality of the Levering oath. As a safeguard against such an argument, the Levering oath was put on the ballot in November 1952 as a constitutional amendment. It passed by a large majority.

Meanwhile, Governor Warren had the opportunity to make several new appointments to the Board of Regents, with the result that the center of gravity of the board shifted toward the Warren faction and away from the faction led by Regent Neylan. At the November 16, 1951, meeting, the regents finally acted by a vote of 12–6 to withdraw their own loyalty oath in favor of the Levering oath. At the same time, the Regents reaffirmed their policy barring Communists from employment and, somewhat curiously, declined to offer reappointment to the nonsigners on condition that they sign the Levering oath. The Supreme Court case would thus continue to go forward, an outcome that was apparently desired both by the regents and the nonsigners. The nonsigners hoped that the Supreme Court would endorse the constitutional arguments (1) and (2) that had been the basis of their victory in the District Court of Appeals, thereby specifically rebuking the regents on principled constitutional grounds.

There were some additional complications concerning the case of John Kelley. After being dismissed in August 1950, he found a temporary visiting position at Tulane University, as mentioned above. But the money for this position ran out after two years, so that in the spring of 1952, Kelley was facing unemployment. It was widely expected that the Supreme Court would rule prior to the summer of 1952, and that under the expected result, the nonsigners could be reinstated for the 1952–1953 academic year. Kelley was being pursued by other universities with offers of permanent employment, but since he preferred to resume his career at Berkeley, he was in a quandary.

Morrey, as department chair, came up with the idea of requesting that Kelley be appointed either as regular associate professor or as acting associate professor as a "new" appointment beginning July 1, 1952. Kelley was prepared to sign the Levering oath. He was also not a party to the lawsuit Tolman vs. Underhill, which challenged the constitutionality of the regents' oath. The regents had acted to bar the employment in any capacity of anyone who was a party to the lawsuit, but had excluded from the ban anyone who was not a party to the suit. Therefore, nothing in principle stood in the way of this (new) appointment for Kelley, and Morrey was moving it forward. But then in May, the regents extended their ban on employment to all nonsigners, so that at the beginning of June, Morrey's proposal was dead, and Kelley had no employment for the following year. Then, in the middle of July, G. Bailey Price, chair of the mathematics department at the University of Kansas, came to the rescue and was able to arrange a visiting position for Kelley at Lawrence.

Kelley recorded his thoughts at this time on the regents' oath, which he had been unwilling to sign, as compared to the Levering oath, which he would ultimately, but with reluctance, sign:

> I do not want to sign the Levering act gizmo, although to me it's quite a different kettle of fish from the ex (oh happy day) loyalty oath. It is not imposed by the Regents, who incidentally, by their rather brave action in withdrawing the old oath, have convinced me that the University has a chance of becoming a decent place again. But it seems to me that the Levering act, as applied to U.C., is unconstitutional in one more way than the old oath, and I think the Regents should have done something about testing the legality. Nevertheless, brawling with the Legislature is a new sort of activity. It is also true that I prefer not to have been fired, and the Court decision should clarify both of my worries. [DM, Kelley file, letter to Charles Morrey, February 11, 1952]

The Supreme Court decision, which was handed down on October 17, 1952, was a victory for the nonsigners, but the grounds for the decision were a serious disappointment. The court ruled that "state legislation [the Levering oath] has fully occupied the field, and that university personnel cannot properly be required to execute any other oath or declaration relating to loyalty other than that prescribed for all state employees." The court ordered the nonsigners reinstated provided that they signed the Levering oath. This point of preemption was never an argument raised by the petitioners either in any hearing or brief, and was in some sense moot, since the regents had already withdrawn their oath in favor of the Levering oath. In other cases decided at the same time, the court upheld the constitutionality of the Levering oath and in so doing indirectly overruled the constitutional grounds (1) and (2) on which the nonsigners had won at the Appeals Court level.

The nonsigners were offered reinstatement effective January 1, 1953, but because of the short notice and prior commitments, some chose to delay their reinstatement until July 1, 1953. John Kelley and Hans Lewy rejoined the faculty at that time. Kelley, as already noted, had spent two years at Tulane and then one year at the University of Kansas. Lewy had visiting positions at Harvard and Stanford. Pauline Sperry chose to retire from the university in 1952, when she was 67. Stefan Peters resigned from the university and remained in New York. Derrick Lehmer signed the oath under duress and had taken leave of absence to serve as director of the Institute for Numerical Analysis of the National Bureau of Standards, located at UCLA, starting in 1951. There had been concern that he might not return if the oath controversy were not settled satisfactorily. After the court decision ordering reinstatement of the nonsigners, he did in fact return, in 1953.

It was assumed that the nonsigners would be provided some back pay, the most reasonable suggestion being back pay equal to the difference of what their earnings would have been at UC minus any earnings during the period, with adjustments for any special expenses incurred. However, the regents dug in their heels at such a request. The regents argued that the Supreme Court had ordered reinstatement, but had not ordered back pay, and they were loath to do this unless specifically ordered to do so by the courts. The regents asked the Supreme Court for clarification on this point, but nothing came down from the court. The debate over this issue put on display the enduring enmity Regent Neylan felt toward the nonsigners. He wrote in a letter to his fellow regents on October 16, 1953, "After dragging the University into a brawl over the signing of a simple noncommunist declaration proposed by their own colleagues, they have lamely signed a drastic loyalty oath. … I think the Regents have learned they cannot buy peace with this small group of conceited little men who still seek to rule or ruin the University" [UA, CU 5, series 4, Box 39, folder 1].

The regents appointed a committee, which was chaired by Neylan, to investigate and make a recommendation on the issue of back pay. The committee report was completed in December 1953 and recommended rejection of the claim for back pay [UA, CU 5, series 4, Box 39, folder 1]. The full Board of Regents subsequently endorsed that stance. Then, in April 1954, a group of nonsigners sued the regents in Sacramento Superior Court for back pay. The case, which became known as Kelley vs. Regents, with John Kelley as lead plaintiff, languished in the court until March 1956, when the regents decided to settle the case for a payment of $162,037.89 in back pay [UA, CU 5, series 4, Box 39, folder 7].

The university and the department survived this crisis and returned to the pathway to growth in stature and excellence, although Charles Stein and Paul Garabedian were permanently lost to Berkeley. The department was fortunate to have been led by the steady hand of Chuck Morrey, who served as chair from 1949 to 1954. But, the controversy left its mark on the university. Clark Kerr comments in his memoirs, "Roger Heyns (Chancellor at Berkeley 1965–1971) once said to me that any time he traced the origins of the problems he endured from the faculty in the second half of the sixties, he was led back to the loyalty oath" [Kerr 2003, p. 28]. As a form of closure, Clark Kerr, as president, nominated Edward Tolman, the leader of the nonsigners for an honorary degree in 1959, a nomination that the regents endorsed, but with some dissension [Kerr 2003, p. 70]. In 1962, Kerr also persuaded the regents to name the new building on the Berkeley campus, which housed the Department of Psychology and the School of Education, Tolman Hall [Kerr 2001, p. 140]. As to legal closure to the controversy, it should be noted that in 1967 the California Supreme Court in Vogel vs. County of Los Angeles voided the disclaimer portion of the Levering oath, thus returning the UC oath requirement to the pre-1949 situation. Also in 1972, the California courts in Karst et al. vs. Regents declared the 1940 regental prohibition on employment of members of the Communist party to be unconstitutional, thus returning matters to the pre-1940 stance.

Finally, the loyalty oath controversy highlighted a major academic personnel policy issue that had been, at least officially, in a state of limbo, namely tenure. For decades it had been the common understanding in the university that full and associate professors were tenured faculty members in the sense that they enjoyed a right to continued employment and could be dismissed only on grounds of moral turpitude or incompetence. However, this understanding was nowhere embedded in the regulations of the university as approved by the regents. Each year, every faculty member received a letter that was in effect a reappointment letter, but was phrased more in terms of specifying the faculty member's salary for the coming year. The loyalty oath controversy put a spotlight on this ambiguity, and demonstrated that everyone's common understanding lacked a clear basis in university regulations. It took some heavy lifting, especially by Clark Kerr, to fix this problem, and it was not until 1958 that regental policy formally recognized tenure [Kerr 2001, p. 140].

College of California, Oakland, ca. 1860

Early Berkeley campus scene, looking west, ca. 1875

William Welcker

Irving Stringham

Mellen Haskell

Derrick N. Lehmer, 1920s

George Edwards, ca. 1920

George Edwards, at a track meet ca. 1920

Pauline Sperry, 1930s

Emma Trotskaya Lehmer, 1928

D. N. Lehmer (middle) and D. H. Lehmer (right) with R. C. Burt of the Burt Scientific Laboratories, Pasadena, where the congruence machine was operating in 1933

Top view of the congruence machine (1933) showing the three series of cog-wheels. The teeth of the gears each have a small hole that can be stopped up. A "solution" occurs when all holes are open.

Sophia Levy McDonald, 1930s

John McDonald, 1936

Benjamin Bernstein, 1936

Bing Wong, 1936

Griffith Evans, ca. 1940

Julia Robinson, ca. 1943

F. N. David, Elizabeth Scott, David Blackwell, and Evelyn Fix, ca. 1962

Jerzy Neyman, ca. 1940

Charles Morrey, 1968

Derrick H. Lehmer, 1968

Hans Lewy, 1975

Alfred Tarski, 1968

John Kelley, 1968

Tosio Kato, 1968

Abraham Taub, 1983

Julia Robinson, 1975

Raphael Robinson, 1968

S. S. Chern, 1968

Chern and Mrs. Chern with President Ford, 1978

Chern receiving National Medal of Science from President Ford, 1978

Rufus Bowen, 1974

Gerard Debreu, 1975

Evans Hall

Wall art in Evans Hall: "Galois"

Mother Functor

Vol. I, No. 3   Nov. 8, 1971

ELITISM IN THE DEPARTMENT

Disquieting thoughts and perturbed sensibilities have been insistently nudging one another in my mind since reading the first issue of our non-establishment newsletter, "Up Against the Blackboard", or whatever it is to be, and bumping straight-on into a bit of rather blatant intellectual snobbery. Some people have been campaigning recently for amity and goodwill between secretaries and academics, but what tortuous and unfathomable routes some are taking to effect it. Let me explain.

I am a secretary - apparently in some minds one of those university subbeings. The secretarial profession had always seemed to me to be an honorable one, requiring a certain amount of skill and brains, and a modicum of commonsense. Unfortunately, of late it has been brought to my attention, in a way that cannot be ignored, that this might not be the case, and that I may have been laboring under a false delusion.

Firstly, our staff columnist, Thomasin Saxe, questions the continuation of our profession at all, on the grounds that it is a dehumanizing one, and that for 8 hours a day we perform menial and servile tasks. Perhaps one should pose the question, where does service end and servility begin? Is there indeed a distinction? I say there is. No matter what one does, service to others is at least a small part of the order of the day. Would you eliminate every form of service? What about maids or janitors? Would you deny them an occupation? And physicians, what of them? Their function is to take care of other people, and it is also desirable that they satisfy and please their patients. We all serve others to some degree. Are we all capable of becoming researchers, or all-day creative thinkers exclusively? Could our society, or any existing society, even make use of us in such numbers?

What is deplorable about someone, say a faculty member, who is inundated with work, who asks for help, and another, who needs and wants to work,

responding to that request? (insofar as it is a request, and not an order or demand.) I hope that "do nothing for nobody" is not the trend of the future; the thought is shudder-provoking.

Secondly, I would like to comment upon the attitude of "Miss Anonymous Nice Girl", and the assumption of some other girl grads that secretaries are on the "lowest rung of the departmental pecking order" and that "we dump on each other".

Who said that secretaries were on the lowest rung? Who says that I am invading taboo territory when I enter the coffee room? Aren't you guilty of a presumption? It has been clearly stated that the coffee room is for the use of all the department, regardless of the position we hold. Hopefully secretaries will be around a lot longer than graduate students, and this is one good reason why this territory should not be off-limits to them. If this is the feeling of the grad students, let's hear more about it! I think we would all be interested. "Nice Girl's" horror at being mistaken for a secretary, heaven forbid, is also interesting and enlightening. I find her stance hypocritical. She deplores mathematical discrimination against women, but seems to take for granted a prejudiced attitude herself.

To summarize, I find my work sometimes enjoyable and often times interesting. I do not appreciate it when someone commiserates with me, and thinks it is a service to tell me I spend 8 hours a day in oppression and servitude. I do not want a faculty member volunteering to share my "menial" duties - or a grad student either for that matter, hoping to show the poor oppressed secretary you care by helping process the barnyard refuse. I am a secretary, and I do not feel that I have to justify that fact. I reject the putdown (unintentional or otherwise).

Doris L. Frederickson,
Secretary

It is strongly rumored that the four color problem
has been solved using a computer program.

Wall art in Evans Hall: Reeb foliation, 8 feet by 12 feet, signed "Zorro 7/22/1074"

MSRI groundbreaking ceremony on January 28, 1984: from left to right, Martin Kruskal, Calvin Moore, Robert Osserman, Hyman Bass, Reese Harvey, S. S. Chern, Irving Kaplansky (with shovel), Is Singer, Dan Quillen, Al Thaler

MSRI as seen from Grizzly Peak looking toward the Golden Gate, 2006

MSRI looking southwest, 2006

Chapter 10

Mathematics: 1950–1955

In this chapter we discuss the development of the mathematics department during the five-year period 1950–1955. This period marked some significant changes for the department. In terms of numbers, there were 28 faculty in professorial ranks as of June 30, 1950, but only 12 of whom would still be in the department on July 1, 1955 because of resignations, retirements and the splitting off of statistics. This period was also marked by a slow recovery from the ravages of the oath controversy, which had culminated in August 1950 just as the 1950–1951 academic year was about to begin. It was not until late 1952 and early 1953 that matters of reemployment were resolved, but the aftermath of the controversy hung for some time. This period was marked by the retirements of the remaining faculty who had been at Berkeley when Evans arrived, as well as the retirement of Evans himself. Altogether, there were eight retirements: Arthur Williams in 1950, Benjamin Bernstein in 1951, Pauline Sperry in 1952, Thomas Buck in 1953, Raymond Sciobereti in 1954, Thomas Swinford in 1954, and Sophia Levy McDonald in 1954. Evans formally retired in 1954 but was recalled for one year in 1954–1955 before he stepped down in 1955. Finally, this period saw the growth of the statistics group and the creation of a separate department of statistics, thus fulfilling Neyman's goal that he had pursued since his arrival in 1938. The creation of the separate department was approved in 1954, to be effective July 1, 1955. This chapter is devoted to appointments and issues other than those in statistics. In the following chapter we will discuss the appointments in statistics and the deliberations and actions that led up to the creation of an independent department of statistics in 1955.

There was a sharp drop in the size of the faculty in all ranks of the professorial series, from 28 in 1949–1950 to 22 in 1950–1951. There were no successful recruitments of new professorial faculty to Berkeley to start in 1950 because of the oath controversy, and six faculty members left: Charles Stein and Paul Garabedian by resignation, and John Kelley, Hans Lewy, and Pauline Sperry (as well as the part-time lecturer Stefan Peters) through action by the regents, and there was one retirement, that of Arthur Williams. A number of temporary instructors and lecturers were employed to ensure that classes were taught. The postwar bulge of enrollments was just ending, and in 1950–1951 enrollment stood at just over 22,000 students, down from over 25,000 during the period from 1946 to 1950. Enrollments from 1951 to 1955 were to remain steady between 18,000 and 19,000. It remained customary for the department to hire new PhDs as instructors, usually for two years. A successful instructor would then be promoted to assistant professor, and the unsuccessful ones would be let go. In some cases, promotion to assistant professor came after one year.

After the oath controversy had largely died down by late 1952, the department was able to begin to recruit from outside to replace the retirements, and so by July 1, 1955, the total size of the faculty in professorial ranks in mathematics and statistics was 28.5, with 18.5 in Mathematics, and 10 in statistics, the fractional position representing a joint appointment, which had come into being for the first time that year. This level of staffing is only marginally higher than the total level of staffing in 1949–1950, which was 28. This number of 28 from 1949–1950 consisted of 6 who could be identified as part of what would become the statistics department upon its formation, and 22 who would have remained in the mathematics department. (For these purposes, Loève is regarded as being in the statistics group, although eventually his appointment was split between mathematics and statistics.) It is evident from these numbers that the statistics wing of the department grew substantially during the period from 1950 to 1955 and achieved the critical mass necessary to become a freestanding department. At the same time, the size of the faculty in all other areas of mathematics actually declined during this period, reaching a level that was only marginally higher than the departmental size in 1934–1935, including those at the instructor rank, which at the time was a more permanent title than it was in 1955.

In 1952, Morrey appointed Ralph Lakness as assistant professor, the first new professorial appointment since 1949 outside of statistics. Ralph Mortimer Lakness was born in Ogden, Utah, on November 22, 1916, and grew up in Oakland, graduating from Castlemont High School in 1935. He then came to UC Berkeley, receiving his BA degree in 1939 in mathematics, and continued his studies in graduate school, receiving his MA in 1940. His graduate study was interrupted when he became an officer in the US Army in December 1941. Returning to graduate study in 1946, he went on to receive his doctoral degree in 1950 with a dissertation entitled "Green's Theorem for Subharmonic Functions for Multiple Spaces," written under the direction of Griffith Evans. He was then hired as an instructor in 1950 at the height of the oath controversy, and after two years was advanced to an assistant professorship.

His principal focus was on teaching: K–12 teacher preparation and the problems of mathematics education in the schools. He was an accredited secondary-school teacher, and served as university representative on committees dealing with credentialing of teachers. According to Professor Sophia McDonald, who had served in these roles for many years, and with whom Lakness worked, he was very knowledgeable about such matters and understood the problems faced by K–12 mathematics teachers. She, in fact, hoped that he would take over her role in K–12 mathematics education after her retirement. The department, however, decided not to replace the expertise and role that Professor McDonald had developed when she retired. This was no doubt a sign of changing priorities, and so Lakness departed the campus in 1955.

Lakness was hired immediately by San Francisco State University (SFSU) and was the first PhD faculty member at SFSU in mathematics. He served as chair for a decade and was very instrumental in building up mathematics at SFSU. The following is a brief summary of his distinguished career, kindly provided by his longtime SFSU colleague Frank Sheehan:

> During World War II, Ralph participated in the early stages of the field now called operations research. Some examples: During a long stretch of duty as an artillery officer in the Aleutians, he analyzed aiming procedures, made improvements, and produced practice firing results that caused Army superiors to visit his remote post (in disbelief). Convinced of the value of his work, the army adopted his improvements as their standard. As his next assignment, he was loaned to the air force, where he planned strategic targeting for bombers in Italy.
>
> Ralph came to San Francisco State in 1955; he was the first PhD mathematician on the faculty. At the time of his arrival, there was no mathematics department (a general education experiment had lumped all departments into divisions). In short order, Ralph brought the curriculum and requirements to an appropriate level for a degree-granting college. He hired other mathematicians, starting with Newman Fisher and Bob Levit and continuing over the ten-year period of his chairmanship. In all, 24 tenure-track appointments were made while he was chair, the majority from UC Berkeley.
>
> Soon after, Ralph moved the mathematics department to develop a graduate program that produced students with master's degrees. Many of them have had successful careers as teachers at community colleges across the state; some have gone on to complete PhD programs; and five have returned to permanent positions on the faculty at San Francisco State.
>
> Ralph served on many committees at San Francisco State. In particular, he was a member of the Academic Senate, helped to plan Thornton Hall (the current home of the mathematics department), and was a co-originator of the engineering program (now the School of Engineering).
>
> Ralph retired in 1995, after 40 years of outstanding service to the university. He died in 1999.

In the spring of 1952, there was an interesting exchange of correspondence and a senate review of a proposal that was a replay of the Courant negotiations in 1949–1950. Indiana University had over several years created a Graduate Institute for Applied Mathematics under the leadership of T. Y. (Tracy Yerkes) Thomas, who had been recruited from UCLA in 1944. The faculty at the institute consisted of Thomas, Vaclav Hlavaty, Eberhard Hopf, Clifford Truesdell, and David Gilbarg. Like Courant's group, they had Office of Naval Research funding as well as support from the army. Thomas was aware from discussions with Courant of the events of 1949–1950, and on March 7, 1952, he wrote directly to President Sproul suggesting that his whole institute be relocated to the University of California. Sproul again referred the proposal to the Academic Senate Committee on Educational Policy for comment. The committee report in April started with exactly the same words as the report two years earlier on the Courant proposal, that "the field of applied mathematics is eminently worthy of cultivation," but then went on to oppose the idea. They wrote that a new administrative structure did not seem to be needed to develop applied mathematics. The committee also reported that the mathematics chairs at Berkeley and UCLA were opposed to the proposal, preferring to build up their own groups gradually. Missing is the opposition in the committee's report on the Courant proposal on the grounds that such institutes distort the proper balance between teaching and research. In the end, the idea of importing this Indiana institute died.

In 1951 and 1952, Neyman had promoted three Berkeley PhD's who serving in temporary faculty ranks to the professorial staff in the statistics wing (Evelyn Fix, Elizabeth Scott, and Haary Hughes). With two separations (Benjamin Bernstein and Robert James), the total faculty in 1952–1953 then stood at 24. In 1953, after the oath controversy had subsided, recruiting was restarted in earnest, and seven new professorial faculty were hired; and with the reinstatement of Kelley and Lewy, the department added nine FTE. As there was one retirement (Buck), the faculty grew from 24 to 32 on July 1, 1953. Three of the new appointments were in statistics—Henry Scheffe, Lucien LeCam, and Terry Jeeves—and these will be discussed in the next chapter. The four appointments in other areas of mathematics were Paul Chambre, Murray Protter, Harley Flanders, and Leon Henkin.

Paul Chambre had been appointed as an instructor in 1951, and after two years was advanced to assistant professor on July 1, 1953. Chambre was born in Kassel, Germany, on August 7, 1918, and became a naturalized US citizen in 1943. After graduating from UC Berkeley with a bachelor's degree in physical chemistry in 1941, he worked in industry as a chemist and theoretician during the war. He then joined a research project on jet propulsion at the Courant Institute at New York University, making use of his expertise in combustion and chemical kinetics. After receiving a master's degree in applied mathematics in 1947 from NYU, he came to UC Berkeley to pursue doctoral studies. He worked in the Rarefied Gas Dynamics Project in the Division of Mechanical Engineering as a research engineer while he pursued his doctoral studies toward a degree in mechanical engineering. As a member of the project, he published a number of technical reports, and his dissertation concerning combustion and rarefied gas dynamics was written under his mentor Samuel Schaaf. Since Schaaf had received his doctoral degree in mathematics in 1944 under Griffith Evans, Chambre was actually a scientific grandson of Evans.

In writing to the dean of letters and sciences recommending the appointment, Morrey stressed again the desire to build up a group in applied mathematics, although this appointment was ultimately to test where the boundaries between applied mathematics and other disciplines lay for purposes of faculty hiring. Chambre was a talented teacher and was particularly successful in Mathematics 220, the graduate-level mathematics methods course for students in physics, chemistry, and engineering. This course had been made famous for many years by Professor Thomas Buck (and there are former students who still comment today on what a splendid course it was). Professor Buck had retired in 1952, and Chambre and others continued the tradition in this course, which was of considerable significance for the relations of the mathematics department with other departments in the physical sciences and engineering. Chambre also taught an equally successful undergraduate course in mathematical methods, Mathematics 120.

Chambre's tenure review in 1956–1957 posed squarely the question of where the boundary of applied mathematics fell. The disposition of the case was that Chambre be given a joint appointment, with 50% in mathematics and 50% in engineering, with the explicit understanding that in personnel reviews, mathematics would assess only the teaching and service aspects of the case as related to the department, while engineering would have full responsibility to assess Chambre's research as well as his teaching and service record in engineering. Chambre was promoted to associate professor in 1957 and to full professor in 1962. Within engineering, his work focused on nuclear engineering, and his appointment was placed in that department. Over many years, his research in this area was productive and very highly regarded, appearing mostly, but not entirely, in journals devoted to nuclear engineering. His work included mathematical analyses of phenomena relating to nuclear reactor design, nuclear reactor safety, and analyses of the long-term performance of geological repositories for radioactive waste. He supervised the doctoral work of two students in mathematics and a number in engineering. Chambre retired from active duty in 1988 at age 70.

A second appointment in 1953 was that of Murray Protter. The appointment was as acting assistant professor, and the "acting" in the title was due to the lateness in the year; it was already July when the appointment was proposed. Within months of his arrival, the department proposed promotion to associate professor for Protter to be effective July 1, 1954. This recommendation was duly approved. Murray Harold Protter was born on February 13, 1918, in New York City. He attended the University of Michigan, graduating at the age of 19 with a BA in 1937 and going on to receive an MA in 1938. He then entered Brown University for doctoral study. This study was interrupted by a year's teaching at Oregon State University and by four years of war-related work, first in Washington, D.C., and then at Vought Aircraft Corporation. He received his doctoral degree from Brown in 1946 with a dissertation entitled "Generalized Spherical Harmonics," written under Lipman Bers. He was the second of Bers's 53 doctoral students, and hence he was a mathematical grandson of Charles Loewner.

Protter stayed on at Brown as an instructor for a year after receiving his degree, and then went to Syracuse University as an assistant professor. Syracuse University had built a mathematics department of some renown in the years following World War II, led by Abe Gelbart and including Charles Loewner, Lipman Bers, and Atle Selberg. Protter joined this group in 1947. This group soon imploded for institutional lack of support, and most of the group eventually departed. Bers and Selberg left in 1949, both to the Institute for Advanced Study, in Princeton, with Selberg remaining there permanently and Bers going on to Courant. Loewner left in 1951 to go to Stanford. Protter left in 1951 to spend time at the Institute for Advanced Study, and in 1953 he was recruited to Berkeley with strong recommendations from Marston Morse, Courant, Bers, and Loewner.

Protter worked in partial differential equations, specializing in the study of equations of mixed type. His work as an analyst was praised, and it was noted that this work was very relevant to the study of transonic flow. Morrey noted in arguing for the appointment that Protter could contribute to ongoing efforts to build applied mathematics. The theme of building in applied mathematics is thus a constant one from the time of Evans's arrival, and it continued through Morrey's chairmanship, and indeed into the present day. Protter was promoted to the rank of full professor in 1958, and his continued work in partial differential equations won him high recognition. His work proving a general uniqueness theorem for a boundary value problem first posed by Tricomi and his work on periodicity problems for pseudo-analytic functions won high praise. He supervised the doctoral work of 16 students during his career. Protter served with distinction as department chair from 1962 to 1965 at a time when the department was growing and was actively recruiting and trying to hold off raids on Berkeley by other universities. He was also active in the affairs of the American Mathematical Society, where he held a number of important offices. Even after his retirement in 1988, Protter has remained actively engaged in the affairs of the department and with professional societies.

The third appointment in 1953 was of Harley Flanders. Flanders had served as an instructor for two years starting in 1951, following a two-year term as a Bateman fellow at Caltech. Harley M. Flanders was born September 13, 1925, in Chicago. He received all his degrees from the University of Chicago—a bachelor's degree in 1947, a master's in 1948, and a PhD in 1949. His dissertation, written under Otto Schilling and André Weil, was entitled "Unification of Class Field Theory." His research interests were wide-ranging, covering differential geometry as well as aspects of number theory and algebra. For a number of years he was the only faculty member at Berkeley doing work in differential geometry. During his time at Berkeley, he supervised the doctoral work of four students. In the department he advocated for higher standards in undergraduate courses and initiated programs for high-school students and for gifted undergraduates. Flanders also had a strong-willed personality, and some of his early interactions with colleagues in the department were not smooth.

He was promoted to tenure as an associate professor in 1958. By this time, the academic market for mathematicians was beginning to heat up in the post-Sputnik years, and universities across the country increased their science, engineering, and mathematics programs and prepared for the baby-boom generation of college students that was soon approaching. Flanders, as well as other faculty members, received offers from other universities at salaries well above what they were receiving at Berkeley. This situation led to conflict between the department and the central campus administration when the administration—or really the Academic Senate Budget Committee, which served as the campus committee reviewing proposals for promotion and salary increases—was not willing to respond to these outside offers in a way that the department thought was appropriate. The situation first came to a head with Flanders, when in the spring of 1960, the administration did not respond to several outside offers, and Flanders resigned. He ended up at Purdue University, where he joined his mentor Otto Schilling, who had moved there recently from the University of Chicago.

The fourth hire in mathematics in 1953 was Leon Henkin, as an assistant professor. Leon Albert Henkin was born April 19, 1921, in Brooklyn, New York. He entered Columbia University in 1937 at age 16 and received his bachelor's degree in mathematics and philosophy in 1941. His strong interest in mathematical logic, which had been stimulated by Professor Ernest Nagel, of the philosophy department, led him to pursue graduate study in mathematics in the fall of 1941 at Princeton University, to which he had been attracted by Alonzo Church. But World War II soon interrupted his studies, and Henkin was involved for nearly four years as a mathematician in classified war work, first on radar, and then with the Manhattan Project, working on the design of isotope separation plants.

After the war, he returned to Princeton and quickly received his doctoral degree under Church in 1947 with a notable dissertation on completeness of first-order logic entitled "The Completeness of Formal Systems." His work featured a new proof of Gödel's theorem and also injected many new ideas into the subject. He remained at Princeton for two years as an instructor and Jewett fellow before assuming an assistant professorship at the University of Southern California in 1949.

With Alfred Tarski, one of the foremost and senior logicians in the world, on the faculty, and the desire to make Berkeley a "center of learning" in this field, it had been the goal at Berkeley for some time to add an up-and-coming outstanding young logician to the faculty. By 1952, attention had focused on Leon Henkin, and in February 1952, Morrey proposed the appointment of Henkin as assistant professor effective July 1, 1952. The appointment was at the top level of assistant professor in light of Henkin's three years of service as assistant professor at USC, and Morrey suggested that the department might wish to recommend promotion to tenure after only one year. The offer was approved but Henkin declined, saying that he wished to await developments in the oath controversy. Recall that the court case of the nonsigners was still before the State Supreme Court and would not be resolved until October 1952. This was another example of fallout from the oath controversy on hiring at Berkeley. In the meantime, USC promoted Henkin to tenure effective July 1, 1952.

In January 1953, Morrey renewed the initiative to bring Henkin to Berkeley. There was considerable correspondence and discussion that spring about whether the offer should now be at the tenured level or at the assistant professor level. In the end, the offer was made at the assistant professor level, which Henkin accepted, and in the process gave up a tenured position at USC. At all events, the department proposed his promotion to tenure within a month or two after his arrival in Berkeley, to be effective July 1, 1954, and this recommendation was approved forthwith. Henkin was promoted to full professor in 1958. His work in logic and foundations of mathematics widened and deepened and includes a massive joint project with Tarski and Monk on cylindric algebras. He supervised the doctoral work of 14 students during his career. Henkin was known as an exceptionally fine teacher, and his scholarly interests expanded further to include mathematics education both at the K–12 and collegiate levels. He was also deeply involved in campus shared governance through work in the Academic Senate, and he served as acting department chair for nearly two years in 1966–1968, helping the department through a governance crisis that will be described in Chapter 19. He served again as chair in 1983–1985. Although he retired in 1991, he has remained an active presence in the department and on campus.

No appointments were made in 1954, the last year of Morrey's chairmanship. It is not clear why there were no appointments in that year, but there is some correspondence in the files indicating that there were some administrative problems. Derrick H. (Dick) Lehmer took over as chair on July 1, 1954, and John Kelley assumed duties as vice chair.

In 1955, three appointments were made in mathematics. First, the department brought in two young mathematicians in the area of functional analysis: William

Bade and Henry Helson. These appointments built on the strong base of expertise in analysis in the department and broadened it into the areas, new to the department, represented by these two young scholars. Their work would complement the research interests of John Kelley and Frank Wolf. The third appointment represented a very new direction, in computing and numerical analysis, with Harry Huskey, who was appointed as an associate professor jointly with electrical engineering.

William Bade was appointed as an assistant professor on July 1, 1955. William George Bade was born in Oakland, California, on May 29, 1924. He attended San Diego High School, graduating in 1942, and then entered Pomona College. The following year he entered the navy for officer training and transferred to Caltech, where he majored in physics. After graduation in 1945, he served until 1946 as a naval officer. He resumed his studies at UCLA as a graduate student in mathematics, earning his MA in 1948 and his doctorate in 1951. His dissertation, "An Operational Calculus for Operators with Spectrum in a Strip," was completed under the direction of Angus Taylor. Taylor then recommended him to Morrey for appointment as an instructor at Berkeley for 1951–1952.

Then, an opportunity arose for him to go to Yale as an instructor to work with Nelson Dunford and Jacob Schwartz. At Yale he continued his research on spectral theory of operators, and also assisted, along with Robert Bartle, with the preparation of the monumental three-volume work by Dunford and Schwartz on linear operators. After three years at Yale, he returned to Berkeley as assistant professor in 1955. Promotion to associate professor (1959) and professor (1964) soon followed. His early work on spectral theory broadened to include work on Boolean algebras of projections, and then jointly with Philip Curtis of UCLA, he began work on a problem of Kaplansky about automatic continuity of homomorphisms of $C(X)$, the spaces of continuous complex valued functions on a compact space X. Their 1960 paper started the ball rolling and made very substantial progress. But, it was nearly 20 years later that the problem was resolved by others in a most unusual and unexpected way; the result is independent of the usual axioms of set theory, and is actually false if one assumes the continuum hypothesis. Bade has continued work on other problems in Banach algebras, even past his formal retirement in 1991. For nearly 10 years, Bade was vice chair of the department for graduate affairs, overseeing graduate admissions and financial awards, monitoring the progress of graduate students, and looking out for their welfare. His work contributed substantially to the health of the graduate program. During his career he supervised the doctoral work of 24 students.

Henry Helson was also appointed as an assistant professor on July 1, 1955, and like Bill Bade, was coming to Berkeley from Yale. Helson was born in Lawrence, Kansas, on June 2, 1927, the son a professor of psychology. After graduation from

Harvard College in 1947, he won a Sheldon traveling fellowship. The following year, he returned to Harvard for graduate work and received his doctorate in 1950 with a dissertation under Lynn Loomis entitled "Fourier Transforms and Spectral Synthesis on Locally Compact Abelian Groups." He had accepted a beginning instructorship at UCLA in 1950, but resigned rather than sign the loyalty oath that had just been imposed by the regents. Instead, he taught for a year in Uppsala, Sweden. Returning to the United States, he accepted an instructorship at Yale in 1951, then received a Jewitt fellowship, which he took at Yale in 1952–1954. He had been corresponding with Frank Wolf and John Kelley about coming to Berkeley in 1954, but this did not work out, and he accepted instead an appointment as assistant professor at Yale in 1954. Negotiations with Berkeley continued and were consummated the following year, when he was appointed as an assistant professor.

Helson's work in harmonic analysis combined both classical themes involving the real line or the circle as well as more abstract themes with results applicable to general locally compact abelian groups. He resolved a celebrated problem of Steinhaus, studied Fourier–Stieltjes transforms with bounded powers, resolved with others the converse of the Wiener–Levy theorem, and with Lowdenslager initiated an important study of prediction theory in several variables. With these achievements he advanced rapidly, winning promotion to tenure in 1958 and promotion to full professor in 1961. He supervised the doctoral work of 20 students during his career. Helson served as vice chair of the department for nearly two years and then as chair for 1965–1966. After he took early retirement in 1993 under the university's voluntary early retirement incentive program, he remained actively engaged with the department and the campus.

The third appointment in 1955 was of Harry Huskey, a renowned expert in digital electronic computer design, the use of computers, and theoretical and experimental numerical analysis. This appointment broke new ground as a joint appointment split 50/50 with the electrical engineering division within the engineering department.

Harry Douglas Huskey was born January 19, 1916, in Whittier, North Carolina, and completed his baccalaureate degree in mathematics at the University of Idaho in 1937. Continuing his studies at the graduate level, first at Ohio University and then at Ohio State University, he received his doctorate in 1942. His dissertation, entitled "Contributions to the Problem of Geocze," was written under Tibor Rado and concerned the calculation of the area of a two-dimensional surface that had been posed by Zoard Geocze some thirty years earlier. He remained at Ohio State for a year as an instructor and published several papers on his thesis subject. In 1943 he went to the University of Pennsylvania as an instructor and there also joined the ENIAC project in the School of Engineering. He was responsible for the "constant transmitter" and the "printer" units and later authored the two-volume manual of operation and maintenance for the entire computer. During 1945–1946, he served on the planning board for the EDVAC, the second computer to be built at the school.

In the spring of 1946, Rado, who feared that Huskey would give up his pure mathematics for computing, persuaded him to accept an assistant professorship at the University of Oklahoma. In doing so Huskey turned down an offer from the University of Pennsylvania as assistant professor. Lehmer, who was stationed at Penn and Aberdeen working on the ENIAC project during 1945–1946, got to know Huskey and was sufficiently impressed by him that he strongly urged his appointment at Berkeley effective July 1, 1946. Huskey's work on surface area was also of interest to Morrey, Morse, and Wolf. Indeed, Morrey had reviewed two of his publications for the *Mathematical Reviews*. Huskey would also be of great help to Lehmer in his efforts to develop electronic computing as a tool in number theory. Evans's view was that it would be easy to get authorization to appoint Huskey as an instructor, even at the highest level of instructor, but that appointment at assistant professor would be more difficult. No action was taken, since it did not seem reasonable to try to appoint Huskey as an instructor.

But then, Huskey received an offer in 1946 from the National Physical Laboratories (NPL) at Teddington, England, to help them with their design of electronic computers. He accepted this offer and ended up never going to Oklahoma. At NPL he had the opportunity to work with Alan Turing, who had sketched out a plan for what he called the ACE computer. Huskey built on Turing's ideas and skillfully implemented a design for this machine. The machine was not completed until 1950, long after Huskey's departure, but he had played a key role in its design. Meanwhile, wheels were turning at Berkeley, and on October 26, 1946, Evans proposed Huskey's appointment as an assistant professor. In his letter to Provost Deutsch, Evans argued that the appointment of Huskey, who he said was an expert on big computing machines, could be used as an inducement for the Bureau of Standards to locate their computing center at Berkeley. Recall that this is exactly the time frame in which the competition for the NBS West Coast computing facility was at its peak as described in Chapter 8. On March 3, 1947, Deutsch wrote back that the review of the proposed appointment was negative and that he was denying Evans's request.

Huskey remained in England through 1947, and upon his return to the United States in January 1948, he accepted a position at the National Bureau of Standards (NBS) in Washington, D.C., as chief of the Machine and Development Laboratory and mathematician in the NBS National Applied Mathematics Laboratories. By this time, NBS had decided on two sites for their computing facilities—one at

UCLA and one in Washington, D.C.—but the bureau was still debating whether to purchase computers, and if so, from whom, or to build them. The decision was to build one at the Institute for Numerical Analysis at UCLA, and in December 1948, Huskey was dispatched with an assignment to build the UCLA computer basically from scratch. The emphasis was to be on speedy construction and low cost.

When Huskey reported for duty on December 6, 1948, he found an empty room with some old resistors, but Huskey and his team did a phenomenal job and had the computer up and running by July 1950. It was dedicated the following month. The machine was substantially smaller than the usual behemoth of the day, and Huskey wanted to name it the "Zephyr," in reaction to the more violent-sounding meteo-rological names of other computers of the day, such as "Whirlwind," "Tornado," and "Hurricane." Some bureau staff in Washington called it "Sirocco," which is the name of the hot, humid wind of southern Italy and the Mediterranean islands. In the end, the cautious bureaucrats won the day, and the machine was simply called the National Bureau of Standards Western Automatic Computer, or SWAC for short. The computer in Washington was similarly called SEAC. The SWAC was a stored-program computer optimized for speed and designed for scientific compu-tation. For a period of time after its completion, it was the fastest machine in the world [Hus].

Huskey stayed on at UCLA with the title of assistant director of the Institute for Numerical Analysis (INA), helping people use the machine. During 1952–1953, he took a one-year leave of absence from INA to go to Wayne State University, where, as founding director of the computational laboratory, he established a first-rate operation. At the same time, he dusted off some old ideas and designed the first "personal computer," which he sold to the Bendix Corporation. Bendix produced and marketed this machine as the G-15 for a price of about $50,000, and it was a commercial success. This machine was the size of a refrigerator, but could be used by one person, hence the designation as a "personal computer." Huskey was clearly an early pioneer in computing as we know it today. He was also an expert in numeri-cal analysis and the logic of design and coding.

Lehmer, of course, knew Huskey both from ENIAC days and from his term as director of INA from 1951 to 1953, and had a very high opinion of him. As Lehmer began his term as chair of the department in July 1954, he set out to attract Huskey to Berkeley from UCLA. This was the time at which funding for INA was going through difficult times. In August he proposed the appointment of Huskey as associate professor, 50% in mathematics and 50% in the Electrical Engineering Division of the School of Engineering. Paul Morton, who was the senior person in electrical engineering working in computing and computers, was also enthusiastic. There was insufficient time to review the proposal for a permanent appointment, so

Huskey was made an acting associate professor in mathematics and engineering for 1954–1955. His appointment survived the usual high level of scrutiny that Sproul made of all initial tenured appointments, and Huskey was appointed as associate professor effective July 1, 1955. It was noted that in the spring of 1954, the Bendix Corporation had made a permanent loan of one of the G-15's to Huskey personally. The machine was installed in his house in Berkeley, and as we shall see, it was described by Lehmer as the most powerful computer in Berkeley in 1956. Huskey supervised the doctoral work of one student in mathematics during his time at Berkeley.

Huskey was promoted to the rank of full professor in 1958, and he made many valuable contributions to both his departments and to computing on the Berkeley campus. His national stature was reflected by election as vice president of the Association for Computing Machinery (ACM), the major national professional organization, for 1958–1960, followed by election as president for 1960–1962. Shortly after the new Santa Cruz campus of the university was formed, Huskey was recruited to that campus in 1968 as founder of its computing program and as director of the computer center. He went on to receive many professional awards recognizing his contributions as a pioneer in computing, and retired from the university in 1985.

The final event in the five-year period 1950–1955 was the fissioning off of the statistics group as a separate department on July 1, 1955, which we will discuss presently. The mathematics department was thereby greatly reduced in size and now had 18.5 FTE in professorial ranks counting Huskey as 50%, for a head-count total of 19 faculty members. In partial differential equations, Morrey, Lewy and Protter provided a strong representation in this field, for which Berkeley was known; in functional analysis and harmonic analysis, Wolf, Kelley, Morse, Bade, and Helson formed the most numerous group; while Pinney, Chambre, and Diliberto worked in applied mathematics and differential equations. Thus, 11 of the 20 faculty were in what one could broadly call analysis of various varieties. Tarski, Henkin, and Foster gave the department strength in logic, while Lehmer, Robinson, and Huskey represented number theory and computing. Geometry was supported by Flanders, and algebra by Seidenberg. Thus, the department had some real distinction but was quite unbalanced. Where it would have been ranked nationally among mathematics departments at this point in time is not clear. Soon a rapid infusion of new positions would transform the department both in balance, overall distinction, and sheer magnitude.

The department had internal discussion during this period about starting a "Program in Applied Mathematics" and hiring several young researchers in this field. This initiative never materialized, and had to await further developments.

The separation from statistics relieved several bones of contention, one being the perception and perhaps the reality that statistics faculty were advanced more rapidly through the ranks. Neyman was a dogged advocate for members of his group and had enough independence in personnel matters within the department that he could push them ahead rapidly. There were also differences in teaching loads, because Neyman felt that the faculty duties as statistical consultants to the campus justified a lower teaching load for statistics faculty. The teaching load for the campus had been modified since 1950 and was 6 hours for full professor, 7.5 for associate professor, and 9 for assistant professors and instructors. The lower teaching load for senior faculty was based on the assumption that they would have more extensive engagement with doctoral students, and the lower teaching load compensated for this differential. This differential was soon to cause problems in the mathematics department. Another change in 1955 was that Dean Alva R. Davis, of the College of Letters and Sciences, stepped down after nine years of service. He had been reasonably well disposed toward mathematics, and his successor, Lincoln Constance, who like Davis was a botanist, required a certain amount of "education" about mathematics and statistics.

Chapter 11

Growth and Separation of Statistics: 1950–1955

During this period, the statistics group within the mathematics department grew substantially, rising from a faculty of six in professorial ranks in 1949–1950 to ten in regular professorial ranks on July 1, 1955. On this latter date, an independent department of statistics was formed, thus realizing the goal that Neyman had set when he arrived in Berkeley in 1938. Altogether in this period there were seven new appointments—five as assistant professor and two at tenure—and three separations. The first separation was Stein, as we have already noted, and then two of the five newly appointed assistant professors during this period separated. Thus, there was a net growth of four. In some sense, Loève might be regarded ultimately as 50% mathematics and 50% statistics, and so statistics could be regarded as having 9.5 FTE faculty at the time of its creation.

The first two assistant professor appointments were of Evelyn Fix and Elizabeth Scott in 1951. Both had been appointed as instructors in 1950, and both were promoted a year later to assistant professor. This was rather faster promotion than was customary for most appointees at this time, but each had spent a number of years as lecturer prior to appointment as instructor.

Evelyn Fix was born January 27, 1904, in Duluth, Minnesota, and received her bachelor's degree in mathematics from the University of Minnesota in 1924, which was followed by a BS in education in 1925. She then taught high-school mathematics in Minnesota from 1925 to 1934, while also receiving an MA in mathematics in 1933. In 1934, she moved to Seattle, Washington, and worked as a high-school mathematics teacher, secretary, and school librarian from 1934 to 1941. She had received a certificate degree in librarianship in 1936 from the University of Washington.

Prompted by a friendship with Evans, developed during a summer-school course she had taken from him in 1931, she attended the summer session at UC Berkeley in 1939 and again in 1940. In 1941 she came to Berkeley to stay, signing on as a research assistant under Neyman in the statistical laboratory. She continued her work during the war supported on Neyman's grants and also taught, first as an associate and then as lecturer in the mathematics department. After the war she completed her work for a doctoral degree under Neyman, which was awarded in 1948. Her dissertation consisted of three parts, two of which were technical reports she had prepared while working in the statistical laboratory, while the third part was entitled "Distributions Which Lead to Linear Regressions." She was appointed as a lecturer for the two years 1948–1950 before being appointed as instructor in 1950 and then as assistant professor in 1951. She was promoted to associate professor in 1957 and to full professor in 1963. She died of a heart attack on December 30, 1965, shortly after returning home from a banquet for the Fifth Berkeley Symposium.

Her research interests ranged over a number of topics from early work on probability in her war work and her thesis, to work with J. L. Hodges on discriminant analysis, to work with Neyman that led to her computation of tables of the power of the chi-squared test and problems of risks, and finally to joint work with F. N. David on statistical problems of biology and health. During her career, she supervised the doctoral work of one student.

She helped with the organization of the periodic Berkeley Symposia on Mathematical Statistics and Probability. Also as her memorial article opines

> Miss Fix participated in the organization of the Statistical Laboratory and then of the Department of Statistics, essentially from the very start. It pleased her to see statistics come alive and she contributed a great deal to the spirit of the Laboratory and Department. In addition to other qualities she had an unusual gift for cooking and many of us will long remember her hospitality, at her apartment and, later, at her home with F. N. David in Kensington.

Perhaps at this point some mention of F. N. (Florence Nightingale) David is appropriate as she was at times a presence in the mathematics and statistics departments at Berkeley, although she was never a regular faculty member. She was born in England in 1909 and received her doctoral degree in 1938 at University College London, under Karl Pearson, the same year that Neyman left University College to come to Berkeley. David also subsequently served as a faculty member at University College London. She began regular summer visits to Berkeley in 1948 where she taught in summer session. She was subsequently recruited to the Riverside campus of University of California, and her memorial article states

> After retiring from UC Riverside in 1977, Dr. David was named professor emeritus and research associate at UC Berkeley where she continued to teach for another decade, and, at the same time, continued her long-term collaboration as a consultant with the United States Forestry Service. She was the author of nine books, two monographs, and over 100 papers in scientific journals. In August of 1992, she received the first Elizabeth L Scott Award at the Joint Statistical Meetings in Boston. She was cited for "her efforts in opening the door to women in statistics; for contributions to the profession over many years; for contributions to education, science,

and public service; for research contributions to combinatorics, statistical methods, applications, and understanding history; and her spirit as a lecturer and a role model. [IM]

Elizabeth Leonard Scott was born on November 23, 1917, in Fort Sill, Oklahoma, where her father, an officer in the U.S. Army, was stationed. After graduating from high school in Oakland, California, she entered UC Berkeley in 1935 and majored in astronomy, graduating in 1939. Shortly after Neyman arrived in Berkeley in 1938, C. D. Shane, director of the Lick Observatory, recommended to Scott that she learn statistics because he felt that applications of modern statistics to astronomy would be important but were not well developed. Scott then entered graduate school in astronomy, but split her time over the next ten years between the astronomy department, the statistics laboratory, and the mathematics department. She was a research assistant in the statistics laboratory for 1939–1941, a teaching assistant in astronomy in 1941–1942, and worked on Neyman's National Defense Research Council contract during the war. She was also a university fellow in astronomy for 1942–1944, and then a teaching assistant in mathematics for 1944–1946. After the war she worked as an associate in astronomy, as a research assistant in the statistical laboratory, and then as a lecturer in mathematics.

During the war she worked on statistical problems concerning the effectiveness of bombing, and then began work on statistical problems in astronomy and bivariate distributions. Her interests began in astronomy, but shifted more and more to statistics. After discussion with her mentors, she decided to submit a dissertation in astronomy in which an astronomical problem is solved by statistical methods. Her dissertation, written under the direction of Robert Trumpler, in astronomy, consisted of two parts: "(I) Contribution to the Problem of Selective Identification of Spectroscopic Binaries, and (II) Note on Consistent Estimates of the Linear Relation between Two Variables." The degree was granted in 1949, and she was appointed as instructor in mathematics on January 1, 1950, and was then promoted to assistant professor 18 months later. Advancement to tenure came in 1957 and to full professor in 1962. She served as department chair of statistics from 1968 to 1973, and her memorial article says that she will be remembered by her deans as a feisty chair of her department and a champion of its students. (The two deans she served under were Walter Knight and the author.) She supervised the dissertation of nine doctoral students during her career.

Throughout her career, Scott contributed both to astronomy and to statistics. There is an observational feature concerning the formula used to estimate the distance to a galactic cluster known as the Scott effect. She began a lifelong collaboration with Neyman that included statistical problems concerning the distribution of galaxies, weather modification (cloud seeding), and carcinogenesis. In other work, singly authored, she explored the effects of ozone depletion and its possible effects. She also undertook statistical studies of career patterns of men and women in academia, work that resulted in several influential reports. One such study, undertaken with Elizabeth Colson, collected and analyzed data on gender disparities and was reprinted by Congress. Another study on salary inequality was done for the Carnegie Commission and was used by institutions to make salary corrections.

She was an effective mentor and role model for many young women in science, and as noted, an award named in her honor was created. She was refreshingly and vigorously outspoken in many venues about discrimination and inequality. Finally, she held a number of important positions in professional societies—president of the Institute of Mathematical Statistics, vice president of the International Statistical Institute, and vice president of the Bernoulli Society—and was elected as an Honorary Fellow of the Royal Statistical Society (London). She retired from active duty July 1, 1988, and died unexpectedly a few months later.

Harry Hughes, the next appointment in statistics, was appointed as an assistant professor of mathematics on July 1, 1952, having served as an instructor for one and a half years beginning in January 1951 and as a lecturer for several years before that. Harry Meacham Hughes was born on August 2, 1917, in Midland, Texas. After graduating from the University of Texas at Austin with a BA in 1937, followed by an MA in 1938, he taught at the University of Tennessee as an instructor for four years. He then served as a naval officer during the war and returned to graduate study in mathematics in 1946 at UC Berkeley. During graduate study, he taught as a teaching assistant and then lecturer in mathematics and worked as a research assistant in the statistical laboratory. He received his PhD at Berkeley under Neyman in 1950 with a dissertation entitled "Estimation of the Variance of the Bivariate Normal Distribution." He was recognized as a devoted and talented teacher, but his research did not develop as hoped, and there is no record of any scientific publications in the literature. Hughes remained at Berkeley for two years as assistant professor and then left for another a position elsewhere on June 30, 1954.

Terry Jeeves was appointed as an assistant professor of mathematics in 1953 after two years of service as an instructor and three years as lecturer. Terry Allen Jeeves was born August 27, 1923, in Berkeley, and after graduating from Berkeley High in 1941, entered UC Berkeley in 1941. He joined the US Naval Reserve and completed college in an accelerated program, graduating in February 1944. His assignment while on active duty in the navy was in radar maintenance, and upon his discharge in 1946, he returned to graduate study in mathematics at Berkeley. His extensive teaching duties apparently slowed his progress to the doctoral degree, which he received in 1952. Written under Neyman, his dissertation was entitled "Identifiability and Almost Sure Estimatability of Linear Structures in n Dimensions." One publication based on his dissertation appeared in the literature in 1954. Jeeves served for two years as an assistant professor and then left for another position on June 30, 1955.

The fifth and final assistant professor appointed in this period was Lucien Le Cam, who was appointed in 1953. Le Cam was born on November 18, 1924, in Croze, Creuse, France, the son a farmer. He graduated from a Catholic boarding school in 1942, and through taking courses and passing special examinations, was awarded a *License es Sciences* by the University of Paris in 1945, which "certified" him as a statistician. This certification allowed him, with the recommendation of G. Darmois, director of the statistical laboratory at the University of Paris, to obtain a position as a statistician for Electricité de France. While so employed, he and other students organized a weekly seminar at the University of Paris. Through this seminar, he came into contact with Neyman, who was visiting in the spring of 1950. Neyman was so taken by Le Cam that he invited him to come to Berkeley as a lecturer. This short visit soon turned into a permanent relocation.

At Neyman's urging, Le Cam enrolled for doctoral study at Berkeley and completed his degree in two years, in 1952. He was appointed as instructor in 1952, and promoted to assistant professor in 1953. His dissertation, "On Some Asymptotic Properties of Maximum Likelihood Estimates and Related Bayes Estimates," was a highly regarded work. In work soon thereafter he reformulated Wald's decision theory and removed restrictions on finiteness of risk that were originally imposed. He won rapid promotion, first to associate professor in 1958 and then to full professor in 1960. Some found him comparable with Charles Stein, who had been lost to Berkeley as a result of the oath controversy. As a further sign of the respect he enjoyed, he was selected to serve as department chair from 1961 to 1965, very soon after his promotion to full professor. In 1973, in order to promote interaction and collaboration between the mathematics and statistics departments, he was appointed a professor of mathematics at 0% time, and he participated in mathematics department affairs.

Le Cam was one of the leaders in the field of theoretical statistics. To quote from his memorial article,

> Le Cam was a principal architect of the modern asymptotic theory of statistics. His life-work was a coherent theory of statistics in which asymptotic approximations played a strong role. His most widely-recognized contributions were the local asymptotic normality condition (LAN) in 1960, the contiguity concept and its consequences, and simple one step constructions of asymptotically optimal estimates. ...While still working on LAN conditions, Le Cam simultaneously tackled a more basic issue in statistical decision theory. He introduced a distance and deficiency between statistical experiments in 1959 and treated insufficiency in 1974. The Le Cam distance and deficiency are fundamentally important in making approximations and asymptotics fit within Wald's (1950) theory of decision functions.

During his career, he supervised the doctoral dissertations of 39 students, a number of whom have gone on to distinguished careers.

Le Cam received many honors during his lifetime, including election to the American Academy of Arts and Sciences, and would have been elected to the US National Academy of Sciences had he become an American citizen. He retained his French citizenship, and according to his memorial article, he always "viewed himself first and foremost as a Breton." But he had certainly come a long way from his origins in rural France.

It is not clear whether one should count Le Cam as a Neyman student from the point of view of inbreeding in the department. Le Cam had extensive training in mathematics and statistics in France, was very much influenced by the Bourbaki tradition, and brought new ideas to bear on statistics. He had also published some notable papers before coming to Berkeley. In fact, his two years of graduate study at Berkeley appeared to be more like a postdoctoral experience than doctoral training.

Two senior full professor appointments were made in this period: Henry Scheffe in 1953 and David Blackwell in 1955. We defer discussion of the Blackwell appointment until after our discussion of the establishment of the statistics group as a freestanding department. As early as 1942, Henry Scheffe had been on Neyman's radar screen as someone he wanted to attract to Berkeley at some point in the future. Neyman was finally able to achieve this goal in 1953, when Scheffe was appointed as professor of mathematics and at the same time as assistant director of the statistics laboratory, so that he became Neyman's deputy. Here also was someone from a very different background in statistics.

Henry Scheffe was born on April 11, 1907, in New York City. He graduated from high school in Islip, Long Island, in 1924 and entered the Cooper Union Free Night School for the Advancement of Science and the Arts to study electrical engineering. The following year he enrolled at the Polytechnic Institute of Brooklyn and at the same time was employed by the Bell Telephone Laboratories. In 1928 he transferred to the University of Wisconsin, Madison, to study mathematics, receiving his bachelor's degree with high honors in 1931. Scheffe remained for graduate study at Wisconsin, receiving his doctoral degree in mathematics in 1935 with a dissertation written under Rudolf Langer entitled "On the Asymptotic Solutions of Certain Differential Equations in Which the Coefficient of the Parameter May Have a Zero." He served as an instructor at Madison for the two years 1935–1937, and then went to Oregon State University, where he was on the faculty from 1937–1939 and 1940–1941. In the intervening year he taught at Reed College, in Portland.

At this point he decided to change his scientific interests from ordinary differential equations to statistics, and to pursue that goal, he took a position as instructor at Princeton University in 1941, working with the statistics group, which included

Samuel Wilks, John Tukey, and others. He served as instructor in 1941–1943 and as lecturer in 1943–1944, and he was employed on a defense contract at Princeton during 1945–1946. During this period he spent a year teaching statistics at Syracuse University. Then, in 1946, L. M. K. Boelter, the new dean of engineering at UCLA, whom we encountered in Chapter 8, recognizing the growing importance of statistics in engineering practice, brought Scheffe to UCLA as an associate professor of engineering. Scheffe in fact spent part of his first year in California at Berkeley on a Guggenheim Fellowship and began joint work with Erich Lehmann. Very shortly thereafter, in 1948, Wald lured Scheffe to Columbia as an associate professor, where he remained until 1953, serving as executive officer of the department starting in 1951 following Wald's untimely death.

Neyman had tried to attract Scheffe in 1950 or 1951, but Scheffe declined because of the oath controversy. Charles Stein, as we have already noted, resigned because of the oath controversy in 1950, and Neyman continued to pursue Stein to return to Berkeley. Even after the controversy had passed, Stein remained uninterested in Berkeley, and so Neyman finally had to admit defeat. But Neyman was persistent with Scheffe, and on September 24, 1952, he was able to write to Dean Davis formally proposing the appointment and arguing that this was a replacement for Stein. The appointment was approved, and Scheffe joined the group on July 1, 1953.

Scheffe continued his long and distinguished career at Berkeley, where he remained until his retirement in 1974. From 1965 to 1968, a difficult time for the university, he served effectively as chair of the statistics department. His memorial article describes his research as follows:

> Scheffe's work can be divided into two rather distinct phases. From 1942 to 1950 his research was devoted to problems of mathematical statistics. The main concern was the investigation of optimum properties of statistical methods which earlier had been obtained on an ad hoc basis. During the remainder of his career (essentially his Berkeley years), he was primarily concerned with the development of new statistical techniques. At least two of the ideas in his papers will be among the few in each generation that prove to have enduring importance: the concept of completeness and the S-method of multiple comparisons.

His book *Analysis of Variance*, written in the 1950s, has remained a classic, and his own contribution to the analysis of variance has been of great significance. Scheffe received a number of professional honors, and after his retirement in 1974, he was appointed a professor at the University of Indiana. He returned to Berkeley in 1977 to continue work on a revision of his book, but he died on July 5, 1977, from injuries sustained in a bicycle accident.

During his long tenure as chair of the mathematics department, Griffith Evans had resisted Neyman's efforts to create an independent department of statistics.

Evans had the support of the campus administration in this stand, but at the same time, he was an ardent advocate for financial support for the statistical laboratory. Evans believed that both statistics and mathematics would benefit from the mutual interactions and intellectual stimulation that would result from the statistics faculty remaining in a unified department. Evans had a broad encompassing view of mathematics and clearly felt that applied mathematics (including mechanics), actuarial science, and numerical analysis/computing should be developed within the mathematics department, as was outlined in Chapter 8.

The new chair in 1949, Chuck Morrey, was willing to take a more relaxed view, and after six months on the job, he wrote on December 20, 1949, to Dean Davis recommending the creation of separate department. The documentation for the narrative in the following eight paragraphs is contained in the University Archives [UA, CU 5, series 3, Box 38, folder 7]. It is interesting to recall that Morrey made this recommendation at exactly the same time as he was negotiating to bring Courant and his entire institute to Berkeley as the Institute of Mathematics and Mechanics to be located within the mathematics department. Morrey closed his letter as follows: "In such a situation, I do not see that there is anything to be gained by keeping Professor Neyman and his group attached to our department against their will." In subsequent discussions, Morrey expressed the view that the relationship between mathematics and statistics would not be altered by the creation of a new department and that a certain relief of the tension would be achieved by separation that should make for even better cooperation. It was also pointed out that creation of a separate statistics department would have no budgetary implications, since the required resources were already in place in the statistics laboratory and in the faculty of the mathematics department.

A particular concern of Neyman's was that as many disciplines realized the importance and significance of modern statistics for their particular field, they were starting to hire dedicated statisticians in departments, schools, and colleges to teach the required statistics courses. Neyman felt that such dispersion of statistics instruction would not be beneficial academically and that an independent statistics department, as opposed to a group within the mathematics department, would be better able to resist this harmful tendency. The response from the campus leadership (the Budget Committee and Dean Davis) was mostly negative to Morrey's proposal, and reflected the long-held views of Evans. But, it was finally agreed in May 1950 that the matter would be revisited the following year.

Following up on this suggestion, on November 6, 1950, the Budget Committee recommended that the entire matter of statistics "be brought under study by a competent and representative committee." President Sproul agreed and set in motion the appointment of a special committee that he charged "to make an analysis of the statistical needs of university in teaching, research, and service, and to recommend organization for achieving ends in this field." The senate Committee on Committees nominated as members of this review committee the following: Ray E. Clausen (genetics) to serve as chair, Percy M Barr (forestry), Roy W. Jastram (business administration), Francis A. Jenkins (physics), Derrick H. Lehmer (mathematics), Paul L. Morton (electrical engineering), and C. Donald Shane (director of Lick Observatory). These members were appointed by Sproul on January 8, 1951. The committee responded to Sproul's charge on June 14, 1951, writing that the committee had met four times, had received documents from Neyman, and had met with Dean Davis and Professors Evans, Morrey, and Neyman. However, Chair Clausen wrote, "the committee has been unable to discover any objective criterion upon which to base a decision as to the desirability of establishing a separate department," and requested more time.

Nearly a year later, on May 19, 1952, the committee submitted its final twelve-page report, in which it recommended the creation of a separate department of statistics within the College of Letters and Sciences. The committee recognized that the creation of a separate department would not solve all the problems confronting statistics, but opined that "the overriding consideration is the magnitude of the task confronting statistics, which should justify the utmost freedom to deal with it, consistent with the higher interests of the university." Professor Lehmer had left the committee, since he was on leave of absence serving as director of the Institute for Numerical Analysis at UCLA for 1951–1953, and was not replaced. The report of the Clausen committee was signed by the remaining six members.

Dean Davis now supported the committee's recommendation and praised the committee, saying that "the committee has done an admirable job whether or not one agrees with its recommendations." He recommended "that a favorable decision be made on the principle and the timing be determined by conditioning circumstances." There was in fact a complication as to timing that arose from a proposal made at the same time for the creation of a freestanding linguistics department in the College of Letters and Sciences. The fact that this new department was linked to a serious retention issue persuaded Davis, Kerr, and Sproul to go forward with that proposal and to hold back the statistics proposal. Sproul feared that presenting the regents with proposals for two new departments might place both in jeopardy.

So, the statistics proposal was laid on the table for the moment. After the establishment of the linguistics department had been approved, to be effective July 1, 1953, the statistics proposal was reactivated when A. R. Davis, now as acting chancellor in Kerr's absence, wrote on August 13, 1953, to the Budget Committee seeking comment on the proposal. The Budget Committee wrote back on September 22, reiterating its longstanding vigorous opposition to the idea. The Budget Committee in effect was ignoring the recommendations of the Clausen committee, a committee that the Budget Committee had recommended be established three years earlier. The Budget Committee report concludes,

> In summary, the Budget Committee believes that neither from the standpoint of research or of instruction can a strong case, or even a plausible case, be made for separating Mathematics and Statistics. If separation is carried out, it appears to the Committee that the grounds can only be those of personality factors and administrative convenience, and these are not apparent from the documentation as it has come to the Budget Committee.

The Academic Senate Committee on Educational Policy, which was also consulted, supported the proposal for a separate department in a memo of October 21, and cited the Clausen committee report as their rationale.

Chancellor Kerr moved judiciously, perhaps in light of the negative recommendation of the Budget Committee, first consulting Dean Davis one last time, who countered the Budget Committee's argument and recommended moving forward in a memo to Kerr of February 23, 1954. Kerr finally recommended to President Sproul on August 13, 1954, the creation of statistics as a separate department effective July 1, 1955, to coincide with the beginning of the 1955–1956 budget year.

President Sproul's staff prepared an eleven-page briefing memo for him on the history and background of the controversy. This memo noted that a separate entity de facto already existed and that those closest to the problem recommend that it be officially recognized. The memo went on to argue that "it does not seem to be practical to attempt to put the egg back into the shell. Under the circumstances, the conclusion seems inescapable that the fait accompli should be recognized and the unit renamed 'Department of Statistics.'" Mixing metaphors, the authors of the report continued,

> Although in this case the horse has escaped, the misgivings of successive committees on Budget regarding recognition of the fact are understandable. Here a willful, persistent, and distinguished director has succeeded, step by step over a fifteen year period, against the original wish of his department chairman and dean, in converting a small "laboratory" or institute into, in terms of numbers of students taught, an enormously expensive unit; and then he argues that the unit should be renamed a "department" because no additional expense will be incurred.

Sproul accepted Kerr's recommendation, and after regental approval, the department was created on July 1, 1955, with Neyman as it founding chair. The statistical laboratory remained as a distinct entity within the new department.

While there are many stories and witticisms about the long battle to create a separate department of statistics, two stand out. The first is frequently attributed to Lincoln Constance, who had served as associate dean of the College of Letters and Sciences under Dean Davis and who was then appointed as the new dean on July 1, 1955. He now had to deal with Neyman as one of his department chairs. At some point he is alleged to have said, "If you had someone [e.g. a department chair] between you and Neyman, why would you ever give it up." Any dean or former dean will immediately resonate with this sentiment. The other was conveyed to the author by Clark Kerr; it addresses jocularly the intellectual rationale for splitting statistics and mathematics. "Mathematics," he said, "seeks ultimate certainty, while Statistics seeks ultimate uncertainty, and so they should go their own ways."

The first appointment in the new department was that of David Blackwell, whom we met before when Neyman tried to bring him to Berkeley in 1942. Blackwell's name had since that time been on Neyman's list of scholars to attract to Berkeley. David Howard Blackwell was born on April 24, 1919, in Centralia, Illinois, the oldest of four children. His father worked for the Illinois Central Railroad, looking after locomotives, and had ambitions for his children to go to college. Blackwell graduated from high school in 1935 and entered the University of Illinois at the age of 16. He graduated in three years in 1938, received an MA in 1939, and then a PhD in 1941. His dissertation, written under Joseph Doob, was on Markov chains, and Blackwell was Doob's third PhD student, following Paul Halmos (1938) and Warren Ambrose (1939). Doob evidently thought very highly of Blackwell, for he was awarded a prestigious one-year appointment as a Rosenwald postdoctoral fellow at the Institute for Advanced Study, where Doob was also spending the year. Blackwell sat in on Sam Wilks's course in statistics along with Henry Scheffe, Jimmy Savage, and others. It was at this point that Neyman tried to bring Blackwell to Berkeley.

According to his own recollection, Blackwell wrote 105 letters of application, to every historically black college or university in the country and landed a job as instructor at Southern University, in Baton Rouge, Louisiana, for 1942–1943, followed by an instructorship at Clark College, in Atlanta, for 1943–1944. The ambition of many black scholars at the time was an appointment at Howard University, in Washington, D.C., and in 1944 Blackwell was appointed as an assistant professor at Howard [DeG, p. 41]. He quickly rose to associate professor in 1946 and to full professor and department head in 1947 at the age of 28. He spent the 1950–1951 academic year as visiting professor at Stanford, and then in the spring of 1954, when Loève finalized his plan to spend a sabbatical year in Paris, Neyman persuaded the campus to make the necessary investment of funds to invite Blackwell as visiting professor of mathematics for the year as Loève's replacement. This was the precursor to his permanent appointment to the faculty as professor in the new Department of Statistics, effective July 1, 1955. He added the title professor of mathematics (0% time) in 1973, and served on the faculty until his retirement in 1989.

Blackwell's research has spanned a number of related fields. He started out in probability theory with his dissertation on Markov chains, which was followed by many other notable papers, including his theorem on renewal, on functions of Markov chains, and on information theory. While in Washington at the end of the war, Blackwell began a collaboration with Abraham Girshick on Wald's theory of statistical decision functions that led to fundamental papers on the subject, especially on Bayesian sequential analysis, and a highly influential 1954 book with Girshick. Their book, *Theory of Games and Statistical Decisions*, is regarded as a classic. Blackwell is generally classed as a Bayesian, and he credits Jimmy Savage as a critical influence on him in that regard [DeG, p. 44]. In addition, Blackwell made a number of contributions to game theory and to the connections of game theory with set theory and logic. Finally, he made contributions to Markovian decision processes and to dynamic programming.

His distinction was recognized by many awards, beginning with an invitation to address the International Congress of Mathematicians in 1954, election to the National Academy of Sciences and the American Academy of Arts and Sciences, faculty research lecturer at Berkeley, Rouse Ball lecturer at Cambridge, and Wald lecturer, among others, and a string of honorary doctorates. He had a phenomenal record of doctoral supervision, with 64 students finishing their dissertations under him at Berkeley.

It is not the author's intent to write a history of the statistics department after its separation from mathematics in 1955. Others will write about the subsequent history and development of the department of statistics. But, there is one point worth noting in connection with David Blackwell. Neyman was naturally the choice to be the first or founding chair of statistics in 1955. But his tenure was short and bumpy. There were conflicts with the new dean, Lincoln Constance, over a variety of issues including teaching loads and the role of statistical consulting. In addition, a breach had opened up between Neyman and some of his oldest colleagues over some internal issues. Neyman suddenly submitted his resignation effective July 1, 1956, after one year as chair. Chancellor Kerr convinced him to withdraw the resignation, but then in December 1956, Neyman again submitted his resignation, and this time it was permanent. The department caucused the next day and decided that the only person who could take over as chair was Blackwell, and so he was suddenly thrust into the job of chair of this new department. His effective and gracious leadership over the next four and a half years set the department on a positive course, and is he credited with a key role in the successful evolution of the department. And now Dean Constance had someone between himself and Neyman! Neyman retained his position as director of the statistical laboratory until his death in 1981. This narrative is based on [Reid 1982, pp. 243–250].

Chapter 12

Change in the Wind: 1955–1957

Beginning in about 1955, the environment and circumstances in which the Mathematics Department found itself both within the campus and in the national mathematics community began to change. The department was also evolving scientifically and organizationally, and these circumstances set the stage for major changes in size and orientation of the department. These changes, which took place in the period 1957–1960, were in many ways comparable to those that had taken place in 1881 and in 1933. In 1957, however, the main stimulus for change came from within the department rather than from outside, as in the previous instances.

These factors included the following. First, the separation of statistics as a separate department brought closure to that longstanding issue, reduced the size of the mathematics department by one-third, and allowed the department to focus more closely and intensely on the challenges and problems that it faced. Second, John Kelley emerged as a leader in the department, first as vice chair in 1954 and then as chair in 1957. He represented a younger generation of mathematicians, and he was able to generate consensus on an academic plan to address the department's problems. Evans had retired in 1954, and the last of the pre-Evans faculty retired in 1954. Third, mathematics as a discipline was making rapid advances in areas that were not represented at Berkeley, generating a sense that Berkeley was in some ways being left behind. Fourth, the job market became very competitive, with considerable shortages, and there was a need for new strategies for recruitment and retention of faculty. Fifth, the new dean, Lincoln Constance, and the chancellor, Clark Kerr, both came to realize the centrality and importance of mathematics as a discipline and had shed their previous conceptions of mathematics simply as a service department, as it had been in the pre-Evans days. Both Constance and Kerr endorsed and implemented, under prodding from the department, a major infusion of resources starting in 1957, under which the department would grow rapidly. Sputnik, which highlighted the need for the nation to expand the training of scientific, mathematical, and engineering personnel, no doubt influenced these decisions as well. Finally,

Kerr recognized the need to relocate both mathematics and statistics from Dwinelle Hall, where they were located in inadequate space alongside the humanities departments, to a location on the mining circle, in proximity to the other science and engineering departments, with which they would naturally interact.

When Dick Lehmer succeeded Chuck Morrey as department chair in 1954, he asked John Kelley, who had just recently been reinstated after the oath controversy, to serve as his vice chair. Kelley played an unusually large role in the department during this time, even though he also served as a part-time assistant to Chancellor Kerr for the academic year 1956–1957. We have already noted that three appointments were made in 1955: Bill Bade and Henry Helson as assistant professors and Harry Huskey as associate professor (jointly with electrical engineering). Kelley played a large role in the appointments of Bade and Helson, especially because they were in his own field of functional analysis. Lehmer played the lead role in the recruitment of Huskey, who was an old friend from the ENIAC project during and after the war.

For the 1956 recruitment cycle, the department had three instructor positions, the result of a budgetary decision that reflected the established and preferred campus practice to recruit at entry level. But, the department was having trouble recruiting at this level, since the instructor salary was becoming increasingly noncompetitive in the mathematics job market. The department prevailed upon the dean to expand the search to include candidates who had completed postdoctoral work elsewhere and might be appointed at the assistant professor level. Three excellent candidates were found: James Eells, Bertram Kostant, and Emery Thomas. Before these candidates could be appointed, a case had to be made that an extensive but unsuccessful search at the instructor level had been made, after which, finally, the dean agreed to upgrade the instructor positions and authorize these appointments. These three appointments strengthened the department in the areas of geometry, topology, and algebra, in which it had been very weak simply because of a lack of faculty in these areas. These appointments established a pattern for future recruitment of junior faculty for the next several years; rather than being restricted to seeking new PhDs for initial appointment as instructors, the department would concentrate on candidates who had a few years of postdoctoral experience and a positive track record. In some cases, in order to be competitive, the department had to appoint some candidates with tenure. To circumvent the university's reluctance to make initial tenure appointments, the department resorted to appointments as visiting associate professors. Acquiescence by the administration in such a recruitment strategy was one of a number of changes that facilitated the department's rapid rise to preeminence during the five-year period 1955–1960, which will be the topic of this chapter and the next one.

James Eells, Jr., was born October 25, 1926, in Cleveland, Ohio, and attended Bowdoin College, graduating with a baccalaureate degree in mathematics in 1947.

He then taught at Amherst for two years and then in Istanbul for a year before entering graduate school in mathematics at Harvard in 1950. There he came under the influence of Hassler Whitney, who supervised his dissertation, which was entitled "Geometric Aspects of Integration Theory." He was awarded a doctorate in 1954, after Whitney had left Harvard in 1952 for a professorship at the Institute for Advanced Study (IAS). Eells followed Whitney to the Institute for Advanced Study (IAS), serving as his assistant for the 1954–1955 academic year. He remained on as a member of IAS for the 1955–1956 academic year. During the previous academic year, he met Charles Morrey, who was spending that year at IAS, and a collaboration began that lead to a joint paper on variational methods in the theory of harmonic integrals. This joint work brought Eells to the attention of the Berkeley department, which found his expertise in topology, functional analysis, calculus of variations, and differential geometry particularly attractive.

Eells was hired at the first step of assistant professor in 1956 and was given an accelerated advancement to the second step the following year. He was also an attractive candidate to other universities: First, the University of Washington tried to recruit him, and then Columbia. Eells was on leave without salary in 1958–1959 at Columbia, and in the spring of 1959, he resigned from Berkeley to accept a permanent position at Columbia. His went on to a distinguished career, first at Columbia, then at Cornell, and then at the University of Warwick, with many contributions to differential geometry, especially concerning harmonic maps. Although he left Berkeley before supervising any Berkeley doctoral students, he has supervised the work of 38 students during his career.

Bertram Kostant was born on May 24, 1928, in Brooklyn, New York, and after graduating from Stuyvesant High School in 1946 enrolled at Purdue University. He was awarded a BS degree in mathematics with distinction in 1950 and went on for doctoral study at the University of Chicago. His dissertation, written under Irving Segal, entitled "Representation of a Lie Algebra and Its Enveloping Algebra on Hilbert Space," was completed in 1954. However, in 1953, Kostant went to the Institute for Advanced Study on an NSF fellowship and remained there until 1955 as an institute member. He was offered an instructorship at Berkeley in 1955, but declined it in favor of a position as Higgins lecturer at Princeton. Berkeley renewed its interest in him the following year with an offer of an assistant professorship, which he then accepted.

It was argued that his appointment would strengthen algebra at Berkeley, which it did, but it achieved a whole lot more. His work on Lie algebras and Lie groups expanded to work in differential geometry and work on the geometry and algebraic structure of homogeneous spaces, representation theory, and cohomology theory, areas that he pioneered. He won very rapid advancement, being promoted to associate professor in 1959, after only three years, and then to full professor in 1961. He was much sought after, and many universities came calling—first Yale, then Chicago with a full professorship in 1960, and then MIT with a full professorship in 1961.

Kostant's case well illustrated the red-hot job market in mathematics and the reluctance of the Berkeley administration to respond to it at the time. There was a raging battle between the department and the administration over the rate of Kostant's advancement. The end result was that Kostant went on leave without salary at MIT for the year 1961–1962, and then finally resigned from Berkeley effective July 1, 1962, remaining at MIT, where today he is a professor emeritus and still active in research. The attractions at MIT included a higher salary than Berkeley was willing to approve and a lower teaching load. Kostant supervised the work of one Berkeley doctoral student, namely James Simons, and has supervised the work of 19 doctoral students overall. The loss of Kostant was a serious blow indeed for Berkeley, and perhaps the only positive outcome of his departure was a more realistic attitude on the part of the administration to the realities of the mathematics job market.

Paul Emery Thomas (known to everyone as Emery) was born on February 15, 1927, in Phoenix, Arizona. His father had been born in North Dakota and his mother in South Dakota, and the family returned to North Dakota, where in 1945 Thomas graduated from Central High School, in Fargo. He entered Oberlin College in 1946 and graduated with a baccalaureate degree in mathematics in 1950. Next came a Rhodes scholarship to Oxford University, where he developed a strong interest in algebraic topology under the influence of J. H. C. Whitehead. Returning to the United States in 1953, he entered graduate school at Princeton and quickly completed his doctoral degree under Norman Steenrod in 1955. His dissertation, "A Generalization of the Pontrjagin Square Cohomology Operation," was a significant piece of work that introduced new cohomology operations. An announcement of it was published in the *Proceedings of the National Academy of Sciences*, and the full version appeared in 1958 in the *Memoirs of the American Mathematical Society*.

Following his degree Thomas served as a research associate at Columbia University for the 1955–1956 academic year, and then he was appointed as an assistant professor at Berkeley starting in 1956. His work branched out into other areas of topology, including characteristic classes, fiber bundles, the structure of classifying spaces, and vector fields on manifolds, and he won rapid promotion, first to associate professor in 1960, and then to full professor in 1963. In his case, Berkeley was able to ward off the outside offers, and Thomas remained at Berkeley for the remainder of his career. His work further broadened to include problems of obstructions to smoothing singularities and embedding problems, but cohomology theory and various cohomology operations remained a central tool in his work. Then, in about 1980, he shifted his interest to number theory and continued publishing results in this area until

2000. He supervised the doctoral work of 30 students during his career. Thomas served as vice chair and chair of the department and as deputy director of Berkeley's Mathematical Sciences Research Institute (MSRI) from 1987 to 1990, and he was a valued citizen of the department and the campus. He retired from active duty in 1991 under the university's voluntary early retirement program, but remained active in research for some years after his retirement. He died in 2005.

These three junior appointments served immediately to broaden the span of coverage of fields in the department even though only one of the three remained at Berkeley in the long term. With the retirement of Evans in 1954, the department had, with the concurrence of the dean, reserved his FTE for the appointment of a senior, established mathematician in the field of geometry or topology. The department had informally offered the position to S. S. Chern (Chicago), N. Steenrod (Princeton), B. Eckman (Zurich), and J. H. C. Whitehead (Oxford). All of these mathematicians had turned down the initiative, and Chern expressed his unwillingness to move unless three or four similar appointments were made at the same time. This reaction led the formulation of a plan that we will see unfolding shortly under John Kelley's leadership. Sproul was aware that the Evans FTE had been reserved for a senior appointment, but he became concerned that the department had not been able to locate a suitable candidate and to bring forward a recommendation. Given these circumstances, he was moving toward withdrawing the FTE, which he in effect did in the spring of 1957, a point that will be discussed presently.

At this point, the department again turned to the perennial challenge of building strength in applied mathematics, and their attention turned to Bernard Friedman, at the Courant Institute, who was one of the large group of mathematicians in the Courant group at NYU that Morrey had tried to attract to Berkeley in 1950. Friedman had been invited to be a visiting professor for the 1955–1956 academic year. During this year, Friedman was a success, creating relationships with faculty in engineering and physics, and he seemed to fit the department's needs very well. The original idea was to move toward a permanent appointment for him in 1957–1958, but in May 1956, both Friedman and the department agreed that an immediate appointment would make their lives much simpler logistically, so that, in particular, the Friedmans would not have to return to New York. In view of the lateness of the proposal, the departmental recommendation was transformed into a proposal for an appointment as visiting professor for 1956–1957 with a change of status to a regular professorship effective on July 1, 1957.

The proposal won campus-level approval, coming up from the department through the dean, the Budget Committee, and Chancellor Kerr, but in August 1956, Sproul again inserted himself, as he had on other initial tenure appointments, questioning Friedman's qualifications. Sproul asked whether Friedman was the best person available at the proposed level, and more basically questioned the programmatic need for the appointment. This stalled matters, and Friedman and his family returned to New York for the 1956–1957 academic year. During the fall of 1956, the department set about to provide responses to the questions raised by Sproul and sent them forward. On January 9, 1957, Chancellor Kerr then renewed his recommendation on behalf of the campus to Sproul for the appointment. Further questions arose concerning the lack of a budgeted provision at the professorial level, and then on March 8, the chair, Dick Lehmer, was summoned to the office of Dean Constance and was effectively given an ultimatum, presumably emanating from President Sproul, concerning the Friedman appointment and the Evans position. Lehmer, who had just been invited by Chancellor Kerr to continue as chair for a fourth year, responded with an angry letter two and a half weeks later. This response deserves to be quoted in its entirety, for it encapsulates a number of issues that had been of concern to the department for some time:

March 27, 1957

Dear Dean Constance:

It was March 8th that you called me into your office to inform me that the Mathematics department had been given the choice of abandoning our quest for the appointment of Professor Friedman or losing the major position left vacant by the retirement of Professor Evans.

That afternoon there was a Department meeting at which it was unanimously decided to press for the appointment of Professor Friedman regardless of the consequences. This fact I transmitted by phone to Miss Annis by phone immediately after the meeting. At the same meeting it was further resolved that a strong protest be prepared and presented to you for possible transmittal to the source of the ultimatum setting forth our views of the situation and our future problems. This document, signed by members of our department is attached.

No American university administration with full confidence in its mathematics department would, in these times, abolish a major position in the department. Yet this has happened twice during the three years I have served as Chairman. During the same period two major appointments of mathematicians have been made in another department. This adds up to a vote of no confidence in the Mathematics Department and in me. For this reason I could not accept for the fourth time my annual appointment to the Chairmanship.

In closing let me express to you my thanks for the sympathetic and very real concern you have exhibited toward the Mathematics Department and my appreciation for the results you have achieved for us.

Sincerely yours,

D. H. Lehmer

The Friedman appointment was finally approved by Sproul in late spring of 1957, thus ending this long battle. Friedman assumed duties as professor of mathematics at Berkeley on July 1, 1957. However, the problems and conflict between the department and the higher administration remained, and had to be addressed.

Bernard Friedman was born on August 15, 1915, in Brooklyn, New York. He enrolled at City College in 1930, and graduated cum laude in 1934. Graduate study in mathematics at MIT followed, where he completed his degree after only two years, at the age of 21. His dissertation, written under Norbert Wiener, was entitled "Analyticity of Equilibrium Figures of Rotation." Courant, who had recently settled at New York University, brought Friedman to NYU as a postdoctoral fellow for the 1936–1937 academic year. Friedman then landed a position as instructor at the University of Wisconsin for three years, and taught at Wilson Junior College in Chicago for another three years. Courant then brought him back to NYU in 1943 as part of his research group doing war-related work.

At the end of the war, Friedman joined the faculty at NYU, initially as associate professor in 1946, and was promoted to full professor in 1948. His work spanned many areas of applied mathematics, including questions of electromagnetic wave propagation, techniques for evaluation of integrals, eigenvalues of matrices, and expansion of functions in series of eigenfunctions of ordinary differential operators. He supervised the doctoral work of 13 students overall, six at Berkley and seven at NYU. His graduate textbook, *Principles and Techniques of Applied Mathematics*, was influential, flowing from Friedman's longstanding interest in teaching. His initial interest was in teaching of applied mathematics at the graduate level, but this later broadened to include the teaching of mathematics at the undergraduate level and in K–12.

Friedman was an active participant in the life of the department and created many links with faculty in other departments. When the department needed someone to step in as chair midyear in 1959–1960, he did so and was then appointed chair to serve through 1962. He presided well over a period of rapid expansion. During the summer of 1966, he was stricken with cancer and died on September 13, 1966, at the age of 51. His death was a tragic loss to his family and colleagues in the department, and it was also a setback for applied mathematics at Berkeley. Several years later, the department established the Friedman prize in his honor, which is awarded each year to a graduate student or students for an outstanding dissertation in applied mathematics. The prize is often awarded to graduate students in other departments, a practice that no doubt would have pleased Friedman.

The department made two further appointments in 1957, both assistant professors in the tradition of the three appointments the year before of mathematicians several years out from the PhD. The first of these was of Jacob Feldman (or Jack as he is known to all). Feldman was born on January 10, 1928, in Philadelphia, Pennsylvania, and graduated with a bachelor's degree in mathematics from the University of Pennsylvania in 1950. He then pursued graduate work at the University of Illinois, but transferred after a year to the University of Chicago. There he came under the influence of Irving Segal and Irving Kaplansky and completed a dissertation in 1954 under Segal entitled "Isomorphisms of Rings of Operators," which concerned rings of operators (subsequently known as von Neumann algebras) and their relationship to the continuous geometries of von Neumann. After completing his doctoral work, he was awarded a National Science Foundation postdoctoral fellowship, which he took in residence for two years at the Institute for Advanced Study. For the 1956–1957 academic year he was a visiting assistant professor at Columbia University working with Richard Kadison.

Feldman's name came to the attention of the department in the fall of 1956, and he was successfully recruited to begin duties on July 1, 1957. Feldman's research broadened and deepened as his career progressed. His work expanded to encompass stochastic processes, especially Markov processes, ergodic theory, and orbit equivalence of actions of groups, and its many connections with operator algebras and statistics. Feldman has a number of important results to his credit. He won quick advancement, first to associate professor in 1960, and then to full professor in 1964. In 1969, the department of statistics voted to make Feldman a 0% professor of statistics as recognition of his work on stochastic processes. Throughout his productive career, he supervised the doctoral work of 21 students altogether and was an active participant in departmental affairs. On January 1, 1993, Feldman retired under the university's voluntary early retirement incentive program at age 64. However, he has remained active in departmental affairs and has continued to publish actively in the years since his retirement.

The second of the assistant professor appointments was of R. Sherman Lehman. The original intent was to make this appointment effective July 1, 1957, but Lehman requested that the start date be delayed for six months until January 1, 1958. Russell Sherman Lehman was born January 20, 1930, in Ames, Iowa, and after completing high school in Dayton, Oregon, enrolled in Stanford University in 1947. He earned a bachelor's degree in 1951, a master's degree in 1952, and his doctorate in 1954. For a period during the oath controversy, Hans Lewy had served as a visiting professor at Stanford, and Lehman came under his influence and in effect wrote his dissertation under Lewy's supervision, even though Lewy was not a permanent member of the Stanford faculty. The dissertation was entitled "Developments in the Neighborhood of the Beach of Surface Waves over an Inclined Bottom." After completing his degree, Lehman spent several months as a consultant at the RAND Corporation, and was then drafted into the army, where he was assigned to the Ballistics Research

Laboratory at Aberdeen. Following his Army service, he received a Fulbright grant to study at Göttingen, and he planned to remain in Europe until the end of 1957. Lehmer, as chair, wrote to him in early 1957 inquiring about his interest in Berkeley. Lehman responded favorably but asked for a starting date of January 1, 1958, and this was how his appointment was structured.

Lehman's interests in mathematics were wide and varied. Early on in his graduate studies he had begun an interest in computational number theory and in particular the Pólya conjecture. This interest plus his experience at Aberdeen no doubt led to Lehmer's interest in bringing Lehman to Berkeley. Lewy supported the initiative to hire Lehman, for he thought very highly of Lehman's dissertation work on the sloping beach problem. In addition, Lehman had developed a longstanding friendship and collaboration with Richard Bellman that led to series of joint papers on dynamic programming. Lehman won promotion to associate professor on January 1, 1962, and then to full professor in 1966. Although Lehman continued his work on water waves, his real interest gravitated toward computational number theory. Using careful error bounds and working on the 7090 computer, he was able to show that the first 250,000 nontrivial zeros of the Riemann zeta function are on the critical line. While this is a tiny number by present-day standards, it represented a nearly eightfold increase over the best previous result of N. A. Meller (1958), and a tenfold increase over the 1956 Lehmer result [Der, p. 258]. He was also able to reduce the previous bound on the first Littlewood violation, where the number of primes less than n exceeds the integrated logarithm function. Slewes had established an upper bound of an unimaginably large number, and then later reduced it to one having 10^{1000} digits. Lehman improved this upper bound to a number with "only" 1166 digits. Subsequent computation over the following forty years by others but starting from Lehman's work has reduced this to a number with 317 digits [Der, p. 236].

Lehman formed a friendship and intellectual partnership with René De Vogelaere (see Chapter 13 for a discussion of De Vogelaere's career) in the development of the numerical analysis curriculum and in establishing computing facilities in the department. Unfortunately, Lehman was victim of a serious automobile accident in 1967, and he suffered a subsequent stoke, which slowed down his scientific output. In the following years, he continued work in numerical analysis, but published only one paper while continuing to teach classes and supervise graduate students. He has supervised the doctoral work of 11 students so far. He took voluntary early retirement in 1994 at age 64.

Although these junior appointments went through the administration smoothly, serious problems between the department and the administration remained, as evidenced by Lehmer's March 28, 1957, letter of resignation to the dean and the protest it contained. The document mentioned in Lehmer's letter setting forth the department's grievances has not been located, but a related and more important document prepared shortly thereafter by Vice Chair Kelley does survive. In this document, which was termed a "white paper," Kelley describes in some eloquence the problems facing the mathematics department and provides a plan for the future development of the department. Kelley was to be the choice of the faculty to assume the chairmanship, and he used this document as a blueprint. His service as an assistant to Kerr in the chancellor's office no doubt helped him to persuade the administration of the wisdom and necessity of implementing this plan. The full document is of historical interest and appears in Appendix 4; some very salient selections from it are reproduced below. One obvious concern expressed by Kelley was the issue of senior appointments, or rather the lack thereof in the past. It was argued that such senior appointments were needed to advance the standing of the department.

This white paper reflects also a signal event in the spring of 1957, in which the junior faculty came together and articulated their grievances in a meeting with department chair Derrick Lehmer. Current faculty who were among the group at the time recall that they all crowded into Lehmer's office in Dwinelle Hall to air their complaints. Among the most important issues for the junior faculty was the differential teaching load of 9 hours for instructors and assistant professors, 7.5 hours for associate professors, and 6 hours for full professors. Issues of salary and rate of advancement were also of concern. In general, it was a fair conclusion that the department was not treating its junior faculty as well as it should have been, particularly in comparison to other universities and in light of the very hot and competitive job market for mathematicians and mathematics faculty. One faculty member recalls from this time a comment by Lehmer to the effect that it did young people good to be mistreated, as he had been at Lehigh years before. It is not clear what Lehmer meant by this. It may have been an expression of Lehmer's well-known wry sense of humor, or he may have been expressing a somewhat classical sentiment that the struggle to overcome adversity builds character.

In addition to the issues concerning teaching load and salary that had been raised by the broad group of junior faculty, the three newly appointed assistant professors in geometry and algebra, Eells, Kostant, and Thomas, pointed out to Lehmer that great advances were being made in their fields and that Berkeley should recruit aggressively in these important and significant areas of mathematics. Lehmer did not respond to this suggestion as these three newcomers had hoped he would [Bertram Kostant, personal communication].

Thus, there was evidently considerable ferment in the department on a number of issues in the spring of 1957. A group of faculty met in Leon Henkin's apartment one evening to discuss the future of the department and the chairmanship. Prominent among the participants in the discussion was Alfred Tarski. Blurred recollections of those attending make it unclear whether this meeting occurred before or after Lehmer had announced his resignation, or was merely contemplating resignation. In any case, a consensus ultimately emerged that, from among current faculty, John Kelley should be the one to assume the leadership of the department. At the same time, the group left the possibility of recruiting a new chair from outside, but nothing came of this immediately. The message that John Kelley was the consensus choice of the faculty to assume the chairmanship after Lehmer's resignation was communicated to Dean Constance and to Chancellor Kerr, and in short order Kelley was offered the chairmanship effective July 1, 1957.

In a sense, the events in the spring of 1957 have a certain similarity to, but also important differences from, two earlier events that were turning points in the evolution of the department. The first, in 1881, was initiated and carried out by the regents and led to the dismissal of Professor Welcker and the hiring of Professor Stringham and a redirection of the department imposed directly by the regents. The second turning point, in 1933, was prompted by the impending retirement of Chairman Mellen Haskell. The change was initiated and carried out by faculty leaders in other departments and by the provost and the budget committee with almost no consultation with the existing faculty in mathematics. A new chairman from outside, Griffith Evans, was recruited and brought in with a mandate for a major change in focus and direction of the department. In 1957, in contrast, the initiating event was the resignation of the chairman in protest to what he considered the administration's shabby treatment of the department. A clearly formulated proposal for change was generated within the department, and the administration was convinced to implement the changes. The results that flowed from this turning point were as dramatic as those that followed from the previous turning points, but, in contrast, the impetus for change came from within the department.

Interestingly, one of the first things Kelley did, even before assuming duties, was to try to escape from them! On May 7, 1957, he wrote to Marshall Stone, at the University of Chicago, as follows [DM, Chairman's File 1957–58]:

Dear Marshall,

You may have heard that Dick Lehmer has resigned as Chairman of the Department of Mathematics here, effective this June. We are in the process of looking for a new Chairman and this letter is simply an inquiry: Are there circumstances under which you would consider coming to Berkeley as Chairman of the Department.

This inquiry is my own, but it is not casual. Dean Constance of the College of Letters and Sciences has asked me to become Chairman and it seems necessary to me that I assume some responsibility for the management of the Department. This responsibility could best be discharged by persuading you to take over. This is not just my opinion. A few weeks ago, a fairly large group within the Department came informally to the conclusion that you would be the best possible Chairman. In fact Alfred Tarski attempted unsuccessfully to telephone you on behalf of this group. Moreover Dean Constance has indicated to me his interest in the possibility of your coming here.

Kelley went on to say that the department was understaffed but that the dean and chancellor recognized this and expected the department to expand greatly in the next five years. Kelley also attached his "white paper" and asked for Stone's comments even if he were to decline the invitation. Stone clearly did decline this initiative, but there is no written response from him in the file.

The white paper that Kelley drafted presumably in April 1957 begins as follows:

> There is an incredible shortage of mathematicians. This shortage is even more severe than in the war years, and there is compelling reason to believe that this situation will get worse. Mathematicians are now quite as difficult to hire as physicists and engineers; they will shortly be scarcer than the latter.

In the text of the white paper, Kelley provides an honest assessment of the standing of the department at the time:

> Our position in the several fields may be summarized: in foundations and metamathematics, very good; in analysis, very good; in geometry and algebraic topology, weak; in algebra, weak; in applied mathematics, weak; in statistics (now a separate department), very good. This is not intended to indicate that the individuals in the Department in the fields labelled "weak" are not good; it is simply that in program, in numbers, and in group effort these fields are weak.. .... The scientific position of our Mathematics Department relative to other mathematics departments in the country may be summarized: we are not in the first three (Chicago, Harvard, and Princeton), and we are in the top ten.

It is worth adding that at the time (1957) there had been for decades a general consensus in the mathematics community that Chicago, Harvard, and Princeton were the top three departments and stood above all others.

The white paper agues for making some senior appointments, for a reduction in teaching load to six hours for all faculty, and for recognition of the job market pressures in advancement and salary decisions. The document summarizes the situation as follows:

> There is now a completely unprecedented shortage of mathematicians. Our department, which has been built on the basis shrewd junior appointments, needs authority to go to a six hour teaching load and sympathetic cooperation on the matter of accelerated promotions in order to maintain its current position. The advance in scientific position which should be expected of us depends in large part on new senior appointments. Three such appointments are needed in pure mathematics, to bolster the fundamental disciplines. One is needed in applied mathematics—the next step in a program for the eventual establishment of an applied mathematics laboratory. One senior appointment is needed in numerical analysis and computing, to serve our

large group of students and to take advantage of our current opportunities. At least three of these five appointments are needed immediately, and the other two within two years.

In conclusion: We have, with meager resources, built a respectable mathematics department. Without some extraordinary assistance, we cannot, in this time of crisis, even maintain our position. We petition for an understanding and help in our effort to make a department of the calibre which this University deserves.

One fairly immediate response from Dean Constance to this white paper concerned the teaching load issue for junior faculty. His view was that teaching loads were set by the departments and that he did not want to know about it. Hence, the reduced teaching load of six hours for everyone went into effect, but implementation was deferred to 1958–1959 because of lack of lead-time and lack of immediate resources to achieve the goal for the coming year 1957–1958.

In his historical article on strengthening the College of Letters and Sciences, Lincoln Constance reflects on his seven years as dean (1955–1962) and the evolution of the mathematics department during this period. He discusses his impressions of mathematics as a discipline and his response to pleas from the department for resources: "Clearly, mathematics showed the most spectacular growth, more than doubling the size of its faculty" [during his period as dean].

But his following comments illustrate how long it apparently took the administration to recognize mathematics as a research discipline.

In the years following World War II, the rapid growth in technology and engineering and the increasing "numerization" of both the biological and the social sciences placed a heavy burden on mathematics instruction. In addition to teaching their own majors and the traditional undergraduate service courses, faculty in the department were moving to create pure mathematics as an independent subject somewhat parallel to and not merely a component of physical sciences and engineering. [Con, pp. 56–57]

This breathtaking statement about the nature and role of mathematics may have reflected the actual state of the department prior to 1933, but did not reflect advances since that time. It also provided an interesting contrast to what J. L. Coolidge had written in 1925 in a retrospective of Benjamin Peirce, who served on the Harvard faculty from 1831 until his death in 1878: before the time of Benjamin Peirce it never occurred to anyone that "mathematical research was one of the things for which a mathematics department existed."

Constance continues:

These changing emphases affected faculty recruitment and stimulated the search for research-oriented experts in numerous specialties within mathematics. The department had for several years been sounding alarms, but the projected staff needs were so great that they were not taken as seriously at first as they should have been.

Strengthening of the department had already begun with the appointment of Griffith Evans as chair in 1934; with his retirement from office in 1949, more aggressive efforts were made by a series of persuasive chairs: Charles Morrey, Derrick Lehmer, John Kelley, and Bernard Freidman. Due to their leadership and the combined support of their colleagues and the administration, Berkeley quickly moved to the top of the national rating charts.

As has been already noted, in 1955, when Constance became dean and after statistics had split off, the department had 18.5 FTE in professorial ranks and was only marginally larger than it had been in 1933. It was to grow explosively, especially starting in 1958. The growth included a number of major tenured appointments, and the department reached 75 FTE by 1968. The department's pleas had certainly been heeded. Clark Kerr, who served as Berkeley's chancellor from 1952 to 1958, provides in his memoirs a basis for this planned growth arising from considerations of academic planning and his philosophical ideas about the structure and mission of a modern university.

Mathematics had quality of faculty under the guidance of Griffith Evans, but not quantity. The sciences and engineering had treated it mostly as a "service" department for training their students. I was convinced, although I was a non-mathematical economist, that mathematics should and would be as central a department in a great research university of the future as philosophy had been in the past. Philosophy, once the "mother" of so many other departments, itself had become one of its more specialized children. Quantitative methods were rapidly getting more emphasis across the board (as they had gradually and intermittently since Pythagoras). Thus, I concluded, if a campus were to have one preeminent department in modern times, it should be mathematics. Also statistics was a new department at Berkeley under the excellent leadership of Jerzy Neyman, and deserving of expansion for related reasons. [Kerr 2001, p. 85]

Kerr does not say at what point in his six-year tenure as chancellor he came to this conclusion about the need to grow and strengthen mathematics. There is some evidence from statements he is quoted as making in 1956 that he was planning for major growth in mathematics, but it was not until 1957–1958 that the resources actually started to flow. And of course, in October 1957, the Soviet Union launched the Sputnik satellite, which among other things galvanized the nation to concentrate resources and attention on science and mathematics education at all levels from elementary school through postgraduate education. This event certainly provided additional justification for the decision to rapidly expand the mathematics department. The statistics department, which was also included in Kerr's text, expanded, too, but not as rapidly as mathematics, perhaps because statistics was already seen as having a very favorable student-faculty ratio when it was formed in 1955.

Before turning to the narrative in the following chapters about the surge in hiring and expansion of the department that started in 1958, it is necessary to provide some context for this growth by describing the campus-level planning issues concerning enrollment trends and the physical development of the campus. Under the leadership of Chancellor Clark Kerr, the campus began in the early-to-mid 1950s a planning process that embraced both academic planning and physical planning for the campus. By 1955, the postwar enrollment boom had passed, and the enrollment on the Berkeley campus stood at about 19,000, down from a peak of 23,000 just after the war, and in fact, just slightly more than the enrollment on the eve of World War II, which was about 17,000. The faculty in professorial ranks in 1955 on the campus numbered just under 800. In 1960, campus enrollment had grown to 25,000, and in 1965 to about 30,000, where it "more or less" stabilized. Faculty in professorial ranks grew to about 1000 in 1960 and to about 1200 in 1965 and with modest growth thereafter. Since overall enrollment and faculty size expanded by about 50%–60% from 1955 to about 1968, the quadrupling of the size of the mathematics faculty from 18.5 to 75.5 represented disproportionate growth, making up in part for past penury, as well as recognizing new circumstances.

One of the first issues of academic planning that were on the table was the ultimate steady-state student enrollment on the campus. The Strayer study of the needs of higher education in California, issued in 1948, recommended a limit of 20,000 students on any UC campus. The baby boom that was already underway and the resulting enrollment pressure that would come as a result of it made it clear that this limit would not provide the access that would be needed. In addition, the campus had already peaked at a 25,000 enrollment in the postwar GI-Bill years, and it was hard to argue that the steady-state size should be limited to something less than 25,000. On the other hand, there were some influential people within the university, especially among the regents, who wanted to concentrate UC enrollment on two megacampuses at Berkeley and UCLA with a number of smaller satellite operations at other locations.

Clark Kerr and campus planners decided on a limit of 25,000 based on two conclusions: First, working backward from an enrollment number using a prescribed student-faculty ratio to arrive at the overall size of the faculty, and then as an exercise in planning, apportioning this faculty complement to departments and schools, it is possible to estimate the size of departments needed to support the stipulated enrollment. Kerr then imagined what would be the consequences for the largest of the departments, such as mathematics and English. How large could a department be and still function effectively as a coherent academic unit and adequately mentor, groom, and advance its nontenured faculty following traditional university academic policy and practice? Kerr thought perhaps something on the order of 50 to 60 would be a desirable limit on department size, and this corresponds by the calculation sketched above to an enrollment limit of about 25,000. [CK, personal communication; see also Kerr 2001, p. 76]. Thus, the mathematics department entered into this calculation in a key way.

Another method of arriving at an enrollment limit is based on physical planning. If one imagines the library at the center of the campus (which is approximately where it is located), then it was Kerr's desideratum that all classrooms, faculty offices, laboratories, and administrative offices directly involved in serving the needs of students be within a (brisk) five-minute walk of the library, and hence that the central campus would have a ten-minute walking diameter. Then, if one prescribes that the campus should maintain as its overarching design feature "buildings in a park" with reasonable limits on height and then quantifies these concepts, one arrives at a limit on the square footage of space that could be built within these urban-design guidelines. Then, if one determines the space needs in square feet per faculty member, and combines this with a student-faculty ratio, one can arrive at an enrollment limit, which interestingly is also about 25,000 [Kerr 2001, p. 75].

The 25,000 enrollment limit has slipped over time owing to political pressures for access, first to 27,500, then to 30,000 (already in 1965), and currently today to about 34,000. The student-faculty ratio, which is at the core of these calculations, has also slipped, which would have the effect of raising the enrollment limits derived from these calculations. In addition, the campus has built some buildings that were not exactly in the mold of "buildings in a park," Evans Hall, for example, and increasingly, buildings have utilized underground space for academic purposes, thus reducing the above-ground build-out. But in President David Gardner's 1988 academic plan for the University, the 25,000 enrollment limit for campuses other than Berkeley and UCLA made its reappearance, inspired by Kerr's original reasoning. Also in the 1988 plan, Berkeley's enrollment limit was to be rolled back to 28,700. But that rollback never occurred.

As a key part of the planning process in the 1950s, Kerr appointed a planning committee charged to develop a formal long-range development plan (LRDP) for the campus. This LRDP was to map out the future physical development of the campus based on academic needs. The LRDP recognized the academic benefit to be gained from locating academically related departments in proximity to each other. The document specified that the "Physical sciences will be grouped in the area east of the Campanile esplanade," and that "Mathematics, Astronomy, and Statistics will be located in a new building north of LeConte Hall adjacent to Physics and Chemistry." Astronomy had been located in the Leuschner Observatory on Observatory Hill, and, as noted before, mathematics and statistics were located in Dwinelle Hall, well away from the other physical sciences. This location was perhaps suggestive of the early (and perhaps some more recent) views of mathematics as a "service" discipline without a research basis.

Not only did the decision to relocate the Departments of Mathematics and Statistics make sense academically at the time, it also reflected one aspect of major physical planning efforts earlier in the century that had culminated in John Galen Howard's 1914 Phoebe Apperson Hearst plan for the Berkeley campus. As recounted in [Wood], these planning efforts had begun in 1896 when the regents, confronted with rapid enrollment growth in the 1890s and the prospect of even more growth in the decades to come, and with a physical plant that was inadequate to accommodate this growth and that was an odd assortment of buildings architecturally, decided to embark on an international competition for a campus architectural plan. The effort was privately funded with lead donor Phoebe Apperson Hearst, the widow of Senator George R. Hearst and the mother of William Randolph Hearst. In the following decades, Mrs. Hearst and her family were to be extraordinarily generous donors to the campus. In 1899, the university and Mrs. Hearst announced the winner of the competition, the French architect Émile Bénard, and the winning plan, which was then called the Phoebe Apperson Hearst Architectural Plan.

After some difficult negotiations and discussions with Bénard, the regents decided to end his involvement with implementation of the plan, and in 1902 they retained John Galen Howard, a New York–based architect, as supervising architect. Howard, who had submitted a plan that was a finalist and ranked fourth in the Hearst competition, moved to California to assume his duties and was also appointed professor of architecture and founding faculty member of the Department of Architecture. While Bénard had already made some modifications to his plan after visiting Berkeley, Howard made a number of additional changes, and in fact, the 1914 plan that the regents approved more resembled Howard's submission in the competition rather than Bénard's original winning plan. Howard's plan called for a broad esplanade with gardens and fountains extending from the western end of the campus to the mining circle near the eastern end of the campus. The axis of this esplanade was oriented directly at the golden gate. On each side of this esplanade would be a series of two- to three-story beaux-arts buildings of white stone (usually limestone) with red tile roofs. The mining circle was so named because one of the very first buildings to be constructed was the Hearst Memorial Mining Building on the north side of this circle. Completed in 1907, this building was funded by a gift from Mrs. Hearst as a memorial to her late husband. The building was an architectural masterpiece and won high praise from all quarters. To quote from [Wood, p. 88], "The mining building stands today as one of the state's, and even the nation's, most distinguished architectural monuments."

Howard's 1914 plan called for additional engineering buildings along the north side of the central axis just west of the mining building, and for a grouping of physical sciences buildings south of the mining circle. This grouping included a building designated for chemistry to the southeast of the circle roughly where the current chemistry complex of buildings is located, a building designated for physics immediately south of the mining circle and facing the mining building, and then a building designated for mathematics to the southeast of the circle roughly where the current LeConte-Birge physics complex is located. Thus, the 1914 plan had laid the basis for the current engineering complex located north and west of the mining circle and had anticipated Kerr's 1956 decision to concentrate the physical sciences in that locale and thus to move mathematics, statistics, and astronomy to this area. But, the rapid growth of physics in the intervening decades and its need for laboratory space meant that physics had already taken over the area that had been reserved for mathematics in the 1914 plan.

Even though the 1914 Howard Plan was subsequently modified in important ways, Howard nevertheless left an indelible mark on the campus up through the termination of his tenure as supervising architect in 1923. The elegant central east–west esplanade never materialized as he planned it, and in fact, in the aftermath of World War II, the central glade or central swale as it was called was filled with a series of ten wooden military barracks-style "temporary" buildings. The last of these was not demolished until the 1990s. While the appearance of the area has improved since the removal of the wooden T-buildings, the presence of the mathematical sciences building, Evans Hall (see Chapter 15), closes off the east end of the central glade, and Moffitt Undergraduate Library is placed right in the middle of the glade, both contrary to the intent of the Howard plan. Both were part of the 1956 LRDP, to which we now return.

The official justification for this new "Physical Sciences and Mathematics–Statistics Building" dated April 21, 1955, stated,

> With recent advances in nuclear studies and the refinement of observational techniques and mathematical procedures, the traditional border-lines between astronomy, physics, and chemistry are growing less distinct. The functional interdependence of these sciences is increasing, as is their mutual reliance on mathematical and statistical methods. Because of this growing interdependence, it is important that Astronomy be given badly needed space facilities. Locating these related departments in adjacent buildings will gain the advantages of shared classrooms and facilities. Great scientific and instructional gains will be derived from increased accessibility of combined libraries and facilitated communications on both the student and faculty level. [UA, CU 5, series 3, Box 4, folder 23]

Funds for this building were appropriated in the 1955–1956 state budget cycle, and construction began some time afterwards. The building was completed and occupied in late 1959.

The building consists of six floors above grade and a basement, and it was unfortunate that from the day this new building was occupied in late 1959, it was too small to accommodate the needs of the three academic departments and the other tenants. Astronomy occupied the sixth floor, mathematics the third and fourth floors, and statistics the fifth floor, the other tenants including a library and the computer center, which together occupied the basement and the first two floors. The building was named Campbell Hall in honor of William Wallace Campbell, a distinguished astronomer who had been the long-time director of Lick Observatory, president of the university, and then president of the National Academy of Sciences.

The mathematics/statistics library, which had been established as a branch library in 1952 and located in Dwinelle Hall, in direct proximity to the two departments, had been growing steadily during the 1950s and was to be moved along with the two departments. The original plan had been to combine the mathematics/statistics library plus the astronomy library, which was moving along with the astronomy department, with the physics library in LeConte Hall to form an integrated physical sciences library. This plan foundered on the lack of sufficient space to accommodate it in LeConte Hall. Consequently, the mathematic/statistics library and the astronomy library were consolidated into what was called the AMS branch library, which would be located on the ground floor of the new building in approximately 4000 square feet of space. This placed further strain on space in the new building. The holdings of the consolidated AMS library at the time consisted of 13,722 volumes, but there is no breakdown between the disciplines. By comparison, the physics branch library had 8451 volumes, and the chemistry branch library 8489.

In addition to the library as a supporting resource for the department, there was a developing need for computing facilities as another increasingly necessary resource. As we recount in Chapter 14, the development of computing resources on the campus had been delayed, but in 1956, a computer center had finally been established in Cory Hall, the campus building housing the Department of Electrical Engineering. It was decided to include space for the computer center in the new building based on a desired proximity to mathematics, statistics, astronomy, and physics. The computer center machine rooms were located in the basement and the administrative offices on the second floor.

The overcrowding in Campbell was exacerbated by the rapid growth of the mathematics department, and it came in some sense as no surprise. Charles Morrey is recorded as saying in a building meeting for the future Campbell Hall on September 19, 1956, "although the Chancellor has stated that mathematics may double in ten years [the statement about planned growth that Kerr made that was noted a few paragraphs above], the rush hasn't started yet" [UA, CU 5, series 3, Box 4, folder 23]. This statement indicates that Kerr may have reached the conclusion about the growth of mathematics quoted from his memoirs but had yet not implemented it. Indeed, the rush was soon to start and mathematics would in fact double in size by 1960. Morrey was rightly worried that the Campbell Hall project would not provide mathematics with the space it needed and went on to say, "if it could be assured that mathematics will be taken care of in the future, his department would go along with the project as planned" [ibid.]. Indeed, the tentative plans were to eventually move mathematics and statistics into a mathematical sciences building to be constructed at some time in the future. The naming of Campbell Hall for a distinguished astronomer suggested that Astronomy would ultimately be the principal occupant of the building. But, before that building was completed, mathematics had vastly outgrown its space in Campbell Hall, and faculty members and graduate students ended up being distributed in spaces in six different buildings on campus. Statistics suffered a similar fate.

One of the most important points of the LRDP was that it recognized that mathematics, as well as statistics and computing, should be located in close proximity with the other sciences and indeed in proximity also to engineering, with which there was increasing intellectual interdependence and synergy. Thus, mathematics, statistics, and computing would be located in the future on the mining circle, if not in Campbell, then in another building facing the mining circle. The LRDP writers anticipated this eventuality, and incorporated it into the LRDP. An early draft of the LRDP projected a major new building at the eastern end of the central glade, that is, just west of the mining circle, that would be for branch library and classroom use, and might be shared by mathematics and statistics. The final version of the LRDP changed this and stipulated that the structure on the west of the mining circle would be an undergraduate library and classroom building and that the alternative plan for mathematics and statistics would be to share a (future) westerly addition of the "biochemistry building" (i.e., Stanley Hall) on the east side of the mining circle.

As it turned out, it was decided to site the undergraduate library (subsequently named Moffitt Library) at the western end of the central glade, just north of California Hall, rather than at the eastern end of the glade, and so the site on the west side of the mining circle was to be assigned to mathematics and statistics, plus classrooms. In fact, the capital project priority list submitted by the campus in 1956 to the president for the period 1957–1962 lists as priority 29 a physical sciences building, which is described as follows:

> At the east end of the Swale (a.k.a. the Central Glade), on the site occupied by Building T-1, it is proposed to locate a building of approximately 40,000 square feet, or more if continuing studies indicate greater site capacity, to house the Departments of Mathematics and Statistics, a possible Computer Center, and additional classrooms which will by 1960 or 1961 be badly needed. The space vacated in the recently funded Astronomy–Mathematics–Statistics Building [Campbell Hall] will by this time be

needed for expansion of the Department of Physics particularly for its offices and seminar rooms. ... Like Physics and Chemistry, mathematics has been and will continue to grow at a rate faster than the campus as large" [UA, CU 5, series 3, Box 4, folder 23].

In retrospect, the estimate of the size of what was to become Evans Hall is a bit amusing, since it ended up not as a building of 40,000 (assignable square feet), but rather a building of 170,000 assignable square feet, and construction was delayed long beyond the 1957–1962 time frame; but more about that building later, in Chapter 15. We now turn to the growth of the department in 1957–1960 under John Kelley's leadership.

Chapter 13

The Kelley Years: 1957–1960

Even though Kelley had long planned to be on leave for the 1957–1958 academic year, he still assumed the position of department chair beginning July 1, 1957, because of the urgent need for new leadership. He in fact went ahead with his leave and remained the leader of the department, guiding its strategic plan for development that he had outlined in the white paper. Clearly he needed someone to handle the day-to-day functions of the position, and consequently he asked former chair Chuck Morrey to serve as acting chair for the year. They were an effective team. A document from 1960 [DM- Academic Plan mathematics, 1953–1967] lays out the strategy that was followed:

> In Fall 1957 Morrey made a trip east to recruit faculty members for the academic year beginning September 1958. As a result informal offers were made to seven young men. Three of these accepted and four refused. The three who accepted were in Analysis and Foundations, fields in which we were already strong. The four who refused were in Algebra or Topology, or both. These refusals, coming after similar failures in preceding years, made it difficult to recruit young people in areas which were not represented by senior people on our staff. To remove this difficulty the Department with the complete support of the Dean and the Chancellor, decided to appoint two senior people in each of areas in which we were deficient, namely, Algebra, Topology, and Applied Mathematics.

Thus began the strategy of hiring in clusters. The first cluster hiring was to be in algebra, where the department successfully recruited Gerhard Hochschild, from the University of Illinois, and Max Rosenlicht, from Northwestern University.

Hochschild has spent the academic year 1955–1956 as visiting professor at Berkeley and so was familiar with Berkeley, and the department thought it had a good chance of attracting him permanently. The department proceeded to recommend his appointment as full professor in January 1958, to be effective on July 1, 1958. Within a week, the department also recommended the appointment of Maxwell Rosenlicht as full professor with the same effective date and made it known to each of them that the other was being recommended for appointment.

Hochschild and Rosenlicht had in fact known each other since 1947 when they were respectively Peirce instructor and graduate student at Harvard. The recommending letters from the department to the dean spoke of the programmatic need for the appointments as follows: "The (almost) simultaneous recommendation for the appointments of Professor Hochschild and Rosenlicht is an attempt to break up the near monopoly of the fields of Algebra and Topology/Geometry which is held by the eastern schools." These recruitment efforts proved successful, and the appointments were readily approved at the highest levels within the university. Hochschild and Rosenlicht accepted the offers.

Gerhard Paul Hochschild was born on April 29, 1915, in Berlin, Germany, where his father was a patent attorney. Soon after Hitler came to power in 1933, Hochschild emigrated to the Union of South Africa, where he continued his education, receiving a Master of Science degree in mathematics from the University of Cape Town in 1938. He then applied and was accepted into the mathematics PhD program at Princeton University. Studying under Claude Chevalley, with whom he shared a connection with South Africa—Chevalley was born in Johannesburg—Hochschild completed his doctoral degree in 1941 with a dissertation entitled "Semi-Simple Algebras and Generalized Derivatives." He became a naturalized citizen in 1942, and after service in the armed forces, was appointed as an instructor at Princeton in 1945. This was followed by appointment as Benjamin Peirce Instructor at Harvard for 1946–1948, and then appointment as assistant professor at the University of Illinois, Urbana-Champaign. Rising quickly through the ranks, he was promoted to full professor in 1952. His work in which he established seminal results in the cohomology of algebras and homological algebra in general, in class field theory, in Lie groups and particular group extensions, and in Lie algebras and their cohomology won him wide recognition by the time of his appointment at Berkeley. He was the first to introduce and study the cohomology groups of associative algebras and their modules, and he went on to show how to use cohomological methods in both local and global class field theory. Today, these are fundamental tools. With Serre he showed how to define a spectral sequence for the cohomology of a group extension—another far-reaching result—and he began a long collaboration with Daniel Mostow in which they have studied many aspects of Lie and algebraic groups and their homogeneous spaces.

After coming to Berkeley, he played a significant role in the development of the Berkeley department, since his presence made it possible to recruit much more effectively in algebra. He exercised intellectual leadership in the department, and continued to develop and expand his research program in several areas of algebra, while supervising the research of 26 doctoral students overall. His contributions to mathematics were recognized by election to the American Academy of Arts and

Sciences and the National Academy of Sciences. In 1985, he retired at age 70 and has pursued a variety of interests since then.

Maxwell Alexander Rosenlicht was born on April 18, 1924, in Brooklyn, New York, and attended New York public schools. His subsequent studies at Columbia University were interrupted by nearly three years of service in the US Army as an infantryman, but after the war he returned to Columbia and graduated in 1947. Doctoral study at Harvard followed, where he completed his research under Oscar Zariski in 1950 with a dissertation entitled "Equivalence Concepts on Algebraic Curves." He then won a National Research Council postdoctoral fellowship for two years, which he took in residence at the University of Chicago and Princeton University. Northwestern University appointed him as an assistant professor in 1952 and promoted him to associate professor in 1955. During 1954–1955 he served as Fulbright research professor at the University of Rome, and was about to be promoted to full professor at Northwestern when Berkeley made its offer of a full professorship in the spring of 1958. Rosenlicht in conjunction with Hochschild decided to come to Berkeley, thereby immediately establishing a major presence of algebra at Berkeley.

Rosenlicht's research had won very considerable recognition, and his work on generalized Jacobian varieties in 1954 and 1957 won especial attention. He was awarded the Cole Prize by the American Mathematical Society in January 1960 for this work. Rosenlicht extended his research to more general algebraic groups and their actions on algebraic varieties, and the structure of quotient varieties, establishing over the years a number of fundamental results. Berkeley had a difficult time retaining Rosenlicht, for he was the object of many attempts to lure him away during the 1960s, but he remained at Berkeley and advanced rapidly. In later years his interests shifted from algebraic groups to differential fields, and he published a number of papers on this topic. Altogether he supervised the doctoral work of 12 students, and he was actively engaged in departmental affairs, serving as chair from 1973 to 1975. He retired from active duty in 1994 under the University's voluntary early retirement plan (VERIP) and died in January 1999 while on a visit to Hawaii.

Although the initiative to hire two senior faculty members in algebra was the centerpiece of the hiring program for 1958, these were only a small part numerically, for the department made a total of ten new appointments to the professorial ranks that year, including three associate professors: Hans Bremermann, René De Vogelaere, and Istvan Fary.

Hans-Joachim Bremermann was born on September 14, 1926, in Bremen, Germany, and received his doctoral degree from the University of Münster in 1951. His dissertation, which was entitled *"Die Charakterisierung von Regularitätsgebieten durch pseudokonvexe Funktionen,"* was a significant step in dealing with a special case of the solution of the Levi problem of characterizing domains of holomorphy in higher dimensions. Although he is not listed as an official Behnke student, he was clearly part of the renowned Behnke School of complex analysis at Münster. In subsequent work published in 1953, he obtained a full solution of the Levi problem. Simultaneously and independently, F. Norguet and K. Oka also published solutions to this famous problem. Bremermann continued his work in several complex variables with additional results on plurisubharmonic functions, on the Dirichlet problem, on holomorphic functions, on infinite-dimensional Banach spaces, and on distributions.

After receiving his degree, Bremermann taught for a year at Münster, then did postdoctoral work with Stephan Bergman at Stanford, followed by a period at Harvard. He returned to Münster, but also spent a period at the Institute for Advanced Study (IAS). In 1957 he was appointed as an assistant professor at the University of Washington, but had an invitation to spend the following year (1958–1959) at the IAS working on application of complex analysis to scattering amplitudes in physics.

He was recruited by Berkeley during the 1957–1958 academic year for an appointment as associate professor to be effective July 1, 1958. He accepted this offer, but because of the prior commitment to IAS, he was permitted to take a leave of absence without salary during 1958–1959 and so joined the Berkeley faculty in person in July 1959. He brought to the department strength in several complex variables, an area spanning geometry and analysis that had not been represented at Berkeley, thereby further broadening the department's intellectual span. He was promoted to full professor in 1966. He was naturalized as a US citizen in 1965.

Early on in his career, Bremermann had shown interests in computation and algorithms. These interests matured and evolved, and the focus of his research moved away from several complex variables in the mid sixties and into the general area of artificial intelligence, computational complexity, neural networks, and evolutionary biology. His work subsequently moved further toward biology, and he worked on population genetics, parasitism and disease, and epidemiology. His work in these areas was highly esteemed, and eventually he moved his university appointment to the Department of Biophysics, while retaining some connection with the mathematics department. As a result of the biology reorganization on campus in the 1980s, when the previous departmental structure was dissolved, his appointment then resided in the new Department of Molecular and Cell Biology. During his years at Berkeley he guided 11 doctoral students to completion in mathematics and more than 15 in the biological sciences. He died on February 21, 1996, at the age of 69.

The second of the three associate professors appointed this year was René De Vogelaere. René Joseph De Vogelaere was born in Etterbeck, Belgium, on May 18, 1926, and he entered the Catholic University of Louvain in 1942 to study civil engineering, obtaining certification as a civil engineer. He then switched to mathematics and proceeded rapidly to obtain his doctoral degree at Louvain in 1948. He then emigrated with his new bride to Canada to take a faculty position as assistant professor of mathematics at Laval University, in Quebec. Notre Dame University, in South Bend, Indiana, invited him as a visiting faculty member in 1953 and then appointed him associate professor of mathematics in 1954.

His research interests lay in the area of nonlinear ordinary differential equations, and their applications to astronomy, mechanics, physics, and meteorology. With the advent and availability of computers, he then began to develop numerical methods suitable for implementation on digital computers. Lehmer knew De Vogelaere and had invited him to the Institute for Numerical Analysis at UCLA in 1951 and had overseen his work there. When the opportunity arose to invite De Vogelaere as visiting associate professor for 1957–1958, Lehmer, as chair, recommended his appointment as a way to strengthen applied mathematics and computing. The following spring, the department under Kelley and Morrey's leadership recommended De Vogelaere for a permanent appointment with 50% in the mathematics department as associate professor and 50% in the newly formed computer center as associate research mathematician. In the following chapter, the development of computing on campus and the establishment of the computer center accompanied by a strategy of making joint appointments in the center and academic departments will be discussed.

The permanent appointment was approved as recommended to be effective July 1, 1958. In 1960, De Vogelaere was promoted to full professor and to research mathematician in the computer center. In 1971 the campus completed conversion of the computer center to a service organization, eliminating any research role, and thus discontinued the practice of employing research scientists in the center. De Vogelaere's appointment was then converted to professor of mathematics 100% time. He took a lead role early on in developing computational facilities for the department, and he devoted himself to developing the curriculum in numerical analysis and in designing computational materials for laboratories that might accompany a number of upper-division courses. During his career he supervised five doctoral students to completion of their degrees.

Although his subsequent formal scientific publications were somewhat sparse, he authored a number of scientific reports and preprints. Moreover, two early papers of his received subsequent recognition and appreciation as having been far ahead of their time. One of these was a 1956 preprint that introduced a precursor of what became known many years later as symplectic integrators, which are very accurate numerical methods for numerical solution of differential equations that have contact structure. The second was a 1955 paper on numerical methods for chemical reaction rates that was again an early precursor of techniques that later became common. In his later years, De Vogelaere also became interested in finite geometries and published several papers in this subject. He died suddenly at home of a heart attack on December 14, 1991, at age 65.

The third appointment in 1958 at the rank of associate professor was Istvan Fary, who was known to everyone simply as "Steve." Fary was born in Gyula, Hungary, on June 30, 1922, and received a master's degree in mathematics from the University of Budapest in 1944, even though his studies had been interrupted by the German army, which had interned him as prisoner of war for two years during the war. He continued his studies and earned a doctoral degree in mathematics from the University of Szeged in 1947. After a brief period of service as an instructor at the University of Szeged, he left Hungary for Paris in December 1948, where he joined the Centre National de la Recherche Scientifique (CNRS). During this period he commenced work on a second doctorate, which was awarded to him by the Sorbonne in 1955. He worked under Jean Leray, and his dissertation was entitled "*Cohomologie d'une Certaine Classe d'Applications.*"

Fary had begun publishing research papers in 1947, and by the time he received his French doctorate, he had compiled a substantial bibliography of 18 items. His early work concerned topological groups and transformation groups, as well as convex bodies, but also included a very notable result in 1949 on the total curvature of a knot, a result now known as the Fary–Milnor theorem. His thesis under Leray was a substantial contribution to algebraic topology and involved very careful analysis of sheaves and of cohomology theories. His results were of importance in algebraic geometry, since they simplified, clarified, and expanded on results of Lefschetz. This work appeared in two highly regarded fifty-page papers in the *Annals of Mathematics* in 1955 and 1957.

After receiving his doctoral degree, Fary moved to Canada, where he was a research fellow supported by the Canadian National Research Council in residence in Montreal. The following year, 1956, he was appointed as an associate professor at the University of Montreal. His celebrated papers in the *Annals* as well as his earlier work on the curvature of knots led to an invitation to come to Berkeley as visiting associate professor for the academic year 1957–1958. Delays in the process of obtaining the necessary visa resulted in Fary not arriving in Berkeley to take up duties until January 1958. Almost immediately, the department decided to try to make the appointment permanent as part of the effort to strengthen geometry and topology. The recommendation for appointment as associate professor of mathematics was approved, Fary accepted, and, as noted, his permanent appointment was effective July 1, 1958.

Fary was advanced to the rank of full professor in 1962. In subsequent years, he continued his research generally in the areas of his previous work, and guided the doctoral dissertations of five students. In the late 1970s he began work on a massive three-volume book, *Stacks and Cohomology*, that was to be published by the Hungarian Academy of Sciences. Work on this book, which would deal with classical and exotic cohomology theories from a unified axiomatic point of view, had progressed, but was cut short by Fary's death on November 2, 1984, at the age of 62.

In addition to these five tenured appointments in 1958, the department made five assistant professor appointments for an amazing total of ten appointments of regular faculty. The five assistant professor appointments were Errett Bishop, Heinz Cordes, François Trèves, Maurice Sion, and Robert Vaught. Of these, Sion left after two years and Trèves after two and a half years, and Bishop left in 1965, but the other two remained at Berkeley for their entire careers.

When Errett Bishop was appointed as an assistant professor, he had already served as an instructor at Berkeley for three years. Errett Albert Bishop was born on July 24, 1928, in Newton, Kansas. His father was a graduate of the United States Military Academy who had served in World War I and then became a professor of mathematics at several universities before settling at Wichita State University. Although his father died when Errett was only five years old, he had substantial influence on his children. Both Errett and his younger sister Mary were mathematical prodigies. He was accepted by the University of Chicago as an undergraduate in 1944 in their special scholarship program. He quickly earned his bachelor's degree in 1947 and proceeded on to graduate study at Chicago, coming under the influence of Paul Halmos, who was to be his dissertation supervisor. His studies were interrupted by two years in the US Army, and he then completed his doctoral degree in 1954 with a dissertation entitled "Spectral Theory for Operations on Banach Spaces."

Upon completion of his degree, he came to Berkeley as an instructor, and in 1958 was appointed as an assistant professor. He developed a strong research program that included, in addition to his dissertation work, basic results on approximation by polynomials and rational functions, extending, for instance, the Mergelyan theorem and the F. and M. Riesz theorem, the general theory of function algebras, and important results in several complex variables, including results about Stein manifolds. Consequently, he won rapid promotion, first to associate professor in 1959 and then to full professor in 1963. In about 1964, he became interested in foundational issues and in particular in constructive mathematics, and devoted the remainder of his career to this subject. His name is best known for his work on constructive mathematics, but he also had many fine results in functional analysis and complex variables to his credit. During the academic year 1964–1965 he became disillusioned with Berkeley, in part owing to the student unrest and perhaps in part to what he perceived to be a lack of receptivity by his colleagues to his work in constructive mathematics. He then made a decision to seek to transfer his appointment to the newly founded San Diego Campus of the University of California. He was one of the early founders of this new department, which has grown to distinction over the intervening years. During his career he supervised nine doctoral students, three at Berkeley and six at UC San Diego. In 1982, he was stricken with cancer and died of the ailment the following year at the age of 54.

As a note, Errett's sister, Mary, who was two years younger, was also admitted to the University of Chicago as an undergraduate under the special scholarship program. She likewise studied mathematics, and earned her doctoral degree under Antoni Zygmund in 1957. She married Guido Weiss, who was another Zygmund student, and she went on to a distinguished career in research, publishing under her married name. Her career was tragically cut short by her untimely death in 1966.

Heinz Otto Cordes was also appointed as an assistant professor in 1958. He was born on March 18, 1925, in Westfalen, Germany, and studied mathematics at Göttingen, receiving his doctoral degree in 1952. The topic of his dissertation, which was written under Franz Rellich, was *"Separation der Variablen in Hilbertsche Räumen."* He continued on as a scientific assistant at Göttingen for four years, during which time he worked on obtaining a priori estimates for certain types of elliptic equations, and he proved the unique continuation theorem for second-order elliptic equations, solving a longstanding open problem. Aronszajn had also solved this problem, but their work was independent of each other's. In 1956, Cordes accepted an assistant professorship at the University of Southern California, and his work came to the attention of faculty at Berkeley, especially Morrey, whose own interests were very similar, and efforts were begun to try to bring him to Berkeley. In response, the University of Southern California, as they did a few years earlier with Leon Henkin, started the process to promote Cordes to tenure as associate professor, but Berkeley prevailed, and Cordes accepted a (high-level) assistant professorship effective July 1, 1958. His appointment added further strength in partial differential equations (PDE) in a younger generation to the department.

Shortly after he arrived at Berkeley, the department began the process of promotion to associate professor with tenure, which took place effective July 1, 1959. Promotion to full professor came quickly, in 1963. Cordes's research deepened and broadened over his career and moved into research on operator algebras, including C^* algebras, Fredholm theory, and pseudodifferential operators. Cordes and Tosio Kato, whose appointment we will discuss shortly, established a longstanding friendship lasting over forty years. Altogether, Cordes supervised 19 doctoral students at Berkeley, including Michael Crandall and Michael Taylor. In 1991, Cordes retired under the university's voluntary early retirement program at age 66, but he has continued his research nearly unabated with numerous publications appearing since his formal retirement.

Maurice Sion was hired as an assistant professor effective July 1, 1958, but resigned after two years to take a position at the University of British Columbia. He was born in 1928 in Skopje, Yugoslavia, and received his bachelor's and master's degrees in mathematics at New York University in 1947 and 1948, respectively. He then came to Berkeley for doctoral work, where he worked under Anthony Morse. His dissertation was entitled "On the Existence of Functions Having Given Partial Derivatives on Whitney's Curves." He spent several years at the Princeton Institute for Advanced Study immediately before returning to Berkeley as a faculty member. His work was in the general area of abstract spaces, analytic sets, the structure of measures, and integration processes. Sion resigned his position at Berkeley on June 30, 1960, to take a position at the University of British Columbia, where he has spent his career.

François Trèves was another one of the five mathematicians appointed as assistant professor on July 1, 1958. Jean François Trèves was born April 13, 1930, in Brussels, Belgium. He was an Italian citizen, and after attending secondary school in Italy through 1950, he enrolled at the Sorbonne to study mathematics. He completed a Doctorat es Sciences in 1958 with work on PDE under the mentorship of Laurent Schwartz. Announcements of his results appeared in a number of notes in the *Comptes rendus de l'Académie des sciences* and were followed by several major papers. His appointment at Berkeley added more strength to an already strong group in PDE. Soon after his arrival at Berkeley, he began to receive attractive offers from several other institutions, and ultimately Berkeley could not retain him. He resigned effective December 31, 1960, just two and a half years after his arrival, to take a position at Purdue. He subsequently moved to Rutgers, and he has been an active and important contributor to the PDE literature over the years.

Robert Vaught is the last of the five assistant professor appointments in 1958. Robert Lawson Vaught was born April 4, 1926, and commenced college studies at Occidental College in 1942, where he met his classmate and future Berkeley faculty colleague Bill Bade. After two years at Occidental, he was inducted into the navy and was assigned to an officer training program at UC Berkeley, where he completed his BA in mathematics and physics in 1945. He served as an ensign in the navy until his discharge in 1946 and then returned to graduate school in mathematics at Berkeley. He soon came under the influence of John Kelley, who had arrived on campus in 1947, and began research under Kelley's supervision. This work led to a joint paper in the *Transactions of the American Mathematical Society*, and Vaught was well on his way to completing a dissertation when Kelley was fired by the regents in 1950 during the loyalty oath controversy. Vaught decided to remain at Berkeley, and therefore had to switch advisors. He selected Alfred Tarski, whom he found inspiring and under whom he completed his dissertation in 1954. The dissertation was entitled "Topics in the Theory of Arithmetical Classes and Boolean Algebras."

Following completion of his studies, Vaught was appointed as instructor at the University of Washington and was subsequently promoted to assistant professor there. When the Berkeley mathematics faculty was rapidly expanding in 1957–1958, Vaught's name came to the forefront, and he was appointed as assistant professor, again following the pattern of hiring young mathematicians with several years of experience rather than newly minted PhDs. Tarski and Henkin were the only members of what they conceived as the logic group on the faculty at the time and had expressed a desire that the logic group in the department represent approximately 10% of the total mathematics faculty. As the size of the mathematics faculty grew past thirty, the case for an additional appointment was natural. This 10% figure was never formal departmental policy, but it was an informal goal desired by Tarski and his successors, and has generally been honored over the years. In addition, several logicians often of a somewhat more philosophical nature were appointed in the department of philosophy, so that the logic group encompassed more than faculty in mathematics. Alfred Foster might be counted as well.

Indeed, in 1957, at the urging and insistence of Tarski, the campus established an interdisciplinary program in logic and the methodology of science. Although students could write a dissertation in logic in either mathematics or philosophy, the program was designed for students who might be more suited to such an interdisciplinary program. "Methodology of science," according to the description of the program, was to be understood to mean "metascience," the study of methods of the sciences by logical and mathematical means. It provided the logic group in the department greater identity, although not to the extent that Neyman had achieved with a separate statistics department, and created links with the Department of Philosophy. Members of the program soon included, in addition to the logic faculty in mathematics, Professors Ernest Adams, William Craig, and Benson Mates, from philosophy; David Blackwell, from statistics; John Harsanyi, from business administration (and a future Nobelist in economics, an honor he shared with John Nash and Reinhard Selten for work in game theory); and finally Yuen Ren Chao, from oriental languages and literature, a renowned scholar of Chinese linguistics, who had earned a doctorate in philosophy at Harvard and had taught mathematics at Cornell in his earlier years.

Vaught's excellent research, which focused especially, but not exclusively, on model theory, won him rapid promotion, first to associate professor in 1960 and then to full professor in 1963. In 1977, he was the recipient of the first Carol Karp prize of the Association of Symbolic Logic for his work on infinitely long expressions. Early in his career Vaught formulated a fundamental conjecture, which came to be called the Vaught conjecture, which is unsolved to this day. During his career he supervised the dissertation work of 18 students, including James Baumgartner, and his future Berkeley colleague Jack Silver. Vaught retired from active service in 1991 under the university's voluntary early retirement incentive program at age 65. He died after an extended illness on April 2, 2002.

After this intense round of hiring, in which ten appointments were made in the professorial ranks, the department had reached a faculty complement in the professorial ranks of 35. Further growth was to come, but the top priority for the 1958–1959 hiring cycle was to find and recruit two senior faculty members in topology and geometry to redress the historical weakness in these areas. The targets ultimately identified were Shiing-Shen Chern and Edwin Spanier, both then at the University of Chicago.

Chern had for some time been someone that Berkeley was interested in recruiting. In his last year as chair, Morrey wrote to Chern in December 1953 proposing an appointment at Berkeley [DM Chern file]. At this time, President Sproul had agreed that the department could make a senior appointment in light of the impending retirement of Evans. Chern in response raised questions about the lack of colleagues in his areas of interest at Berkeley, and after a month's thought declined the initiative. The issue was reopened in the spring of 1958 with a suggestion that Chern spend a semester at Berkeley during the 1958–1959 year as a visiting professor, as a prelude to a permanent appointment. This did not work out, but late in 1958, Kelley arranged for Chern to visit Berkeley for a two-week period in March 1959.

On February 4, 1959, Kelley wrote to Chern indicating departmental plans for building in areas of interest to Chern:

> As you know we were unsuccessful in our attempt to appoint [Raoul] Bott. I still have hopes that [Hans] Samuelson will accept. We have asked John Milnor to visit for the year 1959–60, and he seems somewhat interested. I believe that [Karol] Borsuk will visit us for one semester of next year, and [Roger] Godement will be visiting for the entire year.

Evidently the department had many irons in the fire, and in the end, Samuelson declined interest in the Berkeley initiative and remained at Ann Arbor at this time. Samuelson did subsequently come to California, but to Stanford. Bott moved from Ann Arbor to Harvard in 1959. The letter does not mention the plans to recruit Edwin Spanier, which were already underway.

Chern had previously made plans to spend the academic year 1959–1960 in Europe, and so was not available to accept an offer effective July 1, 1959, but he did indicate interest in an offer from Berkeley under which he would take up duties on July 1, 1960. During February, Kelley secured the approval from Chancellor Glenn Seaborg to negotiate with Chern for such an appointment with a salary that would be competitive. The formal appointment would be subject to the usual academic review. The two-week visit in March went well, and Chern indicated an inclination to accept a permanent position at Berkeley. One key element of that decision was the knowledge that Edwin Spanier was being recruited to Berkeley as well. Chern also inquired about the possibility of bringing Stephen Smale, who was also at the University of Chicago, to Berkeley, and Kelley assured him that plans were already underway to recruit Smale to Berkeley in 1960 (which were ultimately successful).

Kelley added a further incentive to Chern when he wrote on March 26 that the department had established a title of senior professor, which would carry a teaching responsibility of one course per semester rather than the usual two. Kelley wrote that this had been authorized by the senior faculty of the department and had been cleared with the dean. Tarski and Morrey had been appointed senior professors, and Kelley wrote that Chern would likewise hold the rank of senior professor. The administrative wheels turned quickly in this case, and the regents formally approved the appointments of Chern and Spanier on May 18, effective July 1, 1960, and July 1, 1959, respectively.

Shiing-Shen Chern was born on October 26, 1911, in the city of Jiaxing in Zhejiang Province, China, and in 1926 he enrolled as a freshman in Nankai University. After graduation in 1930 he entered Qinghua University in Beijing. In 1932, Blaschke visited China, and Chern took notes of his lectures, which in turn led Chern to seek admission for doctoral study at the University of Hamburg. He was successful, and he enrolled in 1934 with support from a government fellowship. During his studies, he worked under Blaschke, but also had thorough discussions with Erich Kähler. He finished his DSc degree under Blaschke in 1936, and then spent the following year, a decisive one for him, in Paris, working under Elie Cartan. After returning to China in 1937, he was appointed to a faculty position at Qinghua University. This university had moved its location because of the war, and in 1943 he accepted an invitation to visit the Institute for Advanced Study, in Princeton. During this period, Chern made some of his most significant discoveries: the intrinsic proof of the Gauss–Bonnet theorem and his pioneering work on characteristic classes, which were thenceforth known as Chern classes.

In 1946, with the war ended, Chern returned to China, and was asked to organize a new mathematics institute of the Academia Sinica. This institute produced many leaders of Chinese mathematics. But, as the Communist forces advanced toward Nanjing, and the government neared collapse, Chern's work became impossible, and in 1949 he accepted another invitation to visit at the Institute for Advanced Study. The University of Chicago appointed him as professor in 1950, and he remained there until he was recruited by Berkeley in 1959. Chern's appointment at Berkeley was a turning point in the development of topology and geometry at Berkeley, and indeed for the entire department, and his presence and influence was a significant factor in the department's rise to one of the top mathematics departments in the country. Chern created a Berkeley school of differential geometry, and recruited many talented mathematicians and students to Berkeley. He continued his own research unabated for many years and indeed was still working actively into his nineties. He once told the author that the trick is to have enough good ideas when you are young to last you a lifetime. Chern had altogether 45 doctoral students, including three at the Academia Sinica, eight at the University of Chicago, and 34 at Berkeley. The list includes the names of many famous geometers of the twentieth century. The mathematics genealogy project lists 563 descendents, and this number will continue to grow.

Although Chern formally retired in 1979 at age 67 from the university, he remained active and served as founding director of the Mathematical Sciences Research Institute (MSRI) in Berkeley from 1981 to 1984. After his retirement from Berkeley, he frequently visited China and was appointed as the director of a newly created mathematics institute created in his honor at Nankai University, his alma mater. Thus, he was founding director of three mathematics institutes, a record that is unlikely ever to be equaled. After the death in 2000 of his wife of over sixty years, Chern decided to remain in China at the Nankai Institute permanently, where he died on December 3, 2004. There is a forthcoming full-length biography of Chern by Lensey Namioka that will provide additional details on Chern's life and achievements.

Edwin Spanier was the other half of the cluster hire in topology and geometry that Kelley was planning for 1959. Edwin Henry Spanier was born on August 9, 1921, in Washington, D.C. He earned a bachelor's degree in mathematics at the University of Minnesota, and then served for three years as a mathematician in the U.S. Signal Corps. After being discharged from the army, he returned to graduate studies in mathematics at the University of Michigan, where he received his doctorate in 1947 under Norman Steenrod. His dissertation, "Cohomology Theory of General Spaces," was a significant contribution in algebraic topology, and he followed this up with a series of other important results on, for instance, cohomotopy theory, the homology of fiber spaces, and duality in homotopy theory, among many others. His first postdoctoral position was as Frank B. Jewett fellow at the Institute for Advanced Study, followed by appointment in 1948 as assistant professor at the University of Chicago. Promotion to associate professor came in 1954 and to full professor in 1958. His appointment at Berkeley was a major event and helped put Berkeley on the map in algebraic topology and to recruit several young topologists almost immediately.

In subsequent years, Spanier broadened and deepened his work in topology and in 1961 began a collaboration with Seymour Ginsberg, of the University of Southern California, on the structure of formal languages, which led to a long series of papers over a period of 25 years. Later in his career, Spanier's work returned to his original interest in algebraic topology. Over his career, he supervised 17 doctoral students, 4 at the University of Chicago and 13 at Berkeley, some in topology and some in the theory of languages. He retired under the university's early retirement program in 1991, but continued to publish after his retirement. He died of cancer in Scottsdale, Arizona, on October 11, 1996.

The only other appointment in 1959 was that of John W. Woll as assistant professor. Woll was a doctoral student of Salomon Bochner at Princeton. He completed his dissertation in 1956 on "Homogeneous Stochastic Processes," which concerned the structure of stochastic processes that have invariance properties under the action of a topological group. He resigned his faculty position at Berkeley in 1962 after three years of service.

The recruitment in the 1959 cycle set the stage for the 1960 cycle. We have already noted that the arrangements for Chern's appointment were made in 1959 even though the appointment was not to be effective until 1960. The emphasis on senior appointments turned to applied mathematics, the third area of weakness identified in the white paper that needed to be strengthened by addition of such appointments. That effort will be described presently, but first it is opportune to follow up on the immediate impacts of the Chern and Spanier appointments. The department was able successfully to recruit three strong young topologists: Stephen Smale as an associate professor and Morris Hirsch and Glen Bredon as assistant professors.

Stephen Smale was born on July 15, 1930, in Flint, Michigan, and entered the University of Michigan in the fall of 1948 as a freshman. He remained there until 1956, receiving his BA in February 1952, his MA in June 1953, and his PhD in February 1957 after he had already left to take a position as instructor at the University of Chicago. He did his dissertation work under Raoul Bott, with a thesis entitled "Regular Curves on Riemannian Manifolds." He quickly moved on to use the results in his thesis to study immersions of spheres in Euclidean space and established a result that everyone found amazing about turning immersed spheres "inside out." He also established important results on the topological structure of the diffeomorphism group of the two-sphere. This raft of results coming during a short period of time led to considerable recognition, and after two years as instructor at Chicago, Smale went in the fall of 1958 to the Institute for Advanced Study in Princeton as a visiting member. In the spring of 1959, Kelley approached him with an informal offer of an assistant professorship starting in 1960. Smale accepted this offer in April 1959, and according to Kelley, the one factor that influenced Smale was the knowledge that Ed Spanier was going to Berkeley.

In the fall of 1959, the formal offer to Smale was tendered by the chancellor, but Smale's growing achievements and rapidly rising fame made it necessary to revise the offer in December to an associate professorship, a change that was quickly accomplished. Then, in the spring of 1960, as Smale was getting ready to move to Berkeley, he spent some time visiting mathematicians in São Paulo, Brazil, and in the course of this visit found a proof of the Poincaré conjecture in higher dimensions that a homotopy sphere is homeomorphic to a sphere. He first did this in dimension seven and higher, but quickly lowered it to dimension five and higher. Also, he very quickly crystallized from his work a result that was to become known as the *h*-cobordism theorem, which was the key tool in the proof of the Poincaré conjecture and in unlocking the classification of simply connected compact manifolds in dimension five and greater. These results were stunning to the international mathematical community and catapulted his fame even higher. Soon after his arrival at Berkeley, the department initiated the process to promote him to full professor. However, many other universities joined in the competition, and Berkeley lost out to Columbia, when Smale decided to accept a position there effective July 1, 1961, after only one year at Berkeley.

Even before he left Berkeley, Smale had already turned his attention to dynamical systems and laid the groundwork for a program to which he was to devote himself over the next several years. Fortunately, Berkeley was able to lure Smale back to California after only three years at Columbia. He immediately established an important "school" of dynamical systems in Berkeley with many colleagues, post doctorates, and graduate students, and a steady stream of publications. His work on the Poincaré conjecture and the *h*-cobordism theorem was recognized by award of a Fields Medal in 1966, the first awarded to a Berkeley faculty member.

In the early 1970s, Smale's research interests shifted again, now to mathematical economics, which led to a series of papers on economic equilibria and his interaction with the strong group of mathematical economists in the economics department at Berkeley. Smale was appointed as professor at 0% time in the economics department in 1976 in recognition of his work. Then, in the 1980s, Smale's interests shifted once more to the foundations of numerical analysis and computational and algorithmic complexity. Along with many other faculty members he retired in 1994, at age 63, under the university's voluntary early retirement program. Shortly thereafter he accepted a professorship at the City University of Hong Kong, and made yet another switch in his research direction to the theory of learning. During his career he supervised or shared in the supervision of 44 doctoral students, including one at the University of Chicago, one at Princeton, 41 at Berkeley, and one at the City University of Hong Kong. He has 421 mathematical descendents so far. His life and work are the subject of biography by Steve Batterson, *Stephen Smale: The Mathematician Who Broke the Dimension Barrier* (AMS, 2000), which can be consulted for additional information.

Another young topologist hired in the same year, 1960, was Morris (known to all as Moe) Hirsch. Morris W. Hirsch was born on June 28, 1933, in Chicago. He graduated in February 1950 from the Bronx High School of Science, and after periods of study at City College of New York and St. Lawrence College, he transferred to the University of Chicago, where he received a master's degree in 1954 and a doctorate in 1958. His advisor at Chicago was Edwin Spanier, but Stephen Smale, who was an instructor at the time, shared in the advising, and Hirsch was a scientific collaborator with Smale on work on involutions of the three-sphere. Hirsch's dissertation, entitled "Immersions of Manifolds," attracted considerable attention, and was clearly related to research that Smale was doing at the time. Hirsch received a National Science Foundation postdoctoral fellowship, which he took in residence at the Institute for Advanced Study in Princeton for 1958–1960.

During the summer of 1959, Kelley approached Hirsch about coming to Berkeley, and as there was mutual interest, these initial discussions led to his appointment at Berkeley as an assistant professor in 1960. Hirsch's research continued to develop and won him rapid promotion at Berkeley, first to associate professor in 1963, and then to full professor in 1964. In the course of his work on embeddings of manifolds, he resolved in the affirmative an old conjecture of Whitney that every three-manifold can be embedded in the five-sphere. In the early 1970s Hirsch's interests moved more into dynamical systems, but included a mixture of the study of foliations, transformation groups, and affine manifolds. Some years later, his interests evolved further to include the study of neural nets. Hirsch supervised the dissertations of 23 doctoral students at Berkeley. He served as department chair in 1981–1983, and subsequently took advantage of the university's early retirement program and retired on January 1, 1993, at the age of 59. He remained very active in the department after his retirement, supervising graduate students (nine of his doctoral students finished after his formal retirement), teaching courses, and publishing research papers.

The third young topologist hired in the 1960 recruitment cycle was Glen Bredon. Glen Eugene Bredon was born on August 24, 1932, in California. He attended Stanford University, graduating in 1954, and then entered graduate school at Harvard for graduate study in mathematics, where he received his doctorate in 1958. His work focused on topological transformation groups, the study of fixed-point sets, and the use of homology and cohomology theory to describe invariants of the action. Written under the direction of Andrew Gleason, the dissertation was entitled "Some Theorems on Transformation Groups." Following his degree, Bredon was invited to the Institute for Advanced Study for a two-year period, during which time he participated in Armand Borel's seminar on transformation groups in 1958–1959. He was then recruited to Berkeley as an assistant professor in 1960. His work in transformation groups and related topics won him rapid promotion, moving to associate professor in 1965, and to professor in 1967. But in 1968 he was lured away from Berkeley by an offer of a professorship at Rutgers.

Departmental records indicate that the department had actively pursued two other young mathematicians in the general area of topology and algebra in 1959–1960, in the hope of interesting them in Berkeley, but the efforts were not successful. These mathematicians were Michel Kervaire, of New York University, and Richard Swan, at the University of Chicago. The department also approached Paul Cohen and tried to interest him in an assistant professorship at Berkeley, but he declined and accepted a position at Stanford instead. Finally, two more assistant professors were hired during this cycle in different fields: Adam Koranyi in functional analysis and geometry, and Dana Scott in logic and foundations.

Adam Koranyi was born in Szeged, Hungary, on July 13, 1932, and attended the University of Szeged, where he received his diploma in mathematics. He published several papers in which he used Hilbert-space methods to prove and explicate results in complex variable theory, especially interpolation theorems and Loewner's theory of matrix monotone functions. He then took the opportunity to leave Hungary during the 1956 uprisings, and came to the United States, where he enrolled for doctoral studies at the University of Chicago. In his dissertation, "Operator Theoretic Methods Applied to Interpolation Problems for Functions of Several Complex Variables," written under Marshall Stone, he extends these earlier ideas to several complex variables. He completed his work in 1959 and was recruited to Berkeley in 1960 as an assistant professor. Shortly after his arrival he began an extended program of research jointly with Joseph Wolf on bounded symmetric domains, which led to a number of papers, some joint with Wolf and some singly authored. He left Berkeley in 1964 and ultimately joined the faculty at Yeshiva University.

Dana Stewart Scott was born in Berkeley, California, on October 11, 1932, and after graduating from McClatchy High School in Sacramento, matriculated at UC Berkeley in 1950. He became interested in mathematical logic and prospered under the guidance of Alfred Tarski. While still an undergraduate, he obtained some original results that were published as abstracts in the *Bulletin of the American Mathematical Society*. After graduating from Berkeley in 1954 he continued work as a graduate student at Berkeley for one year, but then seeking a change, he decided to transfer to Princeton, where he studied under Alonzo Church. He completed his doctorate under Church in 1958 with a dissertation entitled "Convergent Sequences of Complete Theories." The department made an immediate attempt to bring him back to Berkeley as a faculty member, but this failed both in 1958 and again in 1959, when Scott accepted a position at the University of Chicago. But in 1960, the department's efforts were successful, and Scott was appointed assistant professor effective July 1, 1960. He had also been nominated for a Miller fellowship, a new type of postdoctoral fellowship initiated in 1960 by the Miller Institute for Basic Research in Science at Berkeley. Scott was awarded this prestigious two-year fellowship, but ultimately resigned after the first year in order to take up full-time teaching responsibilities in the department.

Scott had already established numerous results in several areas of logic that had attracted attention, including a solution of the decision problem for the elementary theory of infinite Euclidean spaces, results on automata theory (joint with M. Rabin), and results on ultraproducts and languages with infinitely long expressions. However, soon after his arrival, he proved the remarkable result that under Gödel's axiom of constructibility, measurable cardinals do not exist. This body of work won him rapid promotion to an associate professorship in 1962. However, the following year, Stanford University entered the scene and offered Scott an associate professorship, a position that he accepted. Berkeley tried to lure Scott back on a number of occasions, most notably in 1967 to head up a new Department of Computer Science, which we will discuss presently. Scott went on to a distinguished career, and his loss was a serious and enduring one for the department.

Working jointly with the economics department to develop mathematical economics, the mathematics department appointed Hirofumi Uzawa as an assistant professor in 1960 with a split appointment between the two departments. Uzawa left Berkeley after a year for Stanford, but this appointment was the beginning of a major initiative in mathematical economics that was to begin a few years later.

The final element of this remarkable three-year recruiting effort was to attract to Berkeley two senior faculty members in applied mathematics. For one of these positions, attention came to focus on Tosio Kato, who was then a professor of physics at the University of Tokyo. Kato was familiar to Berkeley, since he had served as a visiting faculty member in the mathematics department from October 1954 to February 1955 as part of a year-long visit to the United States that included visits to New York University and the National Bureau of Standards. Kato had also spent a short time in Berkeley in October 1957. Tosio Kato was born August 25, 1917, in Kanua City, Tochgi-kem, Japan, and received his BS degree in physics at the University of Tokyo in 1941. He received his doctorate in physics, also at the University of Tokyo, in 1951, with a dissertation entitled "On the Convergence of the Perturbation Methods," and immediately joined the physics faculty at the University of Tokyo, winning promotion to full professor in 1958.

Although trained as a physicist, he was in fact a mathematician as well. He studied perturbation theory of linear unbounded operators, convergence issues, methods for estimating eigenvalues, and in addition addressed very specific issues such as solutions of the wave equation for the helium atom. He also did notable work to bring semigroup theory to bear on the Kolmogorov equations for stochastic processes. His work was widely recognized and admired by many US mathematicians, including Einar Hille and Shizuo Kakutani, both of Yale University. Kato's accomplishments and knowledge of both physics and mathematics made him an ideal appointment to build up applied mathematics and establish closer ties between mathematics and physical science and engineering faculty in other departments. His appointment, which was to be effective July 1, 1960, was quickly approved by the university.

Getting the appointment through the UC bureaucracy was the easy part; the hard part was getting the Immigration and Naturalization Service (INS) to grant immigration visas to Kato and his wife. The immigrant quota for Japan was very small and at the time had a long waiting list. The department and the university appealed to Congressman Cohelan and Senator Kuchel to speed the process for Kato, but to no avail. Since Kato was to be a participant on a grant from the Office of Naval Research (ONR), the department appealed to ONR for help in speeding up the immigration process on grounds of national security needs. A change in INS regulations effective in August 1961 then made it possible for the secretary of defense to authorize that certain individuals on visa waiting lists whose services were deemed essential to the government be "paroled" to the United States so that they could begin work while waiting for award of an immigrant visa. This authorization was obtained through ONR, and in spite of the rather off-putting verb "parole" this government regulation employed, arrangements were finally made for Kato and his wife to be paroled to Berkeley. They arrived in March 1962 [DM Kato file].

Thus, after nearly two years and the intervention of the secretary of defense, Kato was finally able to take up duties at Berkeley. Kato prospered at Berkeley, and operator theory and PDEs did likewise. His 1966 book *Perturbation Theory for Linear Operators* has become a classic, and his research continued unabated on operator theory and semigroups, and expanded into the study of nonlinear evolution equations, the equations of hydrodynamics, and dispersive equations. He introduced a key inequality known as Kato's inequality for linear operators, made essential advances in the study of the Navier–Stokes equations and discovered what is called the Kato smoothing effect for the Korteweg–deVries equation, among many other notable contributions. Kato supervised 24 doctoral students altogether—3 at Tokyo and 21 at Berkeley—and retired from active duty in 1988. However, he remained very active in the years following his retirement, publishing prolifically, and indeed at the time of his death in 1999, there were two completed manuscripts still in his computer, which colleagues then published posthumously on his behalf.

The second senior appointment in applied mathematics was intended to be Abraham Taub, but since his appointment was intimately connected with the development of computing and computer science, we will discuss the details in a subsequent chapter devoted to these topics. The original plan was to bring him to

Berkeley in 1960, but this fell through, and instead, the initiative was renewed several years later, and Taub's appointment at Berkeley was consummated on January 1, 1964.

The 20 appointments of faculty to professorial ranks during the three years 1958, 1959, and 1960 of Kelley's leadership had in a real sense remade the department, creating strength in geometry and topology and algebra, and making a start toward building strength in applied mathematics. On July 1, 1960, the department had a faculty headcount of 44 and an FTE count of 41.5 in professorial ranks. Five had half-time appointments in mathematics with the other half in another unit, accounting for the difference between the two numbers. This faculty complement amounted to a doubling in a period of four years—an astonishing rate of growth. In addition to those in professorial ranks, the department had a number (15–20) of lecturers, visiting lecturers, instructors, acting instructors, and visiting assistant professors, as well as some senior visiting professors to fill out the teaching faculty.

The departmental faculty had become much better balanced among subfields, and could be divided up as follows. The geometry and topology group consisted of Chern, Spanier, Smale, Fary, Thomas, Hirsch, and Bredon, together with Kostant and Flanders, who spanned geometry and algebra, plus Bremermann and Koranyi, working in several complex variables and spanning geometry and analysis. The algebra and number theory group consisted of Hochschild, Rosenlicht, Lehmer, Foster, Robinson, Seidenberg, and Lehman, plus Kostant and Flanders. The analysis group consisted first of those who specialized more in PDEs—Lewy, Morrey, Protter, Kato, Cordes, and Trèves—and then those that might be classified more in functional analysis—Kelley, Morse, Wolf, Bade, Bishop, Feldman, Helson, and Woll. The logic and foundations group consisted of four (about 10% of the total): Tarski, Henkin, Vaught, and Scott. Applied mathematics was not so much a group as a number of individuals with varying interests often intersecting with the other four groups—Friedman, Diliberto, Huskey, Kato, Loève, Pinney, De Vogelaere, Chambre, and Uzawa.

In addition to lowering the expected teaching responsibilities for all faculty to six hours, or two courses per semester, Kelley also revised the format of the department's lower-division courses. Prior to this time they had been taught in small (25–35 students) sections by a combination of graduate-student teaching assistants and associates, instructors, lecturers, and some regular faculty. A senior faculty member was assigned to each course to oversee and coordinate the group of instructors. In order to give students more contact with senior faculty, Kelley oversaw the change to the lecture/discussion-section format, whereby a regular faculty member would lecture several times a week to large class (200–300) and then this class would break up into small sections (20–25) that would meet with graduate-student teaching assistants. This format has remained in effect since it was introduced in the late 1950s.

In May of 1959, Kelly had attempted to regularize the appointments of new PhDs similar to the way it had been done in other leading universities. He proposed the creation of Evans lectureships, which would be similar to the Benjamin Peirce instructorships at Harvard and the Moore instructorships at MIT. He proposed a salary of $6500 in order to be competitive in the job market at the time [DM Kelley file]. This proposal ran into several problems. First, the limited-term nature of the appointment was contrary to a longstanding tradition in the university that all qualified nontenured appointees could advance to tenure. Second, the salary level was out of line with that of a beginning instructor ($5232), a normal entry-level position for a new PhD, or even an acting assistant professor ($5916). The proposal never got off the ground, and the department went on as before. However, the administration determined after the large number of tenured appointments during this three-year period that future recruiting would be limited to nontenured appointments. Again, such a restriction was in accord with a longstanding policy and practice of the university to hire at the junior level and promote faculty from within.

Kelley's achievements during these three years shaped the future of the department. They are all the more remarkable in that he was largely absent from the campus during the first year 1957–1958 and operated through Chuck Morrey, who served as acting chair. Kelley served as chair through January 1, 1960, at which time he resigned in order to pursue projects in elementary-school-mathematics education. The position of chair fell at that point to Bernard Friedman, who carried to completion the hiring program that Kelley had started. Friedman served as chair for a total of two and a half years, until July 1962.

During most of this period, the Department of Mathematics and the Department of Statistics, along with the mathematics and statistics library, were located in Dwinelle Hall in the center of the campus along with humanities departments and rather isolated from the other science and engineering departments, which were located in the northeast precinct of the campus. As noted in a previous chapter, the 1956 Long-Range Development Plan sought to remedy this situation through new construction around the mining circle. Planning for Campbell Hall, a new building on the south side of the mining circle, was begun in 1955, and construction was completed in 1959. The departments of astronomy, mathematics, and statistics, a combined astronomy/mathematics/statistics library, and the newly formed computer center moved into Campbell Hall in the fall of 1959. Campbell Hall was far too small for these occupants even on the day it opened. Mathematics had doubled in size since planning had begun. The department overflowed into a variety of other buildings, including most famously the wooden World War II barracks-style structure T-4 (Temporary Building 4) located in the glade northwest of Campbell Hall. The fractionation of the department into several buildings was unhealthy for the academic life of the department. But at least the department was in the science and engineering precinct, with good space for the library and proximity to the computer center. Early planning for a new mathematical sciences building was underway that would resolve the space issue, but completion of this new building, the future Evans Hall, was delayed until 1971. Chapter 15 will relate the details of the planning and construction of Evans Hall.

Chapter 14

Computing and Computer Science 1955–1973

As was recounted in Chapter 8, the Berkeley campus had lost out to UCLA in the 1947 competition to host the National Bureau of Standards Institute for Numerical Analysis (INA). UCLA thus had by 1949 a state-of-the art digital electronic computer designed for scientific computing, the SWAC, or Standard Western Automatic Computer. After much agitation from the Berkeley faculty, Sproul agreed in October 1950 to create a computing facility on the Berkeley campus. This facility was located in Cory Hall and consisted largely of IBM card sorters, card punches, and other equipment for processing data, but with no real computing power to speak of. In 1948, Paul Morton had begun construction of CALDIC, the California Digital Computer. However, faculty who needed a facility like the UCLA facility were very frustrated about the lack of modern computing facilities on campus. Among these was Derrick Lehmer, who had in fact served as director of INA for the two years 1951–1953 while on leave during the loyalty oath period. Another faculty member whose computing needs were not being met was Louis Henyey, of the astronomy department. Other universities were installing modern computing equipment and in addition creating research centers around them. At this juncture, a computer center was envisioned to be more than simply a service bureau and was seen to comprehend a research function as well, a point we shall explore below.

Lehmer expressed his frustration about the absence of any movement to create a modern computing and computing research facility at Berkeley in a response to a letter from Hans Rademacher, at the University of Pennsylvania. On February 6, 1956, Lehmer wrote to Rademacher [DM Lehmer file]:

> I was much interested in the news that the University of Pennsylvania will secure a Univac and set up a research center around it. Such a center located at the very birthplace of electronic computing should have a bright future.
>
> In response to your question as to whether I might be interested in heading such a project, the answer is in the affirmative. The situation of computing at Berkeley is discouraging to say the least. We must have set a record for inactivity. No campus of this size has been provided with so little equipment and staff to do

computing; this in spite of my urging since 1947 that something be done about it. The best computer in Berkeley belongs not to the University but to Harry Huskey and is located in his home [Harry Huskey's Bendix G-15 "personal computer"]. I do my research computing at a distance of 400 miles at UCLA on an informal and inefficient basis.

We do have a faculty committee now under the chairmanship of Teller, "working" on the problem. However there is not much we can do besides recommending over and over the same set up that you are achieving. In case nothing comes of this effort I shall be in a more receptive mood for a definite proposal. It would be quite interesting to be able to work together once more.

The faculty committee mentioned in the letter was chaired by Edward Teller (physics) and was composed of the following members: Louis Henyey (astronomy), William Crum (economics), Derrick Lehmer, Dr. Mills (radiation laboratory or, as it is known today, the Lawrence Berkeley National Laboratory or LBNL), Paul Morton (electrical engineering), Morrough O'Brien (engineering), Glenn Seaborg (chemistry), and Otto Struve (astronomy). Lehmer had circulated copies of his response to Rademacher to the dean, and he received a sympathetic reply from Chancellor Kerr within a week. Teller, who had only in January 1956 taken over as chair of the committee, moved matters forward expeditiously, and on August 27, 1956, Kerr announced that the campus would be acquiring an IBM 701 for the computer center and that it would be located in 173 Cory Hall. The machine, which was expected to be operational in October, would be available for use both on a recharge basis and a no-charge basis with time awarded competitively to proposals. The Teller committee would act as the referee to evaluate proposals from faculty and students to use the machine during the 20 hours per week that the machine would be available for use without charge [UA, CU 153, Box 1, folder 2].

On October 22, 1956, Kerr followed through with an announcement that the computer center was open for business with its IBM 701, and at the same time he announced the creation of the Survey Research Center, which would provide assistance to social scientists in their survey research. The 701 machine, which the campus leased from IBM for $154,000 per year, was capable of performing 27,000 additions or 2000 multiplications per second. This was roughly comparable to the capabilities of the SWAC, which when completed in 1950 could perform 16,000 additions or 2600 multiplications per second. At the same time in 1956, it was announced that UCLA would be creating the Western Data Processing Center (WDPC) to be located in the School of Management, employing an IBM machine worth $2.5 million that had been donated to UCLA by IBM. This center would specialize in computing related to business and social sciences for the entire university, while the Berkeley center would be devoted to scientific computing for the whole university [ibid.].

Professor Louis Henyey, of astronomy, was designated as chair of an operations subcommittee of the Computing Advisory Committee, but this meant that he was de facto running the center. This fact was recognized, and very soon thereafter he was appointed formally as the first director of the computer center. Within a very short period of time, Teller and Henyey succeeded in upgrading the 701 machine and obtained an IBM 704 that was to be shared with the radiation laboratory, with the campus and the rad lab each to get 80 hours per week of machine time. The new machine arrived in early 1959, just as the computer center was getting set to move into new quarters in Campbell Hall, as was described in Chapter 12. After several months of testing, the 704 was ready for use on December 2, 1959. This machine, which could perform 41,700 additions or 4170 multiplications per second, was described as the second-largest computer in the country located on a campus and devoted to scientific computing. By comparison, ten months later, in October 1960, Lawrence Livermore Laboratory announced that their UNIVAC LARC (Livermore Advanced Research Computer) was operational and was capable performing 250,000 additions per second [ibid.].

When in the spring of 1959 Professor Henyey was asked to assume the position of chair of his department effective July 1, 1959, he agreed to take on this departmental assignment. However, he felt that he could not do so while also serving as director of the computer center, but he agreed to remain on as center director for a transitional period. The Computer Advisory Committee, somewhat reconstituted, but still under the chairmanship of Teller, took the opportunity of a change in directorship to also advocate for a fundamental change in the nature and mission of the center. A smaller group consisting of Teller, who was at the time also director of the Lawrence Livermore Laboratory; Carl Helmholtz, chair of physics; John Kelley, chair of mathematics; and Henyey, chair of astronomy, undertook to review the operation of the center and to consult with leading authorities around the country. On December 11, 1959, they wrote Chancellor Seaborg a three-page letter recommending a number of steps. The following are highlights from this letter:

> It is of great interest to many departments and many individuals on the Berkeley campus to have first-rate computing facilities. Such facilities are necessary for the solution of many problems, scientific and non-scientific arising in our modern civilization. The provision of these services to departments and individuals is an important task for the Computer Center.

> However, of even greater importance is research in the computing field. Modern computing is a living and rapidly developing art. In order to maintain a strong position in the field, it is essential that a strong program of research, both in computational techniques and programming and in machine development, be carried on. Such figures as the late John von Neumann have devoted much of their time to this field of research.

Moreover, it is important that the Computer Center take part in a teaching program on computation and design of machines. The current teaching program in Electrical Engineering and Mathematics is expanding rapidly and must be expanded further. The demand for trained personnel in this area is unbelievable; in order to supervise properly the computing machines now on order at IBM alone would, it is estimated to require all of the PhD's in mathematics which will be produced in this country for the next five years. We believe that a strong teaching program in Electrical Engineering, Mathematics, and Physics, centered around the Computer Center and using the computing facilities for student training, is a necessary and natural part of our university program. [DM, Taub file]

The writers envisioned a substantial increase in the budget of the computing center, and that the staffing of the center include key faculty members, who would have 50% appointments in existing departments (e.g., mathematics, physics, electrical engineering, astronomy, and perhaps others) and 50% appointments in the computer center. (Actually, eighteen months earlier, René De Vogelaere had been appointed to such a position with a 50% appointment in mathematics.) Finally, the writers made some specific personnel recommendations, which echoed and supported recommendations from the entire advisory committee. They recommended that Nicholas Metropolis, from the University of Chicago, and Abraham Taub, from the University of Illinois, be brought to Berkeley to lead the computer center, each with such split appointments, with 50% in the center. Metropolis was to be 50% in physics and Taub 50% in mathematics, with Metropolis to serve as director and Taub as associate director. The writers reported that Metropolis and Taub were interested in such an arrangement and that each would find the prospect of coming to Berkeley especially attractive if the other came to Berkeley as well [ibid.].

This 1959 proposal, which would in effect have added a research and teaching function to the service-bureau function of the center, was unusual in terms of university organization. But it was very much in accord with Taub's ideas formulated during the time when he participated in the development of one of the most successful university computer centers at Illinois, and paralleled what UCLA had done at INA and what Rademacher at the University of Pennsylvania had described to Lehmer in 1956. On the other hand, one could view the recommendation as a proposal to create an academic department of computer science, which in addition to its usual academic mission, had a service function to provide computing facilities, consulting, and help to faculty and students from other departments. In a way it paralleled the concept of the statistics department, which in addition to its teaching and research programs had a service function of providing a statistical consulting service to faculty and students in other department.

In spite of the power and influence of the group (which consisted of the chairs of mathematics, physics, and astronomy as well as Teller) that made the recommendation for the appointment of Taub and Metropolis and the reconceptualization of the computer center, the proposal never achieved liftoff, and the ambitious recruitment plans fell through. The center ended up being led by a series of acting or short-term directors for several years, and continued in its service role. Because the computing needs began to exceed the capacity of the existing 704, in the summer of 1962 the center obtained an IBM 7090, a much larger machine. However, the center also ran into financial problems and ran up substantial deficits.

Chancellor Strong became concerned about the state of the computer center in October 1962, when the director resigned amid these financial problems and Strong decided that the campus should seek to appoint permanent leadership for the center. In November, a search committee chaired by Walter Knight, of physics, was appointed and asked to conduct a nationwide search for a new director. Early in the spring of 1963, the committee recommended Taub as their candidate. Murray Protter, the new chair of mathematics, who was a member of the search committee, said that mathematics would provide a natural academic home for Taub with a 50% appointment [UA, CU 149, series 3, Box 50, folder 42]. Taub visited the campus in the spring of 1963 and discussed the future of computing at Berkeley with a number of faculty members and administrators. On March 25, Taub wrote a three-page letter to Chancellor Strong at the request of Vice Chancellor Bressler. In it he wrote:

> The University of California, Berkeley, will not fulfill its potential and play the role it should play in the development of, application of, and training in, the Computer Sciences unless the service computing presently being done at the Computer Center is embedded in an organization with a broader scope than the Center now has. This scope should include a research program and an educational program. The latter must involve an undergraduate and graduate curriculum.

> The new organization may be achieved by expanding the Computer Center and giving it a better charge, or by putting it into a larger organization. The decision as to which course to follow should be made after the various departments and faculty members working in the area of computer sciences are consulted. They should give their views on how their own programs will benefit (or be harmed) by either choice, and under which method of organization the total effort in the Computer Sciences will be most effective. [ibid & DM Taub file]

Discussions continued between Taub and the campus, and they soon matured, first into an informal offer in May, and then, during the summer, into a formal offer to Taub to come to Berkeley. Taub was to be 50% professor of mathematics and 50% director of the computer center. As a condition of his appointment, he sought an agreement and a timeline from the campus leadership to move in the direction he had recommended, which included the creation of a department of computer science. In response to a direct question by Taub, Chancellor Strong wrote to Taub on July 31, 1963,

Let me only add at this point, as concerns matters touched on in your letter, that I would consider a period of five years to be fairly realistic estimate, perhaps, if anything, a conservative one, of the time required to complete the necessary steps in connection with the proposal to move toward creation of a department of Computer Sciences. [DM, Taub File]

With this commitment from the chancellor in hand, Taub accepted the invitation to come to Berkeley. However, because of the lateness of the offer and prior commitments at the University of Illinois, Taub was unable to take up duties until February 1964.

Abraham Haskel Taub was born February 1, 1911, in Chicago and after attending schools in Chicago enrolled at the University of Chicago in 1927. Graduating with a degree in mathematics in 1931, he enrolled at Princeton for graduate work. There he came under the influence of Oswald Veblen and worked with him on spinors, work that led to two publications. His dissertation was completed in 1935 under Howard Robertson and was entitled "Quantum Equations in Cosmological Space." Thus began Taub's lifelong interest in general relativity. After a postdoctoral year at the Institute for Advanced Study, Taub accepted a faculty position as assistant professor of mathematics at the University of Washington, where he moved up the ladder to full professor. During the war he was on leave and served as a research physicist for the National Defense Research Committee, based at Princeton University. During this period he became interested in hydrodynamics and shock waves. In 1948 he was called to the University of Illinois, where he ultimately headed up the Digital Computer Laboratory. This operation was generally regarded as the leader among university-based computer centers, and Taub received considerable recognition for his work as director. Meanwhile, he continued to publish papers on general relativity and hydrodynamics and added a handful of highly regarded publications in numerical analysis. His scholarly focus of interest, however, remained in general relativity.

During the negotiations leading to his appointment at Berkeley, Taub asked that the mathematics department list a graduate course in general relativity that he would like to teach at some point. He also raised the possibility of attracting one of his former doctoral students at Illinois to come to Berkeley with a joint computer center/mathematics department appointment like his own. This was Gene Golub, who was then at Stanford, and unfortunately for Berkeley, he remained there. Professor David Evans held a split appointment 50% in electrical engineering and 50% in the computer center, and it was agreed that Evans would serve as acting director of the center from the summer of 1963 until Taub arrived and that thereafter he would serve as assistant director under Taub. Upon his arrival, Taub threw himself into the job and worked to prepare the way for the creation of an academic unit devoted to computer science.

His first step in early 1964 was to request a change in the campus academic plan that added a section on computer science and spoke of the possibility of a department of computer science. This change was readily approved in the next few months. In August 1964, leaders of the Department of Electrical Engineering, including the new chair Lotfi Zadeh, who had assumed duties in 1963, expressed concern in a meeting with Vice Chancellor Bressler about Taub's efforts and stated that their department was interested in computing and would not want the field "given" to another unit [CU 149, series 3, Box 75, folder 17]. In early 1965, Taub proposed to the College of Letters and Sciences a field major in computer science that included at the upper division about eight semester courses in mathematics plus four in electrical engineering. This proposal was approved expeditiously and was ready for implementation in the fall of 1965 [ibid.; see this same folder for documentation for the following two paragraphs].

Then, in March 1965, Chair Zadeh proposed to the dean of engineering that the Department of Electrical Engineering change its name to the Department of Electrical Engineering and Computer Sciences, citing the developing expertise in computing in electrical engineering and the importance that his department attached to computing. George Maslach, dean of the College of Engineering, sent this proposal on to the chancellor with his strong endorsement. On June 1, Taub countered by proposing, with the support of colleagues from mathematics, the creation of a department of computer science within the College of Letters and Sciences. The college quickly endorsed the proposal. The administration was thus confronted with two proposals, which represented different policy choices. The field of computer science was only in a formative phase at the time, and it was unclear and would remain unclear for some years whether as a discipline (and there was general agreement early on that it was a discipline) it belonged more naturally with other physical sciences in the College of Letters and Sciences or whether it was more naturally an engineering discipline that belonged in the College of Engineering. The administration sat on these two proposals, each of which had a considerable body of influential support behind it, for a year and a half, trying to decide what to do. One possibility was to cut the baby in half, so to speak, and rename electrical engineering as electrical, electronic, and computer engineering, and to name the department in the College of Letters and Sciences the Department of Applied Mathematics and Computer Science. However, this kind of compromise satisfied no one. Perhaps one reason for the delay in acting, among others, was that the administration of the campus was preoccupied at the time with student protests and political controversy on campus.

As the administration thought how to proceed (deny both proposals, approve one and deny the other, or approve both), the tension rose, and word spread both on and off campus of the conflict between Taub and Zadeh. The long delay in making a decision no doubt exacerbated the tensions. Academic Senate committees reviewed the proposals and offered their advice. Issues discussed included the appropriate recognition of the program in computer science that had already been developed in the electrical engineering department, and the possibility of duplication of courses and curricula if a new department were to be created. These were of concern, but the deeper concern from the College of Engineering was that Taub's proposal called essentially for the close integration of the new computer science department and the computer center. This arrangement, it was felt, would give the new department an unwarranted advantage. The Department of Electrical Engineering felt that the computer center should be a service organization for the entire campus and that the director should not be the chair of mathematics, computer science, or electrical engineering, or perhaps even more restrictively, no faculty member in any of these departments should serve as director.

The campus decision came in December 1966 and it was to (1) grant the name change for electrical engineering to become the Department of Electrical Engineering and Computer Science (EECS); (2) to create a Department of Computer Science in the College of Letters and Sciences; and (3) to make the computer center an independent service organization. This outcome was in the long-established Berkeley tradition of "letting a thousand flowers bloom." Although it was initially conceived that the department in the College of Letters and Sciences would concentrate more on theory and that EECS would concentrate more on engineering and hardware, this quickly broke down, and both departments covered the entire spectrum of the field. The nucleus of the new Department of Computer Science was to be formed by faculty from mathematics and electrical engineering who would transfer into the new department. There were differing opinions concerning how many faculty would transfer their appointments. Dean Fretter, of the College of Letters and Sciences, then commenced a search for a founding department chair for the new department. The search committee, which was empowered to conduct a national or international search, recommended as its choice Professor Dana Scott, of Stanford University, and a former Berkeley faculty member. There was universal enthusiasm for this choice, and Scott indicated interest in the position. But just as his appointment was about to go forward in early April 1967, he withdrew his name, indicating on reflection that he did not wish to devote several years of his life at that time to administrative duties [DM Scott file].

The founding faculty in computer science consisted of Professor Martin Graham and Assistant Professors Michael Harrison and Butler Lampson, who transferred their appointments from electrical engineering to computer science, plus Derrick Lehmer, who transferred 50% of his appointment into computer science, and

Assistant Professor Beresford Parlett, who had been 50% in mathematics and 50% in the computer center, and who transferred his 50% mathematics appointment to computer science. Assistant Professor Steven Cook, who had also been 50% mathematics and 50% computer center, transferred his 50% center appointment into the new department in 1969. Taub did not transfer his appointment to computer science, and in July 1967, he resigned as director of the computer center and requested that his appointment be changed to 100% professor of mathematics. He spent the remainder of his career devoted to teaching and to research on general relativity. His achievements were recognized by election to the American Academy of Arts and Sciences. Although he retired from active duty in 1978 at the then mandatory retirement age of 67, he continued publishing actively for many years after his retirement. He died in 1999 at age 88 after a long illness.

With Taub out of the picture as possible computer science department chair, the only full-time tenured faculty member of the new department from the College of Letters and Sciences was Beresford Parlett, who had just been approved for promotion to associate professor effective July 1, 1967. Parlett was to be a key person in the development of this new department. Beresford Neill Parlett was born July 4, 1932, in England and graduated from Oxford University in 1955 with a BA with honors. After a few years working in his family's business, he enrolled in Stanford University to pursue doctoral work in mathematics. Working under the direction of George Forsythe, he completed his degree in 1962 with a dissertation consisting of two parts: "I. Bundles of Matrices and the Linear Independence of their Minors, II. Application of Laguerre's Method to the Matrix Eigenvalue Problem." He spent two postdoctoral years at the Courant Institute at NYU, and then a year teaching at the Stevens Institute of Technology. In the late summer of 1964, Taub approached Parlett about a possible appointment at Berkeley, and shortly thereafter, Parlett's appointment as 50% assistant professor in mathematics and 50% in the computer center was approved, to be effective July 1, 1965. This was first of Taub's joint appointments between the center and academic departments. (Steven Cook was the second such appointment, which was made in 1966.)

Parlett's research on numerical linear algebra and especially on matrix eigenvalue and eigenvector calculations was widely recognized and won him rapid promotion to associate professor in 1967, as already noted. He was promoted to full professor in 1973 and enjoyed a long and successful career working in numerical linear algebra, an area in which he was regarded as a world leader. He received awards for his work and altogether supervised 25 doctoral students to completion. He took advantage of the university's voluntary early retirement incentive program to retire in 1994, and he has continued to publish actively and to organize seminars in the department.

The search for a chair of the new department continued on into the spring and summer of 1967, and a number of outside candidates were contacted, but none indicated interest. One likely factor in these rejections was the well-known dispute over the organization of computer science. Then attention was directed to internal candidates. The College of Letters and Sciences had a brand new dean, Walter Knight, as of July 1, 1967, who inherited from his predecessor, Dean Fretter, the problem of finding a chair for the new Department of Computer Science. After outside possibilities were exhausted, Dean Knight turned to Parlett. When Knight called Parlett to ask him to serve, Parlett tried to beg off, making the very reasonable argument that as a newly promoted associate professor, he really didn't want to take on the job of chairing a brand-new department. The dean then resorted to dire means, and said that if you, Parlett, won't take on the chairmanship, I will dissolve the department [BP, personal communication]. Faced with this threat, Parlett agreed to serve as acting chair of the Department of Computer Sciences for 1967–1968. Subsequently, he agreed to serve three more years as chair, stepping down in 1971.

It was a tall order for a newly promoted associate professor to be thrust into a position of leadership of a new department that had been created under tense and stressful circumstances and surrounded by considerable disagreement between groups of senior and established faculty members on campus. Parlett acquitted himself well in this role, although his administrative duties were clearly a drag on his research. Under his leadership, the department hired William (Velvel) Kahan and Richard Karp as full professors, both of whom went on to distinguished careers at Berkeley. Both Karp and Kahan subsequently received the Turing Award. The department also attempted to hire Donald Knuth (a third K!), but he declined the initiative.

Butler Lampson left the department soon after its founding to work in industry, and he worked with distinction for many years at Xerox's Palo Alto Research Center. He, too, won the Turing Award, and he was also awarded the National Academy of Engineering's Draper Prize in Engineering. Michael Harrison remained at Berkeley, moving back into EECS with the reorganization in 1973, where he had a distinguished career. Steven Cook initiated the idea of NP complexity, and then Richard Karp, first working with Cook and then independently, further developed this circle of ideas, which then led to much fundamental work and to the famous open problem of whether P = NP. Cook subsequently left the university and settled at the University of Toronto, where he has had a distinguished career.

The department made some excellent junior hires, including James Morris, who went on to become dean of the School of Computer Science at Carnegie-Mellon; Jay Early, who became a well-known transformational psychologist; and Susan Graham, who has had a distinguished career at Berkeley. This is a very fine record of hiring, especially under the circumstances of having two competing departments of computer science. The department remained small—seven FTE—and was in an unstable position relative to the much larger EECS department.

Martin Graham assumed the position of chair in 1971, and in the fall of 1972, he appointed a high-level outside review committee consisting of Dana Scott, Alan Perlis, and George Forsythe to visit the department and provide advice and guidance on its future development. The committee's visit and its subsequent report immediately attracted the attention of the new chancellor, Albert Bowker, who was concerned about the viability and academic good sense of having two competing departments of computer science with no visible differentiation in their missions. The campus administration then proceeded to review options for a reorganization of computer science. The option of a single independent department of computer science, with a variety of possible reporting relationships, was rejected in favor of creating a semiautonomous division of computer science within EECS with its own associate chair, reporting to the overall chair of EECS. This option, which had been strongly urged by EECS, was put into effect July 1, 1973. This decision represented a major victory for EECS, and ended nearly a decade and a half of rancorous conflict over the organization of computer science.

Part of the arrangement was that Richard Karp, from the department in the College of Letters and Sciences, was selected to be the first associate chair of the new division. Elwyn Berlekamp, who had a split appointment 50% mathematics and 50% EECS, was the second associate chair. Computer science has thrived at Berkeley under the new structure put in place in 1973. As a further part of the reorganization, Beresford Parlett and Vel Kahan were provided split appointments between mathematics and EECS, and numerical analysis remained an essential part of the mathematics department. A few years later, Richard Karp was given a 0% appointment as professor of mathematics. On at least two occasions, the idea of an intercollege department had been floated as an alternative, once in 1967, by the new dean of the College of Letters and Sciences, Walter Knight, as a possible alternative to the new department in his college, and once again, in 1973, by the dean of physical sciences (the author), as a possible solution to the organizational problem at that time. At neither time did such an idea gain any traction.

While computer science as a discipline has many connections with mathematics and the other physical sciences, its connections with engineering have generally proved stronger and more durable. In addition, an engineering college is perceived as likely to provide greater resources to a computer science unit than a college of arts and sciences would provide. Consequently, engineering has usually prevailed in contests for academic control of computer science departments. When computer science has been located in an engineering college, it has more frequently been a freestanding department, rather than merged with electrical engineering, but both models have been successful. A significant exception to this pattern was for a time the organization of computer science at Stanford, which is generally regarded as one of the top three departments in the country, along with Berkeley and MIT. At Stanford, computer science was organized in the early 1960s as the Division of Computer Science inside the mathematics department. In 1965, this division was converted into a freestanding department in the College of Humanities and Sciences, where it thrived. In about 1985 the decision was made to transfer this computer science department to the College of Engineering, where, it was believed, it would receive more attention and more resources. It has also thrived in this mode. An important part of the reorganization at Berkeley was that EECS would offer, in addition to the undergraduate major in computer science leading to the Bachelor of Science degree in the College of Engineering, an undergraduate major in the College of Letters and Sciences leading to a Bachelor of Arts degree. This major—which must comport with the regulations of the college limiting the number of units that can be required for a major program—requires fewer units than the major in EECS and is viewed as a liberal arts degree rather than a professional degree. It is a popular major in the College of Letters and Sciences with keen competition for entry into it.

Chapter 15

Campus Planning and Evans Hall

In 1960, the Department of Mathematics had completed a period of rapid growth, going from a faculty complement of just over 20 in 1956 to a complement of 44 headcount faculty or 41.5 FTE in four years. These numbers include only tenured and tenure-track faculty in the professorial ranks, and there were a number of instructors, lecturers, and visitors in addition to the regular faculty. But this was only the first phase of the growth of the department. More growth was planned, but of an amount yet to be determined, an amount that would depend on planning, budgetary, and political considerations at the campus, university, and state levels.

This was a period, 1960 through about 1966, during which the state of California invested heavily in education and especially in higher education and the University of California. The Master Plan for Higher Education, which had just been approved by the legislature in the Donohoe Act of 1960, set forth a plan for growth and access on which there was a broad statewide consensus. The Berkeley campus was approaching its planned enrollment limit of 27,500, and the campus academic plan developed in the early to mid 1960s called for a shift in the campus enrollment balance by increasing graduate enrollment and decreasing undergraduate enrollment so that graduate enrollment would reach about 14,000, or about 52% of the total enrollment in 1970 [UCB Academic Planning Committee Progress Report; February 15, 1967]. In 1960, the campus had a total enrollment of 22,000 of which 31.5%, or about 7,000, were graduate students. The campus plan thus represented a major shift in emphasis toward graduate education and in particular a doubling of graduate enrollment by 1970. As is evident today, this plan was never implemented, and indeed the campus took a very different path.

The university enjoyed a favorable student-faculty ratio by mid decade of about 15 to 1 for purposes of state budgets, and in addition, for internal allocations, the university used a weighted formula that placed greater emphasis on graduate and doctoral enrollments. Consequently, the campus expected substantial increases in faculty positions. Finally, the university was moving toward conversion to the quarter system and year-round operation with a state-funded summer quarter, which would result in additional faculty positions. On average, the summer quarter was expected to add 13% to student enrollments and to faculty allocations. As it turned out, events in the late 1960s and early 1970s overtook these bold plans. However, these assumptions were in place and were the foundation for the planning of the new mathematical sciences building, Evans Hall.

Since 1952, the mathematics department had been housed in Dwinelle Hall and was experiencing severe crowding, with many faculty members doubled up in offices and sometimes with even higher occupancy. In addition, the Dwinelle Hall location was remote from the physical sciences and engineering. As already noted in Chapter 12, the mathematics department, along with statistics and astronomy, had moved in the fall of 1959 into Campbell Hall, on the south side of the mining circle. This move placed these departments in proximity to the other physical sciences, as was envisioned in the 1956 Long Range Development Plan (LRDP). It was recognized, even before the completion of Campbell Hall and the move, that the new building would not be large enough to accommodate mathematics and statistics. In Chapter 12 it was noted that the campus in 1957 had determined to construct a new mathematical sciences building on the west side of the mining circle, but at the time the project was rather far down on the list of priorities for campus capital projects. In 1957, the building was planned to contain approximately 40,000 square feet and would replace the World War II temporary building T-1. It was also stipulated that the building might be larger if studies showed greater site capacity.

When in 1962, the campus issued a new LRDP, the construction of a mathematical sciences building on the west side of the mining circle was included as one of the projects in the five-year major capital program. The building was now listed as 180,000 gross square feet and with a projected cost of $5.7 million. Gross square feet, or GSF, is a measure of the total floor area of a building. For a rectangular building, it is the footprint of the building in square feet times the number of floors, including any floors below grade, with an analogous formula if the external geometry of the building is more complex. Assignable square footage, or ASF, is the total number of square feet in all the rooms in the building that can be assigned to units. It excludes the space taken up by walls, all horizontal and vertical circulation (corridors, stairwells, elevator shafts), restrooms, utility closets and shafts, mechanical rooms, and so on. The ratio of ASF to GSF, or the "efficiency" of a building, depends on the specific design, but is generally between 0.55 and 0.6, or expressed as a percentage, between 55% and 60%. The planned mathematical sciences building, at least from the maps included with the LRDP, was to take the place of three temporary buildings, T-1, T-2, and T-3. The 1957 site capacity estimate was surely in ASF, which would correspond to about 65,000 GSF.

It is not clear how planners arrived at the estimate of 180,000 GSF as the new site capacity, but again from the map accompanying the plan, the building is shown conceptually with a footprint of roughly the same size as the Evans Hall building that was ultimately built, 22,500 square feet. Thus, if planners had in mind a footprint of 22,500 square feet, one might therefore conclude that planners had in mind a building of eight floors, or if they had in mind a footprint of 20,000 square feet, the building would have nine floors. In either case, one floor was planned to be below grade, as is known from subsequent planning documents, and the rest above grade. With the larger footprint and two floors below grade, as eventually was the case, the building could have been limited to six floors above grade. Grade here is the grade at the mining circle at the east side of the building, and since the contour of the land slopes down from east to west, grade at the west end of the building is one floor lower. Thus, viewed from the east, the building would be seen as six, seven, or eight floors above grade depending on the footprint and number of floors below grade. Viewed from the west, one additional floor would be visible above grade. The site capacity of 180,000 GSF would translate into an ASF capacity of 108,000, assuming an efficiency of 60%.

This height is consistent with the heights of other buildings recently constructed or about to be constructed at the time: Barrows Hall had eight floors above grade, Latimer Hall had eight floors above grade, and University Hall had seven floors above grade. These buildings, however, were located at some distance from the "classical core" of the campus with its tradition of two to four-story buildings. Although the western face of this mathematical sciences building fronted directly on the classical core, there is no evidence that much thought was given to how this building would fit in with buildings of the classical core.

In October 1960, the Building and Campus Development Committee (BCDC) had discussed in general the space needs of the mathematical sciences and noted that the departments had put forward a case for 121,800 ASF [UA, BCDC minutes]. By 1962, more detailed planning for the mathematical sciences building was underway, and on November 14, 1962, Chancellor Strong submitted to Vice President Morgan a Project Planning Guide (PPG) for the building, which was to accommodate the mathematics and statistics departments, the statistical laboratory, the mathematics and statistics library, and a major classroom complex. The PPG acknowledged an ASF limitation of 107,000 for the site, a number that is consistent with the LRDP site capacity of 180,000 GSF and a building efficiency of just under 60%. The PPG used a projection for the mathematics department of 104 FTE faculty, and a graduate enrollment of 509 for 1968. While not all of these faculty FTE would be used for professorial faculty, perhaps 85%–90% would be, with the rest used for temporary faculty. Thus, a professorial faculty of about 90 was projected, which represented more than a doubling of the 1960 faculty complement. The PPG also projected a more than doubling of the graduate enrollment as well. A doubling of graduate enrollment was of course what was planned for the entire campus as part of the campus academic plan. The PPG explained that the departments felt that these student and faculty projections were too conservative, and that they believed that the actual growth would be even larger than what was projected.

The PPG also incorporated a number of exceptions to the existing space standards, called restudy standards, which were then in place to guide planning for university buildings. These exceptions were called the "Strong exceptions," and have been a part of departmental history ever since, especially when there were discussions about space allocations in Evans Hall. One such exception was to increase the faculty office size from 130 to 140 square feet in order to accommodate a large chalkboard. Another was to increase the research space allotment for applied mathematicians from 60 to 155 square feet, which was the average of the space allotted for mathematicians and physical scientists in the restudy standards. Perhaps the most significant of the Strong exceptions was a generous provision for space for visiting mathematicians and postdoctoral scholars. The vision was that Berkeley, as a distinguished center for mathematical research, would draw many visiting faculty on sabbatical or research leave from their home institutions as well as a number of postdoctoral scholars. These mathematicians would benefit from their visit to Berkeley and would in turn enrich the mathematical research environment. Chancellor Strong endorsed the inclusion of space for more than 60 such visitors, which could almost be seen as a kind of informal mathematics institute within the department. Statistics presented a similar request, which also included a generous allotment of space for the statistical laboratory. Altogether, the PPG presented justification for over 65,000 ASF for mathematics, which was almost five times the amount of space that the department had in Campbell Hall. The PPG justified nearly 27,000 ASF for statistics and the statistical laboratory, more than four times the space allocated in Campbell Hall. The total space allocated in the PPG for all units added up to 114,278 ASF, which was 7000 ASF over the site capacity, but under the 121,800 ASF that the units had originally requested. The discrepancy between site capacity and the request was left unresolved, but the campus said that it would study the disagreement and possibly revise the estimates in the future.

In May 1963, the regents approved construction of the mathematical sciences building, and in July, they appointed Gardner A. Dailey and Associates as executive architect for the project [RM, May 16 and July 11, 1963]. A year later, in July 1964, the schematic plans had been completed, and they were sent to the regents for review and approval. The description of the building provided to the regents at this meeting was as follows:

The structure will be reinforced concrete, nine stories high, located at the east end of the central Glade, west of the mining circle. The building will contain classrooms, class laboratories, offices, a statistics laboratory, and a branch library for the departments of mathematics and statistics. Classrooms occupy the first level with entrance from the Glade. The second level includes the entrance lobby and library, accessible from the mining circle. The third through ninth levels contain office and teaching laboratories with an interior court extending through the top three levels. [RM July 23, 1964]

The size of the building was to be 180,000 GSF and 107,894 ASF, with a footprint of 20,000 square feet, and eight floors above grade as seen from the mining circle. The estimated project cost of $5.2 million, excluding certain equipment, and the timetable presented called for completion of working drawings in May 1965, a construction contract awarded in July 1965, and project completion by July 1967.

The regents' minutes also indicate that

In discussing plans for the building the Committee [on Buildings and Grounds], in general, felt it was quite handsomely[!] designed, but some concern was expressed over the lack of exterior detail shown. Vice President Morgan explained that Architect Dailey has not yet had an opportunity to go into the detail work, but he contemplates using considerable detail on the exterior. He commented that, in the past, Mr. Dailey's work has always included careful attention to detail with most satisfactory results.

The minutes further indicate that "Because of the prominence of the site and the importance of this building, the Committee was reluctant to approve the schematic plans without seeing more of the detail." In the end, the regents approved the schematic design but with the understanding that when the exterior details were developed, they would be sent for review [RM, July 23, 1964].

These details were presented to the regents a year later, at their June 17, 1965, meeting and were approved. The exterior detailing included, among other features, replacing flat exterior walls with what could be described as corrugated exterior walls, in which the exterior wall was divided into 20-foot vertical sections extending the height of the building with alternating sections extending out horizontally five feet beyond the other sections so as to give a pattern of light and shadow on the surface and to satisfy the regents' request for external detailing. This structure affected the internal layout of offices, which were each a little less than ten feet wide but now had to be arranged in an alternating pattern in which two offices were 20 feet long with the next two offices 15 feet long, with this pattern repeated along the exterior of the building. While this design satisfied the desire for external detailing, the end result was that internal space utilization was less than optimal.

In November 1964, the university had included the mathematical sciences building in its 1965–1966 capital budget proposal to the state for construction funding, but in January 1965, the president reported to the regents that the state had deleted this project from the governor's budget because of lack of funds [RM January 21, 1965]. The university pressed the state to restore this item, and ultimately in the fall of 1965 the university was successful, and the project was added back into the state budget through a supplemental appropriation. Therefore, it was not until well into the fall that the university knew that the building had been funded. This was to be the first of several delays in the construction schedule.

Meanwhile, there was ongoing reconsideration of major programmatic elements of the building. First, starting in 1965, the Departments of Mathematics and Statistics renewed their concerns and complaints about the size of the building and argued that the enrollment numbers and faculty FTE used in the 1962 PPG were too small and should be increased. They argued that the size of the building, which at 108,000 ASF was already too small by at least 6000 ASF, should consequently be increased, based on these new projections of enrollment growth and faculty size. Whereas a faculty size of 104 FTE had been used in the 1962 PPG, the mathematics department was now projecting growth beyond that number, and in 1966 the projected size stood at 112 FTE plus an additional allocation for the state-supported summer quarter that was coming on line in 1966. This would amount to a faculty size of about 125. A letter from Professors Spanier, Taub, and Morrey to Dean Knight of November 8, 1967, states that the number 125 was used in the grant applications to the National Science Foundation (NSF) for funding for the building that were filed in September 1966, and a letter from Vice Chair Emery Thomas to Dean Knight of November 13, 1967, reiterates the figure of 112 FTE plus additional provisions for the summer [DM Evans Hall file]. While some of these FTE would be used for temporary faculty appointments of lecturers and visitors, it could reasonably be predicted that the professorial faculty would stand at perhaps 110 if this kind of growth were to occur.

One wonders how a department of this size could possibly have been consistent with Clark Kerr's principle, used to determine campus size, that no department should be so big as to be unmanageable and unable to mentor its junior faculty effectively.

At the time, the department had grown to an FTE count for professorial faculty of over 60, which was already very large, and this projected additional growth represented almost a further doubling. The two departments, which prior to 1955 had been one department, had been located for decades in space that was so totally inadequate that now that they had the opportunity to design a new building, there was a single-minded dedication to ensure that it was sufficiently capacious to accommodate any conceivable future growth plans. Professor Elizabeth (Betty) Scott, of the statistics department, was leading the charge on these issues on behalf of both departments, but especially her own.

In addition, Abe Taub's efforts, which were described in the previous chapter, to redefine the mission of the computer center, as well as his work and that of others leading toward the creation of a department of computer science with close links to mathematics, suggested that it might be wise to bring the computer center into the new building, instead of having it remain in Campbell, as was the original plan. In addition, space in the new building for a future department of computer science was lurking in the background of these considerations. With all of these issues in mind, it was urged by the two departments that the size of the building be increased by 40,000 ASF or more. In a May 24, 1966, letter to Steve Diliberto, chair of the building committee, Robert Steidel, chair of BCDC, reported on his conversations with Chancellor Heyns and Vice Chancellor Connick. Steidel wrote,

> I have discussed the use of this space with Vice Chancellor Connick and again with the Chancellor. Several conclusions have been made. It is in the best interests of the campus that the Mathematical Sciences Building be constructed to the maximum size that the site will allow. The campus architect Mr. DeMonte has estimated this to be 40,000 sq. ft. more than the present program for the building. [DM, Evans Hall file]

Hence, the campus leadership was now persuaded that it was in the best interest of the campus to add an additional 40,000 ASF to the project, for a total of 147,000 ASF.

The first question posed by this proposed increase was how to accommodate architecturally the extra space. One possibility suggested was an underground addition just to the north of the building site, and under a parking lot, but this was rejected, and consideration turned to building additional floors. One strategy was an additional floor or floors underground, which would not have a visual impact, while another would be additional floors above grade. The footprint could be expanded and the interior court reduced. In the end, all of these strategies were used.

The second problem was to find a way to fund the additional space. State funding was not an option, and in fact, in the spring of 1965, when these discussions began, the core state funding for the smaller building was at risk, as has been noted. Planning hinged on external funding, and two such possibilities had been identified. The first was the United States Office of Education, which had funds appropriated under the Higher Education Facilities Act (HEFA). It was determined that the campus might be eligible under Title II of this act, which focused on facilities for graduate education. Second, the NSF was making facility grants at the time to support graduate education.

A related question of how to allocate the additional space was described in the same May 24, 1966, letter. This was relevant because the planned allocation would influence how the proposals to federal agencies were couched. The specification was that 20,000 ASF was to be allocated to the computer center, and the balance to mathematics (14,200) and statistics (5,800). This meant that the computer center was a party to the grant proposals that would be prepared. Again the issue of space for a department of computer science was floating, largely unmentioned, in the background.

The campus had now determined to go forward with the larger project with supplemental funding to be sought from the federal government, and hence took an item to the July 14, 1966, regents' meeting requesting authorization to solicit funds from the two government agencies and requesting, contingent on winning such funding, an expanded mathematical sciences building of 154,460 ASF, an increase of some 47,000 ASF over the original plan. The projected cost of the original project had risen through inflation to $5.9 million, while cost of the enlarged project was estimated to be $8.1 million. The regents approved this request [RM July 23, 1966]. The architects then proceeded to prepare schematic plans for the larger building, which in turn were submitted to the regents for approval at their September 15, 1966, meeting. The written background material for the regents' meeting describes the building as consisting of three floors below grade and nine above, or twelve in all. However, the schematic plans for the building prepared by the architect and dated July 27, 1966, and revised on September 9, 1966, show the building with two floors below grade and ten floors above grade, which was the configuration in which the building was realized. The material for the September meeting lists the gross square footage as 270,000, which essentially agrees with number 270,121 shown on the schematics. The meeting material indicates the building would have 154,460 ASF, for an efficiency of 57%, in agreement with the July 14 agenda item, but the schematics in September show 158,120 ASF, for an efficiency of 59%. One assumes that there were some last-minute changes presented at the meeting that departed from the written materials that had been prepared in advance [RM September 15, 1966]. By the time the building was completed, the expansion by 40,000 ASF had turned into one of over 51,000 ASF. At their April 1966 meeting, the regents had agreed to naming the building Evans Hall, in honor of Griffith C. Evans, an action that fulfilled a recommendation first made by Chairman Kelley in 1959.

At this point, the building was in its final configuration. The changes from the 1964 schematic plans comprised the addition of three floors—one underground and two above grade—plus an expanded footprint from about 20,000 square feet to about 22,500 square feet. The interior courtyard had been reduced from three floors to two. From the east, one would see ten floors above grade, while, from the west, eleven floors would be visible. The detailed allocations of space among the units in this expanded building were not included, but documents from two years later state that the original allocations were 82,088 ASF for mathematics (up from 65,033 ASF in the 1962 PPG), 31,548 for statistics (up from 26,710 ASF in the 1962 PPG), and 20,962 for the computer center (up from 0 in the 1962 PPG). The remainder of the space was devoted to classrooms, the library, and building services [SMCP, Evans Hall Space Allocation file, Garr to Diliberto, August 21, 1968]. No decision had been made about the location of the computer science department, the creation of which had been approved in 1966 and which came into being in 1967, but there was a general expectation that it would be located in Evans Hall in space within the mathematics allocation, since it was expected that many of the faculty in this unit would come from mathematics and especially from the projected expansion in the size of the department.

The schedule for completion of the building was revised and project completion was scheduled for April 1970, a slippage of three years from the original 1963 approval. But the new calendar did not take into account the vagaries of the competition and award process for federal grants. It was stipulated that if the federal grants did not come through, the cost of the building would be cut by elimination of upper floors and not finishing some basement areas. Almost immediately after the regents approved the schematics for the larger building and authorized the grant submission in September 1966, the two departments and the computer center submitted twin grant proposals, one to the Office of Education for funds under the graduate facilities program for $2,331,000 and one to NSF for $3,323,850, with each agency being informed of the proposal to the other. The grant proposals mentioned the proposal for creation of a computer science department in the College of Letters and Sciences (see Chapter 14), and a copy of this proposal was included in the grant proposals.

The government agencies sent site-visiting teams to Berkeley in February 1967, and then began their deliberations. The Office of Education acted promptly, and awarded a grant of $1,366,341 on April 6, 1967. The size of the grant strongly suggested that the two agencies would share in the total amount of federal money that would be provided toward the project. The wheels turned more slowly at NSF, and although the foundation had sent favorable signals earlier, it was not until February 2, 1968, that the NSF awarded their share in the form of a grant of $1,000,000, with a stipulation that the funds could not be used until July 1, 1968. This final piece of federal money arrived just in time, since the state construction funds that had been appropriated in the 1965–1966 budget would lapse on June 30, 1968, unless a construction contract had been signed. The total cost of the project at this time, including equipment, was estimated to be $8.95 million. With $5.85 million on hand from the state (including equipment) and $2.366 million from the federal government, the project was still about $725 thousand short. The university requested this difference from the state in a letter in from Acting Vice President Evans to the state department of finance of February 20, 1968, based on the argument of inflation of building costs since the original appropriation [SMCP, Evans

Hall file]. Fortunately, this extra money was forthcoming, and the project went to bid in time for a contract to be awarded prior to June 30, but only just barely. The completion date had now slipped to early 1971, nearly five years later than the estimated and hoped-for completion date when planning had first begun in the late 1950s. Groundbreaking ceremonies were held September 25, 1968, and the project was underway.

Even as construction progressed, it was becoming increasingly clear that the academic and programmatic assumptions that underlay the planning of the building were not likely to be fulfilled. The new governor who took office in January 1967, Ronald Reagan, was elected on his campaign pledge to "clean up the mess in Berkeley," and he did not share the expansive view of the university and its claim on state funding held by his predecessor, Pat Brown. The student-faculty ratio was soon to slip, and summer-session funding was withdrawn. In addition, the job market for new doctorates in all disciplines tanked in about 1970, and additional funding from the state for growth in graduate enrollments was not to be forthcoming, a situation that has continued to the present day. The plan to become a heavily graduate campus with 52% graduate enrollment, and the plans for continued rapid expansion of the faculty, were dreams that rapidly vanished. The growth in faculty and graduate students in the mathematics department leveled off in 1970 at about 60% of the (very aggressive) projections of growth used to justify the building. The result was a professorial faculty of between 65 and 70 plus some lecturers and visitors and graduate enrollments on the order of 300. Statistics experienced a similar leveling off, and so both departments would be entitled to less space than they had planned on and hoped for.

Inevitably, these events opened up a fierce battle for the extra space that would become available in Evans. The first decision was an obvious one programmatically, and that was to move computer science into Evans. Locating this new unit in proximity to mathematics, statistics, and the computer center made very good sense academically, and comported with the view of computer science as at least in part a mathematical science. The only apparent reason for not making this decision earlier was the concern that there would not be enough room. This small department was allocated 8365 ASF, an increase over their previous allocation of 3700 ASF in one of the WW II temporary buildings. The mathematics department, over its fierce protestations, received an allocation of 52,192 ASF, less than the 65,000 ASF in the 1962 PPG and much less than the 82,000 ASF projected in the 1965–1966 planning exercise. The department would vacate 27,000 ASF in five separate buildings on campus, so the space in Evans was a substantial increase. Statistics received 29,209 ASF, more than the 27,000 in the 1962 PPG, but less than the 31,500 ASF in the 1965–1966 document. Statistics would vacate 11,533 ASF in six separate buildings on campus. These figures highlight just how cramped and how dispersed these departments had been. Mathematics occupied floors 7–10 plus part of 6, while statistics occupied floors 3 and 4, and computer science part of 5. The computer center occupied the second floor plus the basement as computer room. Classrooms were located on the ground floor and the library on the first floor.

After providing for all these needs, some 20,000 ASF of space in Evans was left over, so that in some sense the building was 20,000 ASF (or about one and a half floors) larger than it needed to be to accommodate the programs, including computer science. On the other hand, over 30,000 of the ASF that had been added over and above the 1962 PPG, which had been funded through the federal grants and was needed by the units. The 20,000 extra ASF was assigned to economics, demography, and journalism, and some to Electrical Engineering and Computer Science on a temporary basis.

Evans Hall provided a generous amount of space to mathematics and allowed the department for the first time in many years to be housed in a single location, after being distributed over five different buildings. Having senior faculty, junior faculty, post doctorates, visitors, and graduate students together helped create a better sense of a shared community. The building contained a splendid common room of 1800 square feet on the west side of the tenth floor, which provided an opportunity for all members of the department to meet and interact informally. This nicely furnished space remains a gathering place for faculty and students. When he was chancellor, Clark Kerr encouraged departments that were planning new buildings to include common rooms of this sort, even though they had not been included in the restudy standards and so had to be disguised somewhat in the presentation to the state. Statistics had a similar, but smaller, common room. The offices were functional, and especially on the higher floors had stunning views of the bay and the hills. The classrooms on the ground floor were especially designed with mathematics instruction in mind with the width of the room greater than its depth so that the large chalkboards in the front of the room would be closer to the students. The rooms had copious blackboards extending in many cases on three sides of the room. These were positive features.

However, as Evans Hall was being completed in the summer of 1971, its enormous mass became very evident. It blocked a number of site lines that had been incorporated into the John Galen Howard design of the campus early in the century, and the contrast with the beaux-arts architecture of the classical core of the campus was stark. The poured-concrete exterior of Evans was left unfinished, and one could see the grain marks on the concrete from the wooden forms used, as well as numerous pockmarks and air holes in the surface. It was a building in the "brutalist" tradition, and this, plus its size, immediately gained it the 1960s–1970s nickname "Fortress Evans" or "Fort Evans." The long, narrow, double-loaded interior corridors gave it a stark, forbidding institutional look, almost like a prison, and with little charm. The offices did not have finished ceilings, and occupants looked up into a maze of pipes crossing the twelve-foot ceilings. The office furniture provided was metal and quite sterile. Finally, the elevator system in this twelve-story building, which would have considerable vertical traffic, suffered from some design flaws and has never functioned properly. Thus, while the building was reasonably functional, except for the elevators, all agreed that it was aesthetically challenged.

There is an "urban legend" among the occupants of Evans that the architect of the building committed suicide. Indeed, the regents' minutes of May 16, 1968, noted the "untimely death" of Mr. Dailey and approved the transfer of the executive architect contract to the successor corporation that had been formed by associates in Dailey's office. Gardner Dailey, a well-known San Francisco architect who had opened his practice in 1915, had designed a number of buildings on the Berkeley campus, including Tolman Hall, for which he had won a national award for design excellence. The Graduate School of Education, one of the units occupying Tolman Hall, has the following information on its website:

> In his lifetime Gardner Dailey could stride from one end of campus to another and almost never lose sight of one of his buildings—he also designed Evans, Morrison, Hertz, and Kroeber Halls. The architect did ultimately commit suicide, but his death was related to a terminal illness, not to his professional work.

> Tolman Hall's design, particularly the breezeway, was probably influenced by the groundbreaking urban development by the great Swiss-born architect Le Corbusier, called the Unite d'habitation in Marseille. Finished in 1952, Le Corbusier's structure introduced long wings suspended on pillars, all made of molded concrete, the chief material used in Tolman Hall. As in Le Corbusier's building, the wood grain of the molds can still be seen pressed into Tolman's concrete, part of the aesthetic of the design.

Thus, the legend contains an element of truth, but misses the broader context. Dailey never saw Evans Hall, for he died prior even to the groundbreaking. As to losing sight of Evans Hall, the mass of the building is such that it is in fact visible from essentially everywhere on campus; not only that, it is visible from University Avenue as one approaches the campus from the freeway, and indeed, it is visible from the Bay Bridge and even from downtown San Francisco. In his memoirs, Clark Kerr suggests that the final form of Evans Hall was not what he had had in mind when he was Berkeley's chancellor leading the planning exercise that led to the 1956 LRDP. He writes, "A third unplanned development was the massive size of the math building (Evans hall). It did not loom so large when originally planned, but the federal government kept giving money for additional floors, and the campus kept accepting it all, floor, after floor, after floor" [Kerr 2001, p. 124]. Kerr also outlined his rules for the physical development of the campus, and one of these was "Keeping building heights low—no skyscrapers (the mathematics building escaped this rule)" [ibid., p. 125]. It may be pointed out, though, that he was president of the university and a member of the Board of Regents on the occasions in 1963, 1964, 1965, and 1966 when the plans for Evans Hall came forward to the regents for final approval.

Almost immediately after the mathematics department and other occupants of Evans moved into the completed building in September 1971, the aesthetics and institutional nature of the building prompted a self-generated movement among members of the mathematics department to decorate the walls with wall paintings. At one point, the idea was floated that the wall paintings should be temporary, that they should be painted over each month and new ones added in their place; but this kind of constant renewal soon became impractical. Professor John Rhodes, joined by others, took the lead in these activities, and Rhodes organized a seminar on creativity. In the fall of 1971, Rhodes along with Moe Hirsch and Steve Smale organized a "Colloquium on Social Problems and Mathematics," where one of the topics was the architecture of Evans Hall. Rhodes recounts events in his article "A History of the Evans Hall Murals" [Mathematics Department Newsletter, Fall 2002] as follows:

> Sometime after the seminar [on creativity], I organized a one hour talk in 60 Evans by an assistant professor in the architecture department entitled Fascism and Architecture, being as Evans Hall was despised by many as a dehumanizing workplace. The subject of the talk was the effect of architecture on politics and the spirit. I remember the speaker showed slides of Hitler waving to crowds from the balconies of buildings in Berlin.
>
> After the applause abated at the end of this well received talk, I passed out paintbrushes and paint cans. Even Sarah Hallam (senior administrative assistant of the Department for many, many years) grabbed a paintbrush. We rushed upstairs and painted some of the walls on the seventh and tenth floors.

A seminar announcement for the Seminar on Social Problems Connected with Mathematics indicates that a Professor Anthony Ward of the Department of Architecture spoke in this seminar on November 9. The announced title was "Architecture and Insensitivity in Academia." The campus administration took a dim view of these impromptu painting activities, and workmen from the Department of Physical Plant came over to Evans Hall on November 10 and November 11 to remove the wall paintings. The physical plant then sent a bill to John Rhodes for $433.91 for labor and materials to remove the paintings, a bill that Rhodes declined to pay.

At this point the center of activity passed to graduate students, including especially Richard Bassein and some friends. A few days later Bassein and two friends went into Evans Hall at 3 AM, and with the two fellow students serving as lookouts,

Bassein painted the "Death of Archimedes" on the wall opposite his office (775 Evans). The word got out, and when workmen came to clean it off, a number of students stood in front of the painting to protect it. In light of these circumstances, the workmen appeared to have little interest in cleaning it off, and so it remained. An announcement dated November 19 for the Rhodes, Smale, and Hirsch seminar scheduled for November 30 asked the following questions: "Why did the university bill John Rhodes $433 for cleaning? Why did the police phone Rhodes at 1 AM? Who decides what goes on or stays on the walls?" With respect to the Death of Archimedes mural, the announcement asked, "Who ordered the mural destroyed? Who rescinded the order? Was it legitimate to paint the mural in the first place?" And finally it asked "How was [Dean of Physical Sciences Calvin] Moore involved in the decision to bill Rhodes?"

On December 9. Bassein and other graduate students organized a "paint in" on the seventh floor, and it is reported that as the crowd dwindled, someone from physical plant came to look at the results and called the campus police. Names were taken, but just before the police started to make arrests, someone from the department stepped in and nothing further happened. The wall paintings that resulted from this activity—which carry dates that range between December 9 and December 13, 1971—were, according to the participants, often signed as Zorro so that UC police officers would not be able to identify the artists and UC physical plant would not know whom to bill for the cost of removal. This did not stop John Tronoff, an administrator in physical plant, from sending a bill on December 14, 1971 to John Addison, the mathematics department chair, requesting payment of $1,128.28, an amount that included the unpaid cost already billed to Professor Rhodes, as well as the estimated costs of removing the latest wave of wall paintings. As far as is known, Tronoff never collected one penny [DM, Evans Hall file].

Rhodes continues his narrative in the department newsletter, but he has apparently conflated the events of November 9 with events that took place a month later:

> Voila! That is how the murals got there. Some were painted by famous mathematicians—Thurston and Sullivan. And I know "La Mort de Galois" on the seventh floor was painted by my son's uncle, Jack Knudson. I thought perhaps some of you—young and old—might like to know the history of these paintings. I hope as many of them can be preserved as possible.

These painting, mostly with an anti-establishment, 1970s flavor, including the Galois painting mentioned above as well as the Thurston and Sullivan painting, which represents a curve in the thrice punctured plane along with a description of its homotopy class, survive on the seventh floor lobby walls. The creators of this mural are clearly identified as D.S./B.T. with a date 12/9/71. At the time Thurston was a graduate student and Sullivan was a visitor in the department. The wall mural depicting the death of Archimedes did not survive, and the paintings on the tenth floor were recently painted over.

A few months later in the spring of 1972, the department was provided with a small sum of money, $2000, that could be used to commission artwork for Evans Hall. There was a contest, and the first place winner was a proposal by Charles Pugh to construct a set of models built of wire mesh which showed the series of steps needed to accomplish the inversion of the immersed two-sphere that Steve Smale had shown was possible in 1958. These models were beautifully designed and were a real asset, but to the misfortune of the department and the artist, the models were stolen not long after they were displayed, hanging from the ceiling of the tenth floor common room. The second prize went for a wall mural of starkly geometric form, by a local artist, that is located opposite the eighth floor elevator.

Finally, in 1974 the department sponsored a contest for a wall mural to be painted on the ninth floor area opposite the department chair's office. Richard Bassein won the contest and a $500 commission with a proposal to paint a mural, roughly eight feet by twelve feet, of the Reeb foliation of the three-sphere showing the standard picture of what appears to be snakes swallowing the tails of other snakes. He signed it "Zorro—7/22/74" to keep up the December 1971 tradition. These two murals survive today although the Reeb foliation is in need of some restoration. The author is indebted to Richard Bassein, who is now known as Susan Bassein, for her recollections of these events, which have been used in this narrative about the wall paintings in Evans Hall.

In the late 1970s, six to seven years after Evans Hall was completed, a dispute arose between the campus administration and the Departments of Mathematics and Statistics over space allocation in Evans. The campus had been pressing the central university administration for approval of additional capital projects in order to address space shortages on campus. One response from the administration was to point to Evans Hall and to say that if the campus reduced the space allocated to mathematics and statistics, which was way above what they thought restudy standards prescribed, the campus could address some of their space problems internally without constructing new buildings. This bureaucratic response led the campus administration to open up the issue of these two departments being "overspaced," and to attempt to reduce the space allocated to them. This led to a protracted administrative struggle extending over several years that earned the title "Space Wars," in analogy with the then-popular movie.

The department rolled out the now-famous 1962 "Strong exceptions" to the restudy standards and showed by applying these exceptions that the department was entitled to even more space than it had at the time. In addition, it was pointed out that the departments had sought and won federal money from the Office of

Education and NSF that had funded the enlarged building, suggesting that taking away space might violate the terms of the grant. An official letter from the NSF on the mater was sought. Nevertheless, the dynamic of the situation was that mathematics would have to surrender some of its space, but the goal in negotiations with the administration was to hold the amount lost to a minimum. Shoshichi Kobayashi was the chair of the department at the time, and through subtle and clever diplomacy succeeded in holding the loss to about 5,000 ASF on the sixth and seventh floors, or about ten percent of its total space. This was a victory, since the administration got less than it thought it needed. Unfortunately, the battle also left some scars in the administration that took time to heal. Statistics, which was regarded as more "overspaced" than mathematics was, lost about 6,000 ASF, or about 20% of their total space.

Computer science, which as we saw in the previous chapter had been organized in 1973 as a division within the Department of Electrical Engineering and Computer Science (EECS), continued to occupy space in Evans as well as in the EECS building, Cory Hall. As computer science grew rapidly, space became a serious problem for this unit. Planning started for new building for computer science, and in 1994, construction of Soda Hall, north of Hearst Avenue and very close to Cory Hall, was completed. This building provided some 60,000 ASF for the Division of Computer Science, and because of continued growth, the division is already cramped and is thinking about an addition to Soda. If the campus early on had made a different decision about the organization of computer science and had followed the Stanford model of an independent department in the College of Letters and Sciences, it is not clear that Evans Hall would have been large enough to accommodate this department together with mathematics, statistics, the library, and classrooms, even with all other units (economics and the computer center) decanted out. From this point of view, it is also possible to argue that the enormous growth planned in mathematics for faculty and graduate students in 1965 that did not ultimately occur went into growth in computer science instead.

While it made good sense in 1965 to locate the computer center, as it was then conceptualized, in Evans Hall with the machine room in the basement, modes of access to the machines changed, and this proximity was no longer a priority. In addition, the machine room came over time to be the nerve center of the campus computing and communication network, and its location in the basement of a twelve-story building with a seismic rating of poor (see below) seemed inappropriate. In addition, the machine room was, with annoying frequency, subject to flooding caused by accidents of one sort or another. Consideration of disaster preparedness and campus plans for disaster recovery dictated locating the machine room in a far more secure location. After much debate and discussion, the move was finally made in 2005, and the administrative offices of the center moved at the same time. The end result was that the units for which the building was originally planned (mathematics, statistics, the statistical laboratory, the math/stat library, and a classroom complex) occupied only some 92,000 ASF out of the total of 158,00 ASF. The remaining space is well used, primarily as a home for the economics department and as valuable surge space for other units as their buildings underwent seismic retrofit.

When all campus buildings were initially rated for their ability to withstand earthquakes, Evans Hall had initially received a seismic rating of good. But new evidence of performance of buildings and damage to them during the Kobe and Northridge earthquakes in the early 1990s led to a reassessment of the seismic ratings of many campus buildings. In particular, a number of buildings that had been rated as good were downgraded to lower ratings based on this new evidence. Evans Hall was one of these, and along with many others, received a seismic rating of poor in 1995. The campus then began a program of seismic upgrading, which, given the enormous magnitude of the problem, would probably require at least 25 years to complete. The few buildings with very poor seismic ratings had to be vacated immediately and either demolished or retrofitted, while the buildings rated as only poor, including Evans, waited their turn.

During the winter rainy season the already decrepit-appearing exterior surface of Evans acquired dark stains from water that had seeped into the concrete. While this was unsightly, it was also indicative of a more serious problem. The concrete aggregate used in the construction of Evans was of a lightweight variety that was quite porous and allowed water to seep in. In addition, the steel reinforcing bars (rebar) had been placed closer to the surface of the concrete than was specified by code. Thus, the water easily penetrated to the rebar, causing it to corrode and hence expand. The expansion exerted pressure on the concrete and caused cracking and spalling, whereby chunks of concrete would fall off the building's exterior surface. In some areas, enough concrete had cracked and fallen off so that the steel rebar was actually exposed and visible. This situation was an obvious life and safety hazard. The chair of the mathematics department was serving on a seismic planning committee for the northeast precinct of the campus, and at one meeting, he brought along a one-pound chunk of concrete from the exterior wall that had been found a few days earlier on the patio surrounding the building. He placed this on the conference table and then passed it around the room with obvious effect. The chair of this committee subsequently took the same piece of concrete to a campus-wide meeting with the chancellor and repeated the maneuver.

The result was an instruction from the chancellor to fix the problem immediately. The fix consisted in manual hammering on the surface to check for loose material and then use of a very high-pressure water jet to scour the surface of the building and knock off all loose material. Vertical sections of the building were treated serially, as workmen moved around the perimeter. This work required that all the windows in the section of the building being treated be boarded up. At this point, someone quipped that Evans looked like a boarded-up crack house. At one or two points during the water jetting, the water jet broke through the exterior wall, spraying some water into offices. The problem was that there were some voids in the concrete, indicating that insufficient care had been taken in the process of pouring and settling the concrete aggregate. Next, any exposed rebar was waterproofed, and then the entire surface was patched so that, in particular, the exterior surface was perfectly smooth. The final step was application of an elastomeric coating (polymer paint) to waterproof the surface. The noise, dust, and boarded-up windows disrupted the activities of the occupants, but as the work was done serially, no window was boarded up for more than two weeks. In all, the cost of the project was over $2 million. Finally, it was said that the "fix" being applied was only temporary.

Given the prominence and mass of Evans Hall, the choice of the color of the paint to be applied became an important issue. One concern was to make a selection that would do most to conceal or reduce the apparent mass of the building as seen from a distance, but another was to find a color choice that also made sense when viewed at close range. The campus Design Review Committee's approach used as an analogy how one might select a color that would best conceal a field of petroleum storage tanks as viewed from a nearby freeway. However, the decision ultimately was made by a self-appointed committee of three people: the chair of the mathematics department, the dean of the School of Environmental Design, and the vice chancellor for facilities services. The choice of gray tone with some green that picked up the color of the wooded hills to the east plus orange highlights at the top of the building that reflected the orange tile roofs of other nearby buildings was a fortunate one. Everyone on campus agreed that the appearance of Evans had been improved immensely, even though its massiveness and architectural style are still discordant with the surrounding buildings. A faculty colleague in the arts described the building to the author as a brutalist building with a postmodern paint job.

The future of Evans Hall at this point is uncertain. Many have urged that when its turn comes for seismic retrofit, in say 15 years, instead of being retrofitted, the building be torn down and replaced by a smaller building more in line with the beaux-arts style of the classical core of the campus. It has already been noted that a smaller building would suffice programmatically to house the current configuration of the mathematical sciences units that would occupy it.

Chapter 16

Growth and Stabilization: 1961–1973

The campus planning begun in the 1950s under Clark Kerr that we described in Chapter 12 led to the growth of the mathematics department under John Kelley from 1958 to 1960, but it also encompassed plans for continued growth in the 1960s. The department started out in 1960 with 41.5 FTE regular faculty in professorial ranks, and before the end of the decade, the faculty FTE count was to go to 75. This was truly remarkable growth for a department with a faculty of less than 20 in 1955. However, the new planning process for the campus in the early to mid 1960s, which was described in the previous chapter, made it clear that the department was targeted for even further growth in the period beyond the 1960s. Indeed, according to the campus academic planning exercise described in the previous chapter, the department was to increase to a faculty size of perhaps 110 professorial FTE. In addition, that academic plan called for a doubling of graduate enrollment on the campus and a corresponding cutback in undergraduate enrollment. Graduate enrollment in the department was targeted for more than 500 students.

This additional projected growth in faculty size never occurred, and the shift in enrollments toward the graduate level and away from undergraduates did not take place, especially when budget and political realities intervened. The end result was that by 1973 the department had achieved what was to be a steady-state size and configuration of about 65 FTE, with approximately 300 graduate students. Over time, the total graduate enrollment crept up to 350 or more, an outcome that became an issue in the 1975 external review of the department. In sum, the period 1961 through 1967 was a period of rapid growth, and during the following five-year period, 1968–1973, the department worked to accommodate, absorb, and integrate this rapid growth. The theme of this chapter is the rapid growth during the first part of the period followed by stabilization and integration during the second part.

The years 1961–1973 were also marked by several other events. One was the political turmoil of the 1960s and its aftermath, which are described in the following chapter. By 1973, the turmoil was largely over. Another event was that, in

1973, the struggle over computer science was finally resolved (see Chapter 14), one result of which was that numerical analysis and scientific computing became much more clearly a discipline that belonged squarely to the mathematics department. Furthermore, in its hiring, the department had made considerable gains in applied mathematics, thus addressing a longstanding problem going back several decades. Thoughts or threats that had surfaced at times in the past about the possibility of fissioning off a separate department of applied mathematics were laid to rest, and the department set a course by which pure and applied mathematics would exist and thrive in a single department. Finally, in the fall of 1971, the department had completed their long-delayed move into Evans Hall, which finally provided sufficient space, all in one location, so that the department was no longer physically cramped and fractionated. So, in a real sense, the department had by 1973 stabilized into a steady-state mode. Further change would be relatively slow and evolutionary as opposed to the rapid change during the period described in this chapter.

The previous rapid growth in the last years of the 1950s had involved, at the department's insistence, a substantial number of senior tenured appointments in addition to junior appointments. As the department moved into the next phase of growth, in the 1960s, the administration made it clear that the continued growth would be largely through junior non-tenured appointments, as was the tradition at Berkeley. This pattern held for the initial portion of the period through 1967, but from 1968 through 1972 the pattern changed, with fewer assistant professor appointments and more tenured appointments. In the seven recruitment cycles between 1961 and 1967, the department was to make 54 assistant professor appointments, for an average of nearly eight per year, with no fewer than 12 appointments in 1965 alone. During this same period, the department made only four tenured appointments. By contrast, during the six recruitment cycles in 1968 through 1972, the department made 13 assistant professor appointments and 9 tenured appointments, thus representing a very different mix. No new appointments were made in 1973, but two faculty members in numerical analysis and scientific computing from EECS were added to mathematics with split appointments, and three faculty members in statistics were given 0% appointments in mathematics.

The pool from which the assistant professor appointees during the first part of the period were chosen were new PhDs or very recent ones who had brief post-doctoral experience either at Berkeley or at other institutions. Of the 54 assistant professors appointed between 1961 and 1967, 23 were promoted to tenure, which represented a tenuring rate of about 40%. This stands in sharp contrast to the period prior up to 1960, during which nearly all assistant professors had been promoted to tenure, and in the years after 1970, where again just about every assistant professor won promotion to tenure. Not everyone who was not promoted actually came up for a tenure review. A number of these assistant professors left or were lured away, often to tenured positions elsewhere, prior to a tenure review at Berkeley. Each assistant professor was provided an assessment by the department about prospects for promotion in the fall of the fifth year after the doctorate or the fourth year as an assistant professor at Berkeley, whichever came sooner. In practice, this almost always meant actual promotion to tenure by the end of that year or resignation to accept a position elsewhere. This accelerated tenure review schedule made life for these junior faculty members rather hectic. In addition, most of them could not be accommodated in Campbell Hall because of lack of space, and "temporary" buildings T-2 and T-4 became their home along with the graduate students. One of the 54 assistant professors appointed during this period was to achieve great notoriety. Theodore Kaczynski was appointed as an assistant professor in 1967, but he left after two years, telling the department he wished to pursue other interests. Nearly 30 years later, he was identified as the "Unabomber" and is now serving a life sentence in a federal penitentiary.

As noted, the department peaked in size in 1967–1969 at about 75 FTE in professorial ranks, but then immediately dropped back to below 70 FTE. State budget problems had intervened, and, as noted, the aggressive growth plan that had been put in place did not materialize; the department faculty size stabilized so that it varied slightly from year to year in a range between 62 and 67 FTE, where it remained until 1990. In addition, the College of Letters and Sciences restructured its budgetary process. Vacant faculty positions and salary savings from faculty leaves were no longer left in the department to be used for temporary faculty appointments, but were recaptured by the college and reallocated. The department also restructured its hiring practices in 1968, and recognized again more formally the kind of postdoctoral two-year appointment for new PhDs that was common at other research universities and that had been proposed unsuccessfully nearly ten years earlier by the department as Evans lecturers. The department reduced substantially the number of assistant professor appointments, and started to raise the expectations of what was required for an assistant professor appointment. The result was that in the six recruitment cycles from 1968 to 1973, of thirteen assistant professor appointments that were made, two resigned before ever taking up duties, and another resigned after one year and another after two years, both to accept positions at other universities; of the other nine, seven, or 78%, were promoted to tenure. After 1973, assistant professor appointments became even less common, less than one a year on average, and the promotion rate was over 95%.

In 1967–1968, the department argued to the dean of the need to make more tenured appointments to strengthen certain areas, citing particular needs in topology/geometry, algebra, logic, and applied mathematics. The department cited the almost total lack of tenured appointments over the past seven years, some harmful resignations, and anticipated retirements, and in addition a general need to provide academic leadership. These appeals were successful, and as noted, in the five years from 1968 to 1972, nine tenured appointments were made. The job market during the 1960s in mathematics was white hot, and competition was fierce. The department was able generally to recruit very well and was also able in most cases to fight off raids from other universities and to retain faculty, many of whom were targets of multiple attempts from other universities. The department lost only six tenured faculty members during the whole period. Steve Smale was recruited away by Columbia in 1961, but he returned in 1964, so he is not counted in the six. Bert Kostant was recruited away by MIT in 1962, and Antoni Kosinski, Jerome Levine, Phillip Griffiths, Glen Bredon, and Wilfried Schmid were recruited away between 1966 and 1970.

Of the four tenured appointments made during the first part of this period through 1967, one of these has already been discussed, namely Abraham Taub, who came to Berkeley in 1964 as professor of mathematics (50%) and director of the computer center (50%). The other three were John Addison (1962), David Gale (1966), and John Stallings (1967).

John West Addison, Jr., was born on April 2, 1930, in Washington, D.C., and after graduation from Princeton, where he came under the influence of Alonzo Church, he enrolled for doctoral work at the University of Wisconsin, intending to work under Stephen Kleene in mathematical logic. He received his doctoral degree in 1955 with a dissertation entitled "On Some Points of the Theory of Recursive Functions." After his degree and service as instructor at Princeton, he joined the faculty at the University of Michigan, being appointed as assistant professor in 1957. His work came to the attention of the logic faculty at Berkeley, and he was invited as a visitor at Berkeley for the 1959–1960 academic year. Appointment as associate professor followed in 1962, making him the fifth logician on the faculty, thus keeping the logic group at about 10% of the total faculty. Addison's research focused on the analysis of hierarchies and separation theorems that one finds in descriptive set theory, recursion theory, and model theory. His work demonstrates and highlights interesting similarities between structures in three areas of logic. In particular, the projective hierarchy in descriptive set theory has been a topic of recurring interest to Addison. He was advanced to full professor in 1968 and at the same time became department chair. After serving in difficult times as chair for four years through 1972, he returned to normal duties of research and teaching but served a second four-year term as chair from 1985 to 1989. This placed him second in total service as chair since Griffith Evans's 15 years of service. Addison supervised a total of 13 doctoral students, one at Michigan and the remainder at Berkeley. He retired in in 1994 under the university's early retirement program (VERIP), and has remained active in the department since then.

David Gale was born December 13, 1921, in New York City and received his doctoral degree at Princeton University in 1949 with a dissertation under Alfred Tucker on "Solutions of Finite Two Person Games." After a year as instructor at Princeton, he joined the faculty at Brown University, where he rose through the ranks to full professor. After serving as chair of his department he spent the year 1965–1966 as a Miller visiting professor at Berkeley in the Department of Industrial Engineering and Operations Research (IEOR). Interest in his moving to Berkeley developed in both IEOR and mathematics, and the proposal for a joint appointment with 50% in each department was put forward. As soon as the economics department heard of the initiative to recruit Gale, it offered to join in the effort with a 0% appointment in economics. Gale's research record in game theory, linear programming, mathematical economics, and combinatorics brought strength in areas in which the department was weak, served to strengthen applied mathematics, and created connections with the College of Engineering. The effort was successful, and Gale flourished at Berkeley. His research was recognized by election to the National Academy of Sciences and to the American Academy of Arts and Sciences. In all, he supervised 14 doctoral students, five at Brown and the remainder at Berkeley. Although he retired a bit early (mandatory retirement age had been moved from 67 to 70 in the early 1980s) under the voluntary early retirement program in 1991, he has remained active in the department since that time.

John Stallings was born on July 22, 1935, in Morrilton, Arkansas, and after graduating from the University of Arkansas in 1956, he enrolled in Princeton University for doctoral study. He completed his work in 1959 with a dissertation "Some Topological Proofs and Extensions of Grushko's Theorem" under R. H. Fox. After one year on an NSF postdoctoral fellowship, he joined the Princeton faculty, advancing to assistant professor in 1961 and to associate professor in 1964. Stallings's work concentrates on piecewise linear topology and the group-theoretic aspects of topology, or what became known as geometric group theory. He was widely acclaimed for his solution of the Poincaré conjecture in dimension greater than or equal to five for piecewise linear manifolds, a result that complemented Smale's result for differentiable manifolds, as well as for other results. The department received word that Stallings would be open to an offer from Berkeley, and the arrangements were quickly completed. It was felt that his expertise in piecewise linear topology as well as in geometric group theory would nicely complement those of the topologists already in the department and would compensate for the recent loss in 1966 of two tenured topologists, Jerome Levine and Antoni Kosinski. Stallings was awarded the Cole Prize of the American Mathematical Society in recognition of his achievements, and his kind of topology and algebra has flourished at Berkeley. He has supervised the work of 21 doctoral students in all so far, 17 of these after he came to Berkeley. He took early retirement in 1994, but he has remained very active in teaching, research, and supervision of graduate students in the department, with a number of doctoral students completing their work under him after his retirement.

Of these four appointments, only two, Taub and Gale, were of scholars who had been established for an extended period of time. Addison and Stallings were in the early stages of their careers, six and eight years, respectively, beyond their doctoral degrees, but who by their achievements already commanded a tenured position.

Turning to the assistant professor appointees from 1961 to 1967, we focus attention on the 23 who were promoted to tenure. Four of these young mathematicians left Berkeley soon after their promotion: Antoni Kosinski, Jerome Levine, Phillip Griffiths, and Wilfried Schmid. Antoni Kosinski, a topologist, received his doctoral degree from the Polish Academy of Sciences in 1958 under Karol Borsuk and came to Berkeley as a visitor in 1959. He was subsequently appointed as an assistant professor in 1961, and then won promotion to tenure in 1964, but departed in 1966 to accept a position at Rutgers University. Jerome Levine, another topologist, received his doctoral degree under Norman Steenrod at Princeton in 1962 with a dissertation entitled "Imbeddings and Immersions of Projective Spaces in Euclidean Space." After two years of postdoctoral work, he arrived at Berkeley in 1964 as an assistant professor, and quickly won promotion to tenure in 1965, but then resigned in 1966 to accept a position at Brandeis University. Phillip A. Griffiths received his doctoral degree at Princeton in 1962 under Donald Spencer with a dissertation entitled "On Certain Homogenous Complex Manifolds," and came to Berkeley as a Miller fellow to work under S. S. Chern. At the end of his fellowship in 1964, he was appointed as assistant professor and was promoted to associate professor in 1965, and then to full professor in 1967. But, he then resigned in 1968 to accept a position at Princeton. Wilfried Schmid completed his doctoral degree at Berkeley under Phillip Griffiths in 1967 with a dissertation entitled "Complex Manifolds and Representations of Semi-Simple Lie Groups." The department made an exception to its normal policy of not hiring its own doctoral students immediately after the degree and hired Schmid as an assistant professor. He was promoted to associate professor in 1970, but Berkeley could not hold him against an offer from Columbia. Subsequently, Schmid left Columbia to accept appointment at Harvard. Berkeley also lost Glen Bredon to Rutgers in 1968. It is not clear why the department had such trouble holding on to young faculty in topology and geometry at this time.

Of the 19 who resisted temptations elsewhere, five were in geometry/topology, five were in analysis, three in algebra, three in logic, and three could be classed generally as in areas reaching out to other disciplines. The five in geometry/topology were Shoshichi Kobayashi, Joseph Wolf, Hung Hsi Wu, Charles Pugh, and John Wagoner.

Shoshichi Kobayashi was born January 4, 1932, in Kofu, Japan, and came to the United States for doctoral work. He received his doctoral degree at the University of Washington in 1956 working under Carl Allendorfer in differential geometry with a dissertation entitled "Theory of Connections." During two years at IAS and two years at MIT, his work came to the attention of Chern, who was moving to Berkeley at the time, and he was interested in recruiting Kobayashi to join him. An informal offer of an assistant professorship was extended to Kobayashi by Kelley in 1959, but visa problems emerged that made it impossible for him to accept the offer. The strategy that emerged was for Kobayashi to leave the country for two years, taking a position at the University of British Columbia, and to apply for an immigrant visa for himself and his wife. As with the case of Kato, the waiting list was a long one, but the department, as in the Kato case, persuaded the Department of Defense to declare that his presence in the US working on a DOD grant was essential to national defense, and that he and his wife could be "paroled" to the US while awaiting their emigrant visas. This happened in 1962, and so Kobayashi could come to Berkeley to take up an appointment as assistant professor. His research in differential geometry won him rapid promotion to tenure the following year and to full professor in 1966, and his work over many decades spanned many areas of differential geometry. He supervised the work of 35 doctoral students—a number well near the top in the department—and was selected by his colleagues to serve as department chair for a three-year term from 1978 to 1981 and again for a semester in the fall of 1992. He opted to take early retirement in the VERIP plan in 1994 and has remained active in the department since then.

Joseph Wolf was born October 18, 1936, in Chicago Illinois, and received his doctoral degree at the University of Chicago in 1959 working under S. S. Chern with a dissertation entitled "On the Manifold Covered by a Given Compact Connected Riemannian Homogeneous Space." After three years of postdoctoral work in Paris and at the Princeton Institute, he accepted, at Chern's urging, appointment as assistant professor at Berkeley in 1962. His work won him rapid promotion to tenure in 1964 and to full professor in 1966. His interests in differential geometry expanded to include studies of bounded symmetric domains as well as other topics and expanded further to include representation theory of Lie groups. During his career, he supervised the work of 20 doctoral students. He retired early under the VERIP program in 1994, but has remained very active and engaged in the department.

Hung Hsi Wu was born May 25, 1940, in Hong Kong, and came to the US in 1956. He received his doctoral degree from MIT in 1963 with a dissertation, written under Warren Ambrose, entitled "On the de Rham Decomposition Theorem."

After two years of postdoctoral work, he was appointed as an assistant professor in 1965. He rapidly won promotion to associate professor in 1968 and to full professor in 1973. After coming to Berkeley, he began work on complex differential geometry, holomorphic functions in several variable, and Nevanlinna theory, an area in which Chern shared a special interest. This work won him recognition, and the structure and properties of Kaehler manifolds have been a major theme of his research over the years. He supervised nine doctoral students, and in recent years he has developed an additional interest in mathematics education, a topic on which he has written and spoken frequently. His work in this area has been quite influential in shaping K-12 teacher in-service and pre-service programs.

Charles Pugh was born June 16, 1940, in Philadelphia, Pennsylvania, and received his doctoral degree from Johns Hopkins University in 1965, although all work toward the degree was completed in 1964. He worked under Phillip Hartman, and his dissertation was entitled "The Closing Lemma in Dimension Two and Three." This work, which he soon extended to all dimensions, solved a famous open problem in dynamical systems (or topological ordinary differential equations) that asked whether a flow with recurrent orbits can be modified by an arbitrarily small perturbation so that the new flow has a closed orbit—the answer is yes. Pugh came to Berkeley in 1964 as an instructor, and was advanced to assistant professor in 1965, to associate professor in 1968, and to professor in 1973. His research has ranged widely over dynamical systems and differentiable ergodic theory, and he has supervised the work of 24 doctoral students so far during his career. He retired in 2005 and has taken a faculty position at the University of Toronto.

John Wagoner was born November 19, 1942, in Phoenixville, Pennsylvania, and received his doctoral degree from Princeton University in 1966. His dissertation, written under William Brouwder, was entitled "Surgery on a Map and an Approach to the Hauptvermutung." After a year of postdoctoral work, he was appointed as an assistant professor at Berkeley in 1967. He moved away from his initial work on piecewise linear topology and began work on the study of psuedo-isotopies, publishing a long memoir on the topic, and then began a program of research in algebraic K-Theory, which in recent years has turned to the study of Markov shifts, symbolic dynamics, and the shift equivalence problem. He won promotion to associate professor in 1973 and to full professor in 1978; Wagoner has supervised 15 doctoral students so far in his career. He was selected by his colleagues to serve as chair of the department and did so for three and a half years from January 1993 to June 1996. This was not an easy time to serve as chair, as California was enduring the worst recession since the 1930s and the university's budget was subjected to very deep cuts.

The topology and geometry group was thriving with these new additions to the already distinguished faculty, but the resignations of Griffiths, Schmid,

Levine, Kosinski, and Bredon noted above were damaging and were never really replaced.

The five assistant professors hired in 1961–1967 in analysis who were promoted were Calvin Moore, Donald Sarason, Marc Rieffel, Keith Miller, and Haskell Rosenthal. It is natural to discuss William Arveson together with these five since he was academically their contemporary, even though his initial appointment was in 1969 and was a beginning tenure appointment. He was not recruited from a tenured position outside the university as were the other tenured appointees

Calvin Moore was born November 2, 1936, in New York City and received his doctoral degree at Harvard University under George Mackey in 1960 with a dissertation entitled "Extension and Cohomology of Locally Compact Groups." After a postdoctoral year at the University of Chicago, he was appointed as an assistant professor at Berkeley in 1961. His research interests spanned topological groups, group representations, ergodic theory, and operator algebras, together with an interest in algebra and number theory, especially as they connect with group representations. He was promoted to associate professor in 1965 and to full professor in 1966. His work was recognized by election to the American Academy of Arts, and he also became involved in university administration, serving as Dean of Physical Sciences, as Associate Vice President in the university-wide administration, and as department chair for six years, 1996–2002. Working with S. S. Chern and I. M. Singer, he was a co-founder of the Mathematical Science Research Institute (MSRI). During his career he supervised the work of 13 doctoral students. He retired in 2004, but has remained active in the department and the university.

Donald Sarason was born January 26, 1933, and received his doctoral degree at the University of Michigan in 1963 under Paul Halmos with a dissertation entitled "The H^p Spaces of Annuli." After a year of postdoctoral work, he was appointed as an assistant professor at Berkeley in 1964. He won promotion quickly to associate professor in 1967, and then to full professor in 1970. His research has concentrated on Banach spaces of holomorphic functions in the disc, interpolation problems, and Toeplitz operators. He has supervised so far the work of 39 doctoral students, which placed him near the top of the list in productivity of students, and two of his students were appointed to the Berkeley faculty: Alice Chang and Thomas Wolff.

Marc Rieffel was born December 22, 1937 in New York City and received his doctoral degree from Columbia University in 1963 working under Richard Kadison with a dissertation entitled "A Characterization of Commutative Group Algebras and Measure Algebras." He came to Berkeley as a postdoctoral faculty member with appointment as lecturer, and was appointed as an assistant professor in 1965. He won promotion to associate professor in 1968 and to full professor in 1973. His research has concentrated on the structure and properties of C^* algebras, especially those termed non-commutative tori, and his work has expanded to include issues of deformation quantization and the quantum metric spaces. To date 28 doctoral student have completed their work under him.

William Arveson was born November 22, 1934, in Oakland, California, and received his doctoral degree at UCLA in 1964 with the dissertation "Prediction Theory and Group Representations" written under Henry Dye. His education had been interrupted by four years in the navy after high school. Following his doctoral degree, he spent a year doing mathematical research for the navy and then was appointed as a Benjamin Peirce Instructor at Harvard University for three years, 1965–1968. He then came to Berkeley as a lecturer, and in 1969 he was appointed directly as an associate professor—his work had gained sufficient recognition by this time, which made it necessary and appropriate to appoint him to a tenured position, skipping the level of assistant professorship. Arveson's work has ranged widely over many aspects of operator theory and operator algebras, including such topics as invariant subspaces, nest algebras, questions of positivity in operator algebras, and the dynamics of non-commutative flows. Twenty-six doctoral students have completed their work under him during his career so far. Although he retired in 2003, he has remained active in the department.

Keith Miller was born March 27, 1937, in Monroe, Louisiana, and received his doctoral degree at Rice University under the direction of Jim Douglas, with a dissertation entitled "Three Circle Theorems in Partial Differential Equations and Applications to Improperly Posed Problems." After a year of postdoctoral work, he was appointed as assistant professor at Berkeley in 1965. He subsequently won promotion to associate professor in 1969 and to full professor in 1973. Miller's research has focused on the study of unique continuation problems for solutions of PDEs and on numerical methods, in particular what is called the moving finite element methods. His work has also included safety issues for nuclear reactors. He supervised the doctoral work of nine students, and retired from active service in 2005.

Haskell Rosenthal received his doctoral degree from Stanford University in 1965 with a dissertation entitled "Projection onto Translation Invariant Subspaces of $L^p(G)$" written under Karel DeLeeuw. After a year of postdoctoral work, he was appointed as an assistant professor at Berkeley in 1966. He was promoted to associate professor in 1970, but then was enticed away from Berkeley in 1974 by an attractive offer from the University of Texas, Austin. He early on shifted his research from his thesis topic to the study of the geometry of Banach spaces.

These appointments provided considerable strength in areas of functional analysis, although the resignation of Rosenthal in 1974 left a void in the geometry of Banach spaces that was not to be filled, and Berkeley did not have anyone in this area after 1974. However, appointment of only one person in PDEs did not represent sufficient attention to the challenge of renewal in this field posed by the fast-approaching retirements of Hans Lewy and Charles Morrey, in 1972 and 1974, respectively. It would be some time before Berkeley devoted adequate resources to rebuilding this field.

The three faculty in algebra who were promoted—Andrew Ogg, John Rhodes, and George Bergman—represented a diverse array of interests in algebra. However, in retrospect, it could be said more than three appointments would have been needed to provide sufficient strength for this field. Some others, notably Calvin Moore, had interests in algebra, and in addition there would be more appointments in algebra in the next period 1968–1973, which would redress this imbalance.

Andrew Ogg was born April 9, 1934, in Bowling Green, Ohio, and received his doctoral degree at Harvard University under John Tate with a dissertation entitled "Cohomology of Abelian Varieties over Function Fields." After a year of postdoctoral work, he was appointed as an assistant professor at Berkeley in 1962. His promotion to associate professor came in 1966, and to full professor in 1969. His research spanned the structure and properties of elliptic curves and modular functions, and he supervised the work of seven doctoral students. Taking advantage of the VERIP program, he retired from the university in 1994, but has remained active in the department in the subsequent years.

John Rhodes was born July 16, 1937, in Columbus, Ohio, and received his doctoral degree at MIT in 1962 with a dissertation entitled "Algebraic Theory of Machines" written under Warren Ambrose. After a year of postdoctoral work, he was appointed as an assistant professor at Berkeley in 1963. He won promotion to associate professor in 1966 and to full professor in 1970. His research built on his dissertation in which he showed how to make a prime decomposition of finite semi-groups and translate that into a similar decomposition of finite state machines. Further studies include investigation of the complexity of a profinite semigroup, with applications to finite state machines and languages. Altogether, he supervised the work of 26 doctoral students. He retired in 2001 and moved to Paris, France.

George Bergman was born July 22, 1943 in Brooklyn, New York and received his doctoral degree in 1968 at Harvard University in 1968 under John Tate with a dissertation entitled "Commuting Elements in Free Algebras and Related Topics in Ring Theory." As his work was essentially completed a year earlier, he was appointed directly in 1967 as assistant professor at Berkeley, where he had been an undergraduate and had been the winner of the Departmental Citation in 1964 as the top graduating senior in Mathematics. He won promotion to associate professor in 1972 and to full professor in 1978. His research spanned many areas in algebra and concentrated on general algebraic properties of rings, groups, and algebras. He has to date supervised the work of ten doctoral students.

During this period the logic group added three new faculty members who won tenure: Robert Solovay, Jack Silver, and Ralph McKenzie. These appointments brought the logic group to seven faculty altogether, so that the logic group kept pace with the growth of the department (to about 70) with logic remaining again at about 10% of the total.

Robert Solovay was born December 15, 1938, in Brooklyn, New York, and received his doctoral degree in 1964 at the University of Chicago under Saunders Mac Lane with a dissertation entitled "A Functorial Form of the Differentiable Riemann-Roch Theorem." He had effectively completed the requirements somewhat earlier, and had been appointed as Higgins Lecturer at Princeton for the period 1962–1964. While still a graduate student he had, in addition to his dissertation research, obtained significant results in set theory. A year at the Institute for Advanced Study followed, and then in 1965 he came to Berkeley as an assistant professor. He won rapid advancement to tenure as associate professor in 1966 and as full professor in 1968. His results in set theory built on and vastly extended the work of Cohen on forcing, and his work was widely recognized. His research placed him in a preeminent position in logic, and led to his election to the National Academy of Sciences and the American Academy of Arts and Sciences. He supervised the doctoral work of 14 students including his future Berkeley colleague Hugh Woodin. He took early retirement in 1994. Since his retirement, he has shifted his scientific interests into different areas.

Jack Silver was born April 23, 1942, in Missoula, Montana, and received his doctoral degree in 1966 at Berkeley with a dissertation entitled "Some Applications of Model Theory to Set Theory" written under Robert Vaught. After a year of postdoctoral work, he was appointed as an assistant professor at Berkeley in 1967. He was advanced to associate professor in 1970 and to full professor in 1975. His research has focused on questions of the properties and implications of various large cardinal hypotheses, most notably the hypothesis of the existence of measurable cardinals. He has supervised the work of 18 doctoral students to date.

Ralph McKenzie was born October 20, 1941, in Cisco, Texas, and received his doctoral degree at the University of Colorado under Donald Monk, with a dissertation entitled "The Representation of Relation Algebras." He served a postdoctoral year as an instructor at Berkeley for 1966–1967, and was then appointed as an assistant professor in 1967. Promotion to associate professor came in 1971 and to full professor in 1978. McKenzie's research is in the area of general algebra, an area in which Tarski had worked but which was not then represented at Berkeley. McKenzie published actively in this field. He took advantage of the university's early retirement program in 1994. He then accepted appointment at Vanderbilt University to the chair that Bjarni Jonsson had held until his retirement. Mckenzie has supervised the doctoral work of 24 students, 21 at Berkeley and 3 at Vanderbilt.

The remaining three faculty members hired in this period and promoted were mathematicians whose interests reached out to other fields, and who could be considered to work in applied areas. One of these, Beresford Parlett, who works in numerical linear algebra, has already been discussed in the chapter on computing and computer science. The other two were Lester Dubins, a probabilist whose appointment was 50% mathematics and 50% statistics, and Oscar Lanford, who worked in mathematical physics.

Lester Dubins was born April 20, 1920, in Washington, D.C., and received his doctoral degree at the University of Chicago in 1955, working under Irving Segal with a dissertation entitled "Generalized Random Variables." He had a postdoctoral appointment at Carnegie-Mellon for two years, followed by two years at the Institute for Advanced Study and then two years as a visitor at Berkeley. In 1961 he was appointed as an assistant professor with a split appointment: 50% mathematics and 50% statistics. He was quickly promoted to associate professor in 1963 and then to full professor in 1966. His research ranged over many topics in probability theory, including the probability underlying gambling, the theory of martingales, the role of finitely additive measures in probability theory, and many others. From the point of view of his work, he could well be counted among the faculty in analysis rather than applied mathematics. He supervised the work of eight doctoral students, and he retired in 1991 when he reached the then mandatory age of retirement for faculty. However, he and another UC faculty member challenged the legality of UC's mandatory retirement policy under state law, and prevailed in court. Consequently, Dubins was reinstated as professor in 1997. Meanwhile, the university had eliminated mandatory retirement in 1992, a year before federal law eliminated it throughout the country. Dubins continued to serve as professor until he retired voluntarily in 2004.

Oscar Lanford was born January 6, 1940, in New York City, and received his doctoral degree in physics at Princeton in 1966, working under Arthur Wightman, with a dissertation entitled "Construction of Quantum Fields Interacting by a Cutoff Yukawa Coupling." He served as an instructor in physics at Princeton during his last year there and was then appointed as an assistant professor at Berkeley in 1966. He won promotion to associate professor in 1970 and to full professor in 1975. His research focused on statistical mechanics of systems of particles and the time evolution of such systems. This led to work on dynamical systems in which he established the Feigenbaum conjecture by computational methods. In 1982 Lanford accepted appointment as professor at the Institut des Hautes Études Scientifique (IHES) in the suburbs of Paris. He was permitted to take leave without salary from Berkeley, since it was thought that he would return after a brief period in France. This turned out not to be the case, and Lanford remained in Europe, but moved from IHES to a faculty position at the Swiss Federal Institute of Technology (ETH) in Zurich. Although he did not officially resign for some years, he had effectively departed from Berkeley in 1982. Overall, he has supervised the work of seven doctoral students, five at Berkeley and two at ETH.

As the department moved into the next phase of development, stabilization and consolidation during 1968 to 1973, the emphasis turned toward tenured appointments. After the rapid growth through junior appointments, the department had convinced the dean and the chancellor of the need to add tenured appointments. Three were made in 1968—Wu-Yi Hsiang, Ichiro Satake, and Rainer Sachs—which provided strength in topology, algebra, and applied mathematics.

Wu-Yi Hsiang was born February 3, 1937, in Lianh-Chi, Anhwei Province, China, and received his doctoral degree from Princeton University in 1964 under John Moore with a dissertation entitled "On the Classification of Differentiable Actions of the Classical Groups on Pi-Manifolds." Immediately on completion of his degree, he was appointed as an assistant professor at Brown University; three years later he moved to the University of Chicago as an associate professor. In 1968 he was recruited to Berkeley as an associate professor, and shortly thereafter, in 1970, was promoted to full professor. In arguing the case for this appointment, the department cited three recent losses at the tenured level in topology—Jerome Levine, Antoni Kosinski, and Glen Bredon—and Hsiang indeed strengthened the topology group. His work initially continued to focus on differentiable transformation groups and later shifted somewhat to the study of minimal immersions and embeddings. Much later, in the 1990s, his attention turned to sphere packing and the Kepler problem. Altogether he supervised the work of 21 doctoral students, one at Brown, three at Chicago, and 17 at Berkeley. He retired in 1998 and spent the next several years as a faculty member in Taiwan, but he recently returned to Berkeley.

Ichiro Satake was born on December 25, 1927, in Tokyo, Japan, and received his doctoral degree in 1958 from the University of Tokyo. He served on the faculty of the University of Tokyo from 1952 until 1963 as lecturer, assistant professor, and professor. In 1963 he came to the United States as professor at the University of Chicago. Satake had served as visiting professor at Berkeley in the fall of 1966, and the department began its recruitment of Satake shortly thereafter, in 1967. These efforts were successful, and Satake joined the Berkeley faculty permanently in 1968. The department argued for this appointment on the basis of the need to strengthen algebra, and Satake, whose work lay in the crosscurrent of number theory, several complex variables, and Lie groups, brought important breadth and depth to the department. Satake overall supervised the work of eight doctoral students, with five of these at Berkeley. Taking early retirement in 1983, he took a professorial position in Japan, but continued to spend a considerable amount of time in Berkeley.

Rainer Sachs was born June 13, 1932, in Frankfurt am Main, Germany, and received his doctoral degree in physics from Syracuse University in 1958, working under Peter Bergman with a dissertation entitled "The Structure of Particles in a Linearized Gravitational Field." After three years of postdoctoral work, a year in the US Army, and a year at Stevens Institute of Technology, he was appointed in 1963 as an associate professor at the University of Texas, Austin. He was advanced to full professor there in 1965, and then at Abe Taub's urging, he visited Berkeley during the 1966–1967 academic year. Taub was very eager to bring a younger scholar working in general relativity theory to Berkeley, and he promoted Sachs's appointment as a joint appointment in physics and mathematics. Sachs's appointment would strengthen applied mathematics generally, a long-term and recurring goal of the department. One obstacle to the Sachs appointment was that the Berkeley physics department had been skeptical of theoretical work in general relativity because of its apparent remoteness from experiment. However, recent experimental developments had brought about a change of heart, and the Sachs appointment was warmly endorsed by both departments as a 50% mathematics–50% physics split appointment. Altogether he has supervised three doctoral students. Later in his career Sachs switched his field of interest to mathematical biology and has published actively in this area. He retired early in 1994, but has remained very active in research.

Alfred Tarski had reached the age of mandatory retirement in 1969, and the logic group made the case for a new tenured appointment in logic. A number of possibilities were pursued, but the choice devolved onto Ronald Jensen, a leading scholar in set theory. Ronald Jensen was born on April 1, 1936, in Bonn, Germany, and he received his doctoral degree from the University of Bonn in 1964 under Gisbert Hasenjaeger. He was appointed as professor in the Berkeley mathematics department in 1970, but in 1972 he took leave without salary for two years before resigning in 1974 to return to Europe. He has made many contributions to modern set theory, and has had a peripatetic career, supervising the work of 12 doctoral students in seven different institutions, including two students at Berkeley.

The department then turned to the problem of adding further strength in applied mathematics, a longstanding challenge. Investigation of candidates in the fields of combinatorics, game theory, and information theory who might also strengthen links with computer science led to the possibility of recruiting Elwyn Berlekamp, then on the staff of Bell Labs. Elwyn Berlekamp was born September 6, 1940, in Dover, Ohio, and received his doctoral degree in electrical engineering at MIT in 1964 working under Robert Gallager with a dissertation entitled "Block Coding with Noiseless Feedback." Immediately upon receipt of his degree, he was recruited to Berkeley as an assistant professor of electrical engineering, where he remained until February 1967, when he resigned to join the research staff at Bell Labs. The initiative to bring him back was to be a joint one between the mathematics department and the Department of Electrical Engineering and Computer Science, and it came to fruition in 1971 when Berlekamp returned to Berkeley as full professor with an appointment split 50/50 between the two departments. His work on algebraic coding theory and error-correcting codes won wide recognition scientifically and laid the basis for a firm, Cyclotomics, that he established in Berkeley in 1973. The firm was sold to Eastman Kodak in 1985. Error-correcting codes developed by Berlekamp are widely used in compact discs and in the transmission of data from deep-space satellites. Berlekamp was granted industrial leave and ultimately cut back on his appointment so that it became only 50%, all in mathematics. He also developed fast, efficient algorithms for factoring polynomials over finite fields, which were of use in algebraic coding theory, but which have had impact and use more generally, specifically in factoring polynomials over the integers. His research interest subsequently shifted to combinatorial game theory, where his work is also well known. His research has been recognized by election to the National Academy of Sciences, the National Academy of Engineering, and the American Academy of Arts and Sciences. He has supervised the work of 13 doctoral students at Berkeley. He retired in 2002 but has remained active in the department.

At the same time, the department investigated candidates in the field of what one would call today scientific computation, focusing on numerical studies of partial differential equations. The name of Alexandre Chorin, then at the Courant Institute at NYU, came to the fore. Alexandre Chorin was born June 25, 1938, in Warsaw, Poland, and when the war erupted, his family fled through Russia and finally to Israel. They subsequently moved to Switzerland, and Chorin completed his baccalaureate degree at the University of Lausanne, where he studied under de Rham. He then came to the United States for doctoral work at the Courant Institute, and completed his work under Peter Lax, with a dissertation entitled "Numerical Study of Thermal Convection in a Fluid Heated from Below." He joined the faculty at NYU, and then in 1970 the Berkeley mathematics department successfully nominated him for a visiting Miller professorship for the year 1971–1972. A permanent appointment as associate professor followed in 1972 with advancement to full professor in 1974. Chorin's research on numerical methods in fluid mechanics and turbulence won him wide acclaim, prizes, and awards. He invented the projection method and the vortex method, both of which are of fundamental importance and everyday use in the numerical solution of fluid mechanics and aero-dynamical problems. His work with Grigory Barenblatt in 2000 led to fundamental revisions in the laws governing turbulence, and finally his work on optimal prediction has opened up new areas of research.

Chorin was elected to the National Academy of Sciences and the American Academy of Arts and Sciences in 1991, and in 2001 he was appointed as a university professor, one of only 13 such faculty on active duty so honored in the entire UC system, and the first full-time member of a mathematics department to receive this honor. Richard Karp and Gerard Debreu, who hold or held 0% appointments in the Berkeley mathematics department, have also been so honored. Chorin has attracted a large number of graduate students, and he has supervised the work of 47 doctoral students so far, which ties him for the record in the department with Rob Kirby. Shiing-Shen Chern with 45 and Steve Smale with 44 are close behind. However, counting those with 0% appointments, they all trail David Blackwell (64) and Irving Kaplansky (55).

The appointments of Sachs, Berlekamp, and Chorin immensely strengthened applied mathematics. To build further strength in mathematical physics, the department made an attempt jointly with physics in 1972 to attract Barry Simon to Berkeley with a tenured appointment. Unfortunately, that effort failed when Simon declined the initiative.

The department still needed to add strength in topology beyond the Hsiang appointment, especially with the resignations of topologists in the 1960s, and also in algebra, in addition to the Satake appointment. Opportunities soon presented themselves to attract two mathematicians who would meet these needs: Rob Kirby in 1971 and Robin Hartshorne in 1972.

Robin Kirby was born February 25, 1938, in Chicago, Illinois, and received his doctoral degree at the University of Chicago in 1965 working under Eldon Dyer with a dissertation entitled "Smoothing Locally Flat Embeddings." He joined the mathematics faculty at UCLA as an assistant professor immediately after receiving his degree. In 1968, Kirby found a proof of the stable homeomorphism conjecture, which was the key step in unwinding the basic structure of topological manifolds. In joint work with Laurence Siebenmann, the annulus conjecture was resolved, as was the triangulation problem for topological manifolds, with the exact obstructions to existence and uniqueness of triangulations clearly identified. These results brought considerable recognition to Kirby, and shortly thereafter he responded positively to an initiative to bring him to Berkeley, and he was appointed as professor in 1971; the actual transaction was an intercampus transfer. Kirby and low-dimensional topology have thrived at Berkeley, and he has attracted a large number of doctoral students; as noted above, 47 students have completed work under his direction so far. His work has been honored with the Veblen Prize from the American Mathematical Society and by the Award for Scientific Reviewing from the National Academy of Sciences, and he has been elected to the National Academy of Sciences. In the department he served for many years as vice chairman for graduate affairs, and he served as deputy director for several years.

Robin Hartshorne was born March 15, 1938, in Boston, Massachusetts, and received his doctoral degree from Princeton University working under John Moore (and Oscar Zariski, at a distance) with a dissertation entitled "Connectedness of the Hilbert Scheme." He then returned to Harvard University, where he had been an undergraduate, as a junior fellow in the Society of Fellows. In 1966, he was appointed as an assistant professor at Harvard, and in 1969 he was advanced to associate professor. In 1971 the opportunity arose to recruit him to Berkeley, and this was successfully completed with his appointment as associate professor in 1972. Advancement to full professor came in 1974. Hartshorne's research in algebraic geometry had won wide attention, and he was the first faculty member at Berkeley immersed in the Grothendieck school of algebraic geometry. His Harvard doctoral student Arthur Ogus soon joined him at Berkeley, further strengthening algebraic geometry at Berkeley. Hartshorne supervised the doctoral work of 26 students, 22 of them at Berkeley. In 1996, he solved the Zeuthen problem, a prize problem with a monetary award for its solution that had been posed in 1901 by the Royal Danish Academy of Science, and on which a number of mathematicians had worked. The original prize had carried a monetary award, but it was time limited. Hence no money was forthcoming to Hartshorne, although he did receive a certificate acknowledging the solution. Hartshorne retired from active duty in 2005.

In addition to the tenured appointments that we have just discussed, the department made thirteen assistant professor appointments during the period 1968–1973, four of whom either never took up duties or served only briefly. Of the other nine, seven were promoted to tenure. These were Tsit-Yuen Lam, Jerrold Marsden, and Alan Weinstein, who were appointed to the faculty in 1969; Robert (Rufus) Bowen and Blaine Lawson, who appointed in 1970; and Paul Chernoff and David Goldschmidt, who were appointed in 1971. Four of these were in various aspects of geometry and topology; two were in algebra, both in fields of algebra not then represented at Berkeley; and one was in analysis.

Tsit-Yuen Lam was born February 6, 1942, in Hong Kong, and came to the United States for graduate study in mathematics in 1963. He received his doctoral degree at Columbia University in 1967 with the dissertation "On Grothendieck Groups" written under Hyman Bass. He spent year at the University of Chicago as instructor, then came to Berkeley as a lecturer the following year, and in 1969 he was appointed as assistant professor. His appointment further strengthened algebra, and since his research interests, which focused in algebraic K-theory, were in a field that had not been hitherto represented at Berkeley, his appointment added significantly to the coverage of algebra. His dissertation work on Grothendieck groups and K-theory for the group algebras of finite groups led to a series of papers jointly with Irving Reiner on this topic. Lam expanded his interests to include K-theory of fields, the study of quadratic forms over various kinds of fields, and a variety of topics in ring theory, quaternion algebras, and divisional algebras. His achievement won him rapid advancement to associate professor in 1972 and to full professor in 1976. His work was recognized by the award of a Steele Prize, and his extensive publication record includes eight solely authored books or monographs, plus some edited volumes and several more monographs in preparation. He has supervised the dissertation work of 17 graduate students so far. Lam has served as vice chair of the department on four different occasions, and served a term as deputy director of the Mathematical Sciences Research Institute (MSRI).

Jerrold Marsden was born August 17, 1942, in Ocean Falls, British Columbia, and received his doctoral degree at Princeton University working under Arthur Wightman with a dissertation entitled "Hamiltonian One-Parameter Groups and Generalized Hamiltonian Mechanics." He came to Berkeley as a lecturer in 1968 and was appointed as an assistant professor the following year. He began an extended program of research, applying methods of infinite-dimensional Lie groups, such as diffeomorphism groups, to various partial differential equations of mathematical physics to obtain existence and uniqueness theorems in various contexts. The results applied the equations of fluid flow, relativity, and elasticity. He won promotion to associate professor in 1972 and then to professor in 1976. He has published extensively, with over 300 items listed in *Mathematical Reviews*, and his work has been recognized by election to the American Academy of Arts and Sciences. In 1997, he took early retirement from Berkeley and accepted a faculty position at Caltech. Altogether, he has guided the doctoral work of 34 students, 24 at Berkeley, 1 at Cornell, and 9 at Caltech so far.

Alan Weinstein was born June 17, 1943, in New York City, and received his doctoral degree at UC Berkeley in 1967 working under S. S. Chern with a dissertation entitled "The Cut Locus and Conjugate Locus of a Riemannian Manifold." After a postdoctoral year at MIT and another at the University of Bonn, he was appointed as an assistant professor at Berkeley in 1969 as one of the few post-1940s Berkeley doctorates that have been hired into faculty positions at Berkeley. He was quickly advanced to associate professor in 1971 and to full professor in 1976. His research focus moved early on from Riemannian geometry to symplectic and Poisson geometry, a field in which he was one of the pioneers. His work won wide recognition and led to his election to the American Academy of Arts and Sciences. He has supervised the work of 25 doctoral students to date.

Blaine Lawson was born January 4, 1942, in Norristown, Pennsylvania, and received his doctoral degree in 1969 from Stanford University working under Robert Osserman with a dissertation entitled "Minimal Varieties in Constant Curvature Manifolds." He came to Berkeley as lecturer immediately after his degree and was appointed as an assistant professor the following year, in 1970. He was advanced very rapidly to associate professor in 1971 and to full professor in 1974. Lawson's research has ranged widely over many aspects of differential geometry, and has won him wide recognition and election to the National Academy of Sciences. In 1978 he took leave from Berkeley to assume a faculty position at SUNY Stony Brook, but he did not finally resign from his Berkeley position until 1982. He has supervised the work of 30 doctoral students overall, 6 at Berkeley and 24 at Stony Brook.

Robert Bowen (known to all as Rufus) was born February 23, 1947, in Vallejo, California. He entered Berkeley as an undergraduate in 1964 and graduated after three years in 1967. At graduation, he was selected as university medalist, an award given to the top graduating senior. He remained on for graduate study at Berkeley and came under the influence of Steve Smale, with whom he worked on dynamical systems. Although the mathematical results that became his dissertation were presented at the 1968 AMS Summer Institute on Global Analysis, his doctoral degree was not awarded until 1970. In this work he showed that, for an Axiom A diffeomorphism, the topological entropy is equal to the radius of convergence of its zeta function. He spent the academic year 1969–1970 at the University of Warwick working with the group there in dynamical systems. This remarkable young mathematician was hired immediately as an assistant professor in 1970, promoted to associate professor in 1973 and to full professor in 1977. Then, in the summer of 1978, Rufus Bowen died suddenly and unexpectedly of a massive stroke, an event of enormous sadness for his family, for Berkeley, and for mathematics. He had amassed a prodigious publication record of some 50 papers of exceptional quality and had already supervised the work of four doctoral students. Had his life not been tragically cut short, his achievements would have been even more spectacular. The department established a yearly series of lectures, the Bowen lectures, in his memory. The list of Bowen lecturers, beginning in 1980, is an impressive one indeed.

David Goldschmidt was born May 21, 1942, in New York City, and received his doctoral degree in 1969 at the University of Chicago under John Thompson with a dissertation entitled "On the 2-Exponent of a Finite Group." After two years of postdoctoral work at Yale University under Walter Feit, he was recruited to Berkeley as an assistant professor in 1971. Finite group theory had begun a renaissance in the 1960s, and Goldschmidt was the first representative of this reinvigorated field to come to Berkeley. His work on finite group theory won him promotion to associate professor in 1973 and to full professor in 1979. His interests subsequently shifted to include aspects of graph theory, and in 1991 he resigned from his faculty position at Berkeley to accept the directorship of the Center for Communication Research, in Princeton, New Jersey. While at Berkeley he supervised the work of six doctoral students.

Paul Chernoff was born June 21, 1942, in Philadelphia, Pennsylvania, and received his doctoral degree at Harvard University in 1968 working under George

Mackey, with a dissertation entitled "Semi-Group Product Formulas and Addition of Unbounded Operators." He came to Berkeley for three years of postdoctoral work combining an NSF postdoctoral fellowship and a lecturer position. In 1971 he was appointed as an assistant professor, and was advanced to associate professor in 1974 and to full professor in 1980. The focus of his research has been on properties of various classes of unbounded operators and the semigroups that they generate. Overall, he has supervised the work of four doctoral students. He retired from active duty in 2005.

We mention briefly at this point two unaccepted offers of assistant professorships: Karen Uhlenbeck and Michele Vergne. Uhlenbeck received her doctoral degree in 1968 at Brandeis University under Richard Palais with a dissertation entitled "The Calculus of Variations and Global Analysis," and then came to Berkeley as a lecturer in 1969 with a two-year appointment. In 1971 she was offered an assistant professorship at Berkeley but declined the offer. Michele Vergne received her doctorat d'État at the University of Paris in 1971 under Jacques Dixmier with a dissertation entitled *"Recherches sur les Groupes et les Algèbres de Lie."* She came to Berkeley in 1971 as lecturer and was then offered an assistant professorship in 1972. Although she accepted this offer, she was uncertain when she could take up duties because she had to return to France for the following year. Subsequently, she resigned without ever taking up duties. These were the first two women offered professorial positions at Berkeley in many years, and unfortunately these recruitment efforts were unsuccessful. As is evident, both Uhlenbeck, who now holds the Sid W. Richardson Foundation Regents Chair at the University of Texas, Austin, and Vergne, who is on the faculty at the Ecole Polytechnique, went on to very distinguished careers as mathematicians.

After the intense period of hiring ending in 1972, the department made no appointments in 1973, although an offer of an assistant professorship was made to William Thurston. Thurston, who was a Berkeley PhD in 1972 under Moe Hirsch, declined the offer, but that was not the end of the matter regarding the department's effort to attract Thurston back to Berkeley.

However, there were other additions to the faculty in 1973. First, as previously noted, the reorganization of computer science in the spring of 1973 as a division in the Department of Electrical Engineering and Computer Science (EECS) led to the request of Beresford Parlett to split his appointment between EECS and mathematics, with 40% in EECS and 60% in mathematics. Parlett had been appointed and promoted to tenure in mathematics and had then shifted his appointment to the computer science department in the College of Letters and Sciences when it was formed in 1967. Vel Kahan, who had been hired in that computer science department in 1969, requested that his appointment also be split between EECS and mathematics, 50/50 in his case. These additions to the mathematics department further strengthened and solidified numerical analysis and scientific computing as an area in which the department would prosper.

William (Vel or Velvel) Kahan was born June 3, 1933, in Toronto, Canada, and received his doctoral degree at the University of Toronto in 1958 working under Byron Griffiths with a dissertation entitled "Gauss–Seidel Methods of Solving Large Systems of Linear Equations." He then joined the faculty at the University of Toronto, rising to the rank of full professor, and was recruited in 1969 as full professor to the fledgling Department of Computer Science at Berkeley. His research focused on questions of numerical linear algebra, although he goes far beyond these confines and has addressed issues of standards for floating-point computation, reliability, and error analysis. His work is well known and has been honored by the Turing Prize, the Emmanuel Piore award, and election to the American Academy of Arts and Sciences and the National Academy of Engineering. He has supervised the doctoral work of five students, including his future colleague James Demmel, and was awarded the Distinguished Mentoring Award by the campus.

In an effort of reaching out to a broader community of mathematical scientists on campus, the department also added three members of the statistics department as 0% professors effective July 1, 1973. These included David Blackwell and Lucien LeCam, who had been in the mathematics department before the 1955 split and who have been discussed in earlier chapters. The third faculty member added was David Freedman. Freedman, who is a probabilist, had come to Berkeley in the statistics department in 1961 after completing his degree in mathematics with Willi Feller at Princeton.

As noted at the beginning of this chapter, the department underwent extraordinary growth during the first part of this period (through 1968) and then stabilized and absorbed this growth during the second part of the period. Change since 1973 has been evolutionary and has built on what had been accomplished during these earlier periods. Although the department has matured in many ways since 1973, it does not look fundamentally different today in terms of size and overall complexion from what it looked like in 1973. Of course, during the period covered in this chapter, there were many social and political upheavals going on in he country, for which Berkeley was a focal point. The following chapter is devoted to brief description of some of these events and their effect on the department.

Chapter 17

Departmental Life in the "Sixties": 1964–1973

The general period of the 1960s, during which the mathematics department grew rapidly and the other events described in the last three chapters took place, was a time of political change and social turmoil, with Berkeley serving as a notable epicenter. This chapter will try to describe the impact of this turmoil on the department and how it responded socially and organizationally. The "sixties" as a political and social phenomenon can be said to have begun in 1964 with the free-speech movement at Berkeley. It is a matter of some debate when the "sixties" ended. The student protests ended effectively in 1970, but the social and political tensions that generated these protests remained: concerns about civil rights, the Vietnam War, opposition to and resentment of the so-called military–industrial complex; the university's relationship to it lingered, and the entire counterculture movement remained active after 1970. The narrative of the previous chapter ended in 1973, at which time the academic and structural changes in the department had stabilized. Although 1973 would be a convenient time to end most of the narrative of this chapter, since many of the political and social tensions had wound down toward more normal levels by this time, it is natural to continue the narrative a few years beyond 1973 on certain points. One notable exception to the winding down of ferment was the issue of representation of women and minorities in the student body and on the faculty. This was a continuing issue that was to grow in importance; indeed, it has remained a lively and important issue to this day and will also be discussed in subsequent chapters.

The topic of this chapter is how the internal structure and governance of the mathematics department evolved in response to outside forces and other factors during this period. By 1964, the transition of the department from a good department to one of the very top departments in the country had been completed, and this excellence was recognized nationally. This transformation began with the very successful hiring in the 1958–1960 period that we have already described, and in the following years, the department continued to recruit well and was able to fight off most of the raids from the outside. The administration was becoming cognizant of the need to offer competitive salaries in the then highly competitive job market. The losses of Steve Smale in 1961 to Columbia and Bert Kostant in 1962 to MIT were serious blows, but the department's ability to lure Smale back in 1964 was a boost to morale. There were other losses later in the decade that were also damaging which we discussed in the previous chapter.

In 1964, the American Council on Education (ACE) conducted an evaluation of leading American universities as judged by the scholarly reputation of the faculties and the effectiveness of their PhD programs, discipline by discipline. The evaluation concluded that Berkeley was "the best balanced distinguished university in the country." Clark Kerr took this as an indication of the full recovery of Berkeley from the oath controversy in the early 1950s and a refutation of a view by some at the time that Berkeley would be permanently damaged and drop out of the top tier of universities [Kerr 2001, p. 57]. In this 1964 evaluation, the Berkeley mathematics department was ranked a close second behind Harvard and just ahead of Princeton in quality of graduate faculty. This was the first time that the mathematics faculty at Berkeley had been ranked at this high a level, and it was formal recognition of the status of the department as one the very best in the country. In fact, the rankings by junior scholars gave Berkeley a first-place tie with Princeton ahead of Harvard. However, in the rankings by rated effectiveness of graduate programs, Berkeley was third, well behind Harvard and quite far behind Princeton, and just barely ahead of Stanford. This pattern of slightly lower ranking for the effectiveness of the graduate program than for faculty quality was to continue in future rankings, and posed a continuing challenge to the department. A further sign of distinction was the award to Steve Smale of a Fields Medal in 1966. Over time, the number of faculty who received prizes and awards for their research grew, as did the number of faculty who were elected to the National Academy of Sciences and the American Academy of Arts and Sciences.

The department had an unusual age distribution because, from 1955 until 1972, there was only one scheduled retirement: Tarski in 1969. In 1955, Tarski was the only faculty member in the department over 50, and in 1960 he was the only faculty member over 55. So at this time, the "senior" faculty in the department were comparatively young. Of course they grew older, and by 1970 the age distribution of the tenured faculty was more evenly distributed. Of the 44 headcount faculty in 1960, seven were assistant professors. This ratio changed substantially during the decade with the addition of many assistant professors, and by 1968, the faculty headcount had risen to 79, with 28 assistant professors. The department had a very different feel to it with this growth and the infusion of nontenured faculty coupled with the dispersion of the department physically into five different buildings on campus. By 1973, the department had declined to approximately 70 headcount faculty and 2 assistant professors. The growth in size and complexity of the department in turn made the job of departmental chair much more difficult, demanding, and time-consuming, and as we shall see, these circumstances led to a leadership crisis in 1967, whose solution required restructuring of departmental administration and how department chairs were selected. This problem was not fully resolved until 1973.

In 1955, with a faculty of 20, the entire department with spouses could gather for a departmental social event in a member's house. By 1960 this was no longer possible, and further growth led to fractionation of the department both by field of specialization and by age group, and also, as the decade wore on, by political orientation as well. Through the activities of some of its members, notably Steve Smale, Moe Hirsch, and John Kelley, the department achieved a reputation as a locus of political activity. The department has generally been without the internal feuds or longstanding disputes between individuals or groups that have troubled other departments. There have been sharp disagreements on specific issues, but in general they have not been lasting. Perhaps the sheer size of the department reduced the potential for such quarrels, and, in any case, the absence of such conflicts and an underlying mutual respect among the faculty has always been one of the strengths of the department.

As noted, the signal event that marked the beginning of the "sixties" as we understand the term was the free-speech movement. By way of background, the university had for decades, as a matter of policy, sealed itself off from the partisan and political activities and life of the state. This was virtually in the form of a contract or at lest a tacit agreement between the state government and the university. As Neil Smelser puts it in his foreword to Clark Kerr's memoir,

> A second consideration arises from the fact that the state in mandating freedom from partisan and political influences, is in fact enunciating a two-way contractual understanding between the state and the university, though the terms of the contract usually remain implicit. That contract is, in essence, that in exchange for their freedom (including academic freedom), the various constituencies within the university—administration, faculty, staff, and students—will reciprocate by holding to standards of civility, dispassion, and political neutrality in carrying out the university's mission. These constituencies, however—particularly students and faculty—are in reality not apolitical. From time to time, they engage in partisan activities, sometimes launching them from university campuses. Most often they do so in the belief that they are (and ought to be) acting with impunity, under the cloak of academic freedom. From the outside, however, their activity is regarded though not always articulated as a breach of the contract, and it invites political intervention. [Kerr 2003, p. xvii]

Kerr himself wrote on this same theme,

> As a graduate student at Berkeley beginning in 1933, I somehow got the impression of an evolving understanding that the university would stay out of politics and in turn that politicians would stay out of the university, and that the university took this understanding very seriously. The constitution of the state so implied; and President

Sproul, with his long experience in Sacramento, knew, I assumed how destructive political involvement would be in terms of public opinion and support. It was better to erect a wall between politics and the university. [Kerr 2003, pp. 125–126]

One result of this policy choice was a ban on political speakers on the campuses of the university. Kerr also writes of his efforts as president beginning in 1958 to relax these restrictions somewhat, for instance by liberalizing the rules as to what outside speakers could be invited to speak on campus, although Communist speakers were still excluded. The ban on Communist speakers was finally lifted in 1963, to the great annoyance of many political conservatives. One could in fact interpret the oath controversy, which was described in Chapter 9, as an example of the political consequences of perceived breaches in this wall between the university and politics. The issue was to recur in 1964.

But the issue in 1964 concerned students (and faculty) individually or in groups launching advocacy campaigns from the campus. This kind of activity conflicted with existing university policies, and the university's attempts to limit such activity was cast by the students and their supporters as a free-speech issue, hence the emergence of the free-speech movement. As a result of confusion and misinformation about the ownership of a small piece of land adjoining the campus that had been a venue for student activism, and missteps by the campus administration, conflicts between campus authorities and students in September grew into a full-fledged confrontation by December, with the occupation of Sproul Hall by protestors, followed by hundreds of arrests by campus police supplemented by police from neighboring communities and Alameda County sheriff's deputies. Chancellor Edward Strong, who had been in office since 1961, resigned under pressure at the end of December, and after an interregnum, Roger Heyns, from the University of Michigan, was appointed chancellor starting in the fall of 1965.

It is not the author's intent to write a history of the free-speech movement or the New Left/counterculture political movement, since that has been done elsewhere, nor to write in detail about subsequent campus protest movements, but some brief words are in order to set the context. The issues raised by the free-speech movement won wide support among the faculty, and matters were settled in the spring of 1965 by negotiations giving the students most of what they wanted. The new rules that were promulgated granted substantially more freedom for student advocacy. These new rules resulted in an increased level of political activity and agitation on campus and changed the nature of campus life in many subtle ways. Moreover, the protest demonstrations in the fall of 1964 set a pattern for the future, in which there would be nearly constant campus demonstrations and police actions in response. The twin issues of the Vietnam War and civil rights drove the agendas for activist groups of students and faculty. One early activity in which Steve Smale was involved, along with many others, was a protest and demonstration against the passage of troop trains through Berkeley. It was described in the press as trying actually to stop the trains. Antiwar activists staged antiwar marches and other events that gained public attention and that were linked with the campus.

Public reaction against the events at Berkeley, of the kind that Sproul might have predicted, set in. Among other subtle signs, undergraduate applications and enrollment declined. This was before the days when Berkeley had more applicants than it could accommodate, and all applicants who met minimum UC eligibility requirements were admitted. Consequently, the shortfall in enrollment could not be made up by increasing the percentage of applicants who were admitted. In the minds of many parents and college-age young men and women, Berkeley became known as "Berzerkeley," and students stayed away. However, for other students of a different mindset, Berkeley was perhaps a magnet. Then, Ronald Reagan came on the scene in 1966 and staged a run for governor, using as one of his main campaign themes, "I will clean up the mess at Berkeley." This very overt message resonated with the voters, and he swamped his moderate Republican opponent in the June primary and then went on to defeat the incumbent Pat Brown in the general election in November by a million votes. The argument that in a premier university, such as Berkeley, Harvard, or any of a number of others, there will always be intellectual and social ferment, as well as likely dissent from established norms, as part of the intellectual environment, was not being effectively made. To the extent it was made, it had little impact on public opinion.

In his memoirs, Clark Kerr describes some events at the November 16–17, 1966, regents' meeting, a meeting that fell immediately after Reagan's election and prior to the momentous January 1967 meeting after Reagan had been inaugurated. One of these related directly to the Berkeley mathematics department and says something about the nature of a premier university:

> The regents meeting brought two important developments. One revolved around the case of Stephen Smale, a Berkeley faculty member. Smale had just won the Field prize in mathematics, the equivalent of the Nobel Prize. The Berkeley campus had quite properly nominated him for an over-scale salary. But Smale had also been in the headlines for trying to stop a troop train on its way through Berkeley. I discussed his salary with the chancellors, most of whom thought I should take the item off the agenda as being a very divisive issue at that moment. I decided not to do so and said that if I were not willing to support the Smale nomination, I did not deserve to be president. In the end, the board did support my nomination of Smale by a vote of 9 to 6 with 3 abstentions; too close for comfort. On the merits of my recommendations, I still had the support of the board, even when they were highly controversial. I had been working with the board for fourteen years as chancellor and president, and it knew my record well. [Kerr 2003, p. 295]

A minority group of regents had for some time been pressing for Kerr's dismissal, but the election of Reagan shifted the dynamics inside the regents so that this group was now in the majority, and at the next regularly-scheduled regents meeting in January 1967, Kerr was fired. In between the two meetings there had been an emergency meeting called by regent Pauley over incidents on the Berkeley campus concerning a navy recruiting table and student demonstrations, followed by police action and arrests, and then a mass meeting of students calling for a strike.

In the fall of 1966, the university had submitted its 1967–1968 budget request to the state, proposing an increase of about 16% in order to meet increasing enrollment and other needs. One of Reagan's first acts as governor was to deny this increase and propose instead a 10% decrease in state funding. Ultimately, the governor backed off on the 10% decrease, and he also relented on his campaign promise to appoint a special commission to be headed by former CIA director John McCone to "investigate the charges of communism and blatant sexual misbehavior on the Berkeley campus" [Kerr 2003, p. 288]. The university did not do well with its budget requests during the years when Reagan was governor, and it is not clear that the governor believed that it was part of the state's mission or responsibility to support a public preeminent research university that could compete with and surpass the best private universities.

Protest demonstrations and police responses followed with great regularity for several years, peaking in intensity in 1969 and 1970. In May 1969, conflict arose over a 2.8-acre parcel of land that the university had purchased for student housing, but which some campus and community advocates felt should be a park instead: "People's Park." The campus had purchased the land over several years, evicting the residents and demolishing the existing buildings, but construction of the student housing had been delayed for a variety of reasons, including the drop in enrollment. This sequence of events set the stage for conflict. In a street battle for "possession" of the land between police and demonstrators, one person was killed and another blinded by gunfire. The National Guard, armed with rifles and fixed bayonets, was called in by the governor, and these military units patrolled the campus and surrounding areas for several weeks A few days after this, on a Tuesday afternoon, a helicopter sprayed CS gas on a crowd in Sproul plaza, and the gas quickly spread over campus, creating a major disruption. There was significant disruption of the academic routine.

In May 1970, when the United States invaded Cambodia, campuses across the country erupted in protest, and at Kent State University, Army National Guard troops opened fire on unarmed students, killing several and wounding many more. The outcry at Berkeley was sustained and passionate, and there were calls that classes be "reconstituted," which could mean that normal academic content was dropped and antiwar activities substituted; or alternatively, that the curriculum or class meetings and assignments was modified, or in some cases that the course was discontinued. What exactly happened in each of the thousands of courses that spring was never clear, but anecdotal information was sufficient to suggest that all of the above did occur in some form and that the disruption of normal instructional activities was substantial. As in 1969, the publicity was intense, and the regents evidently soon lost confidence in the leadership of the campus. Berkeley's reputation as "Berzerkeley" in the public mind was secure and would last for several years. Chancellor Heyns suffered a heart attack during that summer, and although he fully recovered from it, he submitted his resignation as the new school year began, to be effective July 1, 1971.

These two years were the peak of the "sixties" at Berkeley, and they were stressful years indeed. In the first year, the physical environment deteriorated, while in the second, the instructional environment was compromised. The challenge was to recover from these events, and it would be some time before the recovery was anywhere near complete. These disruptive events evidently had no impact on faculty recruitment in mathematics, for as we have recounted in the previous chapter, recruitment during these two years and in the immediately succeeding years was very successful. There was no notable disruption in the research activity of the faculty, and it is at least possible that the "political" events provided the promise of a certain spark and spice in life at Berkeley that may have been attractive to many young faculty.

The new chancellor who assumed duties on July 1, 1971, was Albert Bowker. He was a shrewd and effective leader of the campus for the next nine years, working hard to restore the credibility of the campus with its many constituencies in the state. Bowker was also a member of the mathematical sciences community. He had graduated from MIT in 1941 with a major in mathematics, and during and after the war he moved into statistical work. He pursued graduate studies in statistics, receiving his doctorate in statistics from Columbia University in 1949. While still a graduate student, he came to Stanford in 1948 as a faculty member in mathematics, but he quickly moved over to the newly formed statistics department as its founding chair and assistant professor, while still technically a student. He presided over the development and growth of this department to one of the two leading statistics departments in the country. Following his work in department building, Bowker gravitated toward university administration, first as graduate dean at Stanford and then as chancellor of the City University of New York, before coming to Berkeley as chancellor. Toward the end of his term as Berkeley chancellor, he was also to play a key role in the establishment of the Mathematical Sciences Research Institute (MSRI), as described in Chapter 19.

The only campus-level event to have an impact on the mathematics department subsequently was a strike of the buildings-trades unions against the campus in the spring of 1972. Picket lines were thrown up around campus, and union members respected these picket lines and staged a sympathy strike. Among those not working were the custodial staff, so that the campus and Evans Hall in particular were without custodial service to clean the restrooms, empty trash, and hence Evans and other campus buildings became a bit grubby over the two-month period. Some non-academic staff walked out in sympathy, and some faculty took their classes off campus so as to not cross picket lines or stopped teaching altogether, all of which brought back memories of the spring of 1970. The striking workers failed to attract public sympathy, and when the campus held firm, the strike collapsed. This strike also had a more indirect effect on the department. There had been a tradition in the department whereby many faculty members would lunch together at large tables in the campus dining commons. Faculty would simply walk down and be assured that there would be a table of colleagues one could join. The dining commons was shut down during the strike because the truckers who delivered food and supplies would not cross the picket line. The faculty scattered to many small restaurants off campus for lunch, and the tradition of lunches in the dining common never revived. Not too long after the strike, the campus dining commons went out of business for lack of clientele.

In the fall of 1971, just after the department moved into Evans Hall, Steve Smale and others created an informal biweekly department newsletter to serve as a vehicle for communication among faculty, graduate students, and staff. Articles were to be solicited from all members of these groups. The newsletter had a degree of countercultural irreverence, most vividly evident in its name: the *Mother Functor*. It lasted for only nine issues, drawing many articles from faculty, students, and staff members. It was discontinued somewhat before the building-trades strike began in the spring of 1972, but it remains in the minds of all who were members of the department at the time. A complete set of all nine issues resides in the University Archives, and one of the illustrations in this book shows a sample article from the newsletter. The term "Mother Functor" had actually not originated at Berkeley, but had been in use at the 1969 Summer Institute in Category Theory at Bowdoin College, and had appeared in the mathematics literature in 1970 in a brief humorous article reporting on the Bowdoin institute by the pseudonymous author "Phreilambud" [RMCS, IV]. (The author is indebted to George Bergman for this reference.) For the non-mathematician reader, a functor is a technical term for a particular variety of function that arises in a field of mathematics called category theory. Common use of the term goes back to at least 1942, and in fact the *Oxford English Dictionary* traces it use back to the 1930s in logic. One could envision a kind of universal functor from which all other functors might be derived, a "mother of all functors," so to speak, and hence the term.

A theme running throughout this period was the allegation by various groups that the university was too much a part of the "military–industrial" complex, and this feeling resulted in the department passing a motion by a nearly unanimous vote in the spring of 1971 that it was the sense of the department not to accept Department of Defense contract and grant funds. However, there is no evidence that this motion resulted in anyone in the department doing anything differently as a result. At the same time, there was a feeling among a number of groups of students and staff—and perhaps some faculty—that the university was an "oppressor." What this meant exactly is not clear, although this kind of voice emerges from the *Mother Functor* newsletter, and there was a general feeling among some that the university was "exploiting" or mistreating its employees, specifically nonacademic staff and graduate students in their role as teaching assistants. The feeling of exploitation led many to support unionization of staff employees and of the graduate-student teaching assistants. Although the campus had collective bargaining agreements with small groups of employees, notably in the building trades, large-scale unionization had to await a change in state law.

In 1978, the California legislature passed and Governor Jerry Brown signed the Higher Education Employer–Employee Relation Act (HEERA), which authorized collective bargaining for university employees for the first time and created a Public Employees Relations Board (PERB) to oversee and implement the act (sort of a small version of the National Labor Relations Board). When unions representing staff won collective bargaining rights in representation elections held under HEERA, department chairs became "management" when dealing with employees in the department, a position that was uncomfortable to some faculty members, especially if they had grown up in families that were philosophically committed to organized labor. This added a disincentive for many faculty members to serve as department chair. The university participated in the drafting of HEERA and lobbied for provisions that it thought would preclude teaching assistants, or as they were later called, graduate student instructors (GSIs), from qualifying for unionization. What resulted was a twenty-five-year struggle through PERB and the courts on this issue. In 2000, under political pressure from the legislature, the university caved in and dropped its opposition to unionization. The United Auto Workers (UAW) won a representation election by an overwhelming majority. Mathematics graduate students were generally very strong supporters of the union and held leadership positions, including the presidency, of the UAW local on the campus.

The ongoing concern over civil rights throughout this period led to the formation in the mid 1960s of the Mathematics Opportunity Committee (MOC) in the department, whose charge was to attract and help support women, underrepresented minorities, and students from disadvantaged backgrounds to apply, enroll, and succeed in graduate school at Berkeley. Leon Henkin, among other faculty members, took a lead role in establishing the MOC, and was also influential in campus-level activities along these same lines. This committee was allowed to use a certain percentage of the graduate admissions target and graduate student support to advance its agenda. The guidelines for its operation had to be fine tuned to make sure they complied with applicable federal and state laws as they evolved over time.

The growth in the size and complexity of the department as well as the turmoil on campus transformed the nature of the duties of department chair, and a number of difficulties resulted. As noted, Bernard Friedman took over as chair in January 1960 and served for two and a half years until 1962. Murray Protter then assumed the duties and served a three-year term until 1965. The established method of selecting department chairs when a new one was needed was for the dean to ask each ladder-rank faculty member to write a confidential letter to the dean discussing from that faculty member's perspective the state of the department—the problems, challenges, strengths, and weaknesses—and then to suggest some names of colleagues who would be good choices for department chair. The dean communed with these responses, which were of varying degrees of helpfulness, and then selected the "victim" and had the task to convince that person to take the job.

In the spring of 1965, Henry Helson was the person who was selected by the dean to serve as the next chair, which he did during the 1965–1966 academic year. Helson, however, had a sabbatical leave for 1966–1967, which he had planned to take in Europe. Since he was unwilling to forgo these plans, Leon Henkin agreed to serve for the 1966–1967 academic year as acting chair during Helson's sabbatical leave. By this time, the department had grown to over 70 faculty members, and the chair's job had become a full-time job, especially with all the turmoil on campus and its spillover into the department. Matters were also complicated by the ongoing planning for expansion of the department described earlier; the conflict over the organization of computer science, already described in Chapter 14; and the complex physical and space planning issues for Evans Hall. Both Helson and Henkin felt strongly that the nonacademic staff needed to be reorganized with the addition of a high-level staff member serving as assistant to the chair in order to relieve some of the burden of the job. Both Helson and Henkin articulated this point to the dean, but without effect. Helson wrote to the dean in early 1967 that he would not resume duties as chair upon his return in the summer unless such a reorganization was implemented. Henkin also made clear his unwillingness to continue beyond his one-year term absent this reorganization.

It is well to recall that, prior to 1936, the department had no staff support at all. In 1936, the administration authorized the department to hire a half-time secretary/stenographer to assist Griffith Evans with his correspondence as chair. The department hired Sarah Hallam into this position, who was also a graduate student in the department. Over time, she received a master's degree, and her position in the department evolved into a full-time staff position. As the department grew, additional staff members were added to keep pace with the growth and to handle the increased complexity of the work of the department. Not all of the staff positions were state-funded, because federal grants at the time were able to support some staff positions. By 1967, the staff had grown to about 18. Sarah Hallam was in charge, and there was a pool of secretaries who provided clerical and technical typing support to the faculty. Each faculty member was assigned to one member of this pool. The chair's office required two secretaries to handle the sizable correspondence as well as the paperwork for the large number of new appointments and to prepare faculty dossiers for campus review. There were staff who managed the paperwork for the budget of the department as well as the paperwork for the many faculty government contracts and grants, and finally there were some student service staff, and the group in logic and methodology of science had a secretary. As a result of her long career at Berkeley, beginning in the 1930s, Sarah Hallam was very friendly with the senior faculty, and she worked closely with the department chair. In fact, many junior faculty and graduate students believed that Miss Hallam, as she was called, was the one who really "ran" the department. She was a formidable personality, and although a number of staff members were unhappy serving under her, she was effective in her role in administration of the department.

The reorganization that was sought would more formally recognize Sarah Hallam's role in working closely with the chair and sharing some of the workload by changing her title from administrative assistant to assistant to the chair, and would allow her to shed some of the more routine duties with regard to overseeing the staff. The dean's office was dragging its feet, using the so-called "one over n" argument that "we cannot do this for you unless we do it for everyone else." There were other demands presented by Helson and Henkin, including recognition for purposes of merit advancement that the chair's position in a department of this size was a full-time job and resource requests for allocation of a substantial number of tenured positions for recruitment and allocation of additional teaching assistants, which the dean had agreed to. Finally, a new organized research unit, the Center for Pure and Applied Mathematics (CPAM), had been approved but not funded. This unit in the short term would provide resources for grant administration and relieve the chair of some of the administrative burden and in the long term could grow into an organization to promote applied mathematics within the department and related departments. However, it could do neither without funding.

So, as June 1967 came to a close, the department was without a chair for the coming year. In addition, Dean Fretter had completed his term, and as noted in an earlier chapter, on July 1 there was a new dean, Walter Knight. Dean Knight finally persuaded Helson to reconsider his refusal to assume the job, but within weeks, when the dean's office continued its foot-dragging on the department's requests, Helson resigned in protest. At this point, the department was without a chair, and no one was willing to take the position. Emery Thomas, who had signed on to be vice chair, continued on in that position and valiantly kept essential business going. The dean continued to meet with faculty groups, but could not resolve the basic issues dividing the faculty from the dean's office. It was of some embarrassment to him that he was so at odds with one of his largest and most visible departments that no one would agree to serve as chair. Finally, in December, the dean and Leon Henkin entered into discussions that bore fruit. The position of assistant to the chair was created and Sarah Hallam was appointed. Core funding for CPAM as requested was finally provided by the dean, and the other issues around resources were worked out. Henkin agreed to serve as acting chair for the remaining half year only, and to try to help work out better arrangements for the succession.

When the dean polled the department in the spring of 1968 in the usual manner for selection of a new chair, none of the candidates that emerged from the polling was willing to serve, and matters were once more at an impasse. Henkin conceived the idea of exploring whether some members of a younger generation of faculty might be willing to serve. He approached John Addison, whose promotion to full professor was in progress, Joe Wolf, and Calvin Moore, both of whom had been promoted to full professor less than two years earlier. Only Addison was willing to consider the possibility of taking on this job, and after some discussions and negotiations, he was appointed chair. Wolf and Moore agreed to serve as vice chairs, so leadership of the department had indeed passed to a younger generation. Addison served as chair with distinction for four years.

When time came to find a new chair in 1972, the College of Letters and Sciences had undergone a reorganization in which the dean was no longer directly responsible for the over 40 different departments and their chairs. Instead, four divisional deans (for humanities, social sciences, biological sciences, and physical sciences) had been appointed who had direct responsibility for the departments in their division and who worked under the overall direction of the dean of the college. Calvin Moore had been appointed in 1971 as the Dean of Physical Sciences, and so it fell to him to select a replacement for Addison. The new divisional dean learned firsthand how generally useless and scattershot were the faculty responses to the usual polling process. Dean Moore approached over ten faculty members in the department to be the new chair, and all of them refused, thus bringing on a possible repeat of the previous problems. Finally, Emery Thomas was convinced by Moore to reconsider his refusal and to take on the position for a year with a specific charge to find a permanent structural solution to the recurring problem of finding a department chair.

Thomas was as good as his word, and the department under his leadership designed a new method of selecting a candidate for the department chair. What emerged was an election system inside the department with a complex set of rules, which has worked well and withstood the test of time for over 30 years. Although the inclusion of graduate students and staff in the electoral system has a certain "sixties" flavor to it, the students and staff have very much appreciated their inclusion in the selection process.

Briefly, the election procedure was as follows: When a chair was due to step down, a six or seven-person election committee came into being, consisting of the three immediate past chairs (exclusive of the sitting chair, who generally recused him- or herself from the entire process), plus one or two representatives of the junior faculty, one graduate student, and one nonacademic staff member. This committee canvassed the department regarding potential candidates, and came up with list of names of candidates who had substantial support, and for whom it might be regarded as being their "turn" to serve as chair. The committee then tried to persuade the candidates they had identified to "stand" for election, with the idea that the number of candidates should be large. Those that agreed then formed the slate of candidates, and soon a tradition developed of department meetings open to all faculty, students, and staff, where the candidates offered brief statements and then responded to questions. The election took place by secret ballot with a weighting under which the total votes cast by faculty counted for two-thirds of the total, the votes cast by graduate students counted for one-sixth, and the votes cast by the nonacademic staff counted for one-sixth. The winner was announced to the department (but without the vote tally), and the name was sent to the dean. Some years later, the rules were modified so that the top two candidates (in order) were announced and the two names sent to the dean.

The dean maintained the option, before making a decision, of invoking the traditional method of polling the faculty for their comments on the state of the department and their recommended choice(s) for chair. The comments were always useful, and although the unanimous or near-unanimous response to the poll was to designate the winner of the election, this process would alert the dean if there was any substantial undercurrent of opposition among the faculty to the winner of the election.

The department gained in this process by having a far more definitive and organized say in who the next chair would be. The candidates benefited, since it is far easier for someone who may be reluctant to agree to serve as chair to agree to stand for election than to actually agree to do the job, especially if the slate of candidates is a large one. Thus, candidates could hope that they would lose and so be "off the hook." The winner came out of the process with a real vote of confidence from colleagues, and usually a decreased level of apprehension about the job. The dean got a person willing to serve who had proven support from the department. On the other hand, the dean's discretion in picking someone he or she wanted was seriously limited. The department and the winning candidate, however, lost on one score, and that was the ability to negotiate for additional resources, a strategy that had been very successful in 1967. But, in general, the dean's cupboard has been pretty bare in terms what could be offered, and so not much was lost. Although the departmental election specified a nominal term of three years, the dean has always had the formal authority to appoint department chairs, and appointments are in fact for one year, renewable. A chair who is willing to serve substantially more than three years stands for reelection, in which the full election process is repeated. The author is not aware of any other department on campus that has an election system of any kind.

The first election took place in the spring of 1973. All the candidates except Max Rosenlicht announced that they would serve only one year. Although Rosenlicht initially said he would consider serving two years, he subsequently said that he would serve only a year. Max Rosenlicht was elected and appointed chair starting in the summer of 1973. His relationship with staff and graduate students during the 1973–1974 academic year was a bit bumpy. Toward the end of his first year, an informal poll by the dean indicated that in spite of reservations by staff, students, and some faculty, there was sufficient support to justify his continuation for a second year. However, at the end of the second year he stepped down and a new election was scheduled. His two-year term was somewhat contentious not only because of his relations with staff and students, but also as a result of the controversy over appointment decisions during his term of office that will be described in the following chapter.

The refusal of candidates to consider more than a one-year term and then the resignation of the first elected chair after two years of service caused some concern over the chair election system, particularly if it were to mean elections every year or perhaps every two years. The election in the spring of 1975 resulted in the election of John Kelley, who had served previously as chair with distinction in 1957–1960. He agreed at the start to serve out a three-year term, a change from the first election. His election had a soothing effect on the department, and a decade of contention largely came to an end, along with the general relaxation of tensions in academia and society.

The chair election system has worked well over the years and is still in place. Subsequent chairs were Shoshichi Kobayashi (1978–1981); Moe Hirsch (1981–1983); Leon Henkin (1983–1985), following his prior service as acting chair; John

Addison (1985–1989), serving his second term following service in 1968–1972; Alberto Grünbaum (1989–1992); John Wagoner (January 1993–1996), with Kobayashi serving for six months in the fall of 1992 because of Wagoner's planned sabbatical leave; Calvin Moore (1996–2002), who was reelected for a second three-year term in 1999; a team consisting of Hugh Woodin (2002–2003) and Ted Slaman (2003–2006), an arrangement necessitated by Slaman's planned leave abroad in 2002–2003; and finally Alan Weinstein (2006–).

Chapter 18

Recovery: 1974–1985

The period from 1974 to 1985 is the last period in the evolution and development of the department that we will examine in detail. For the campus as a whole, it was a period of recovery from the "sixties" accompanied by efforts and attention on rebuilding damaged relationships with the people of the state. However, when the new governor, Jerry Brown, known for his phrase "smaller is better," assumed office in 1975, the university's budget problems continued for another eight years. In the early 1980s, the nation's severe economic problems, which were reflected in the California economy, created additional serious fiscal problems for the university and the campus. Faculty salaries lagged far behind competitive levels, and state support for other critical resources for the university were constrained. It was not until a new governor, George Deukmejian, and a new president of the university, David Gardner, took office that the university's budget situation was dramatically reversed in the 1984–1985 budget cycle. At that point, the university had regained its financial health and momentum from which it had been partially derailed nearly twenty years earlier.

The mathematics department was affected directly and indirectly in many ways by these cycles of state and public support and neglect. There were 22 successful faculty recruitments into ladder-rank positions during this twelve-year period, but because of retirements, resignations, and deaths, there was very little net growth. Nine of these appointments were at the tenure level and thirteen were at the non-tenure level, a proportion that was the same as in the prior period 1968–1973. Of the nine tenured appointments, only four were of mathematicians who were more than six years beyond their doctoral degrees. Hence 18 out of the 22 were of mathematicians early in their careers. There was some attrition even among the new appointees, and seven of these appointees resigned or retired or were no longer active in the department by the end of the period. In addition, there were also a number of damaging resignations of senior faculty who had been at Berkeley prior to 1974.

In addition to the ladder-rank faculty appointments, the department brought 8 to 10 postdoctoral faculty each year for two-year appointments, a practice that was customary in other major departments. The department was able to secure the appointments of one or two mathematics candidates as Miller fellows, a prestigious two-year research position open to candidates in the sciences. There were Miller visiting professorships, which allowed the departments to invite senior distinguished visitors for a yearlong appointment with no teaching duties. The department hosted a substantial number of other visiting faculty, many of whom came on their own resources, usually sabbatical leaves or other fellowships; some visitors also received stipends for part-time teaching in the department. The department was now able to accommodate larger numbers of these visiting and postdoctoral mathematicians in far better office space and in one location because of the move into Evans Hall in 1971. The general level of mathematical activity in the department was thereby enriched and increased.

During a few years in the early 1980s, there were an unusually large number of failed recruitments, in which the department was unsuccessful in recruiting targeted faculty away from other universities. In the three years 1980, 1981, and 1982 there were no successful recruitments at all, and at the same time there were losses of faculty to other universities. This lack of success in recruitment and retention during the early years of the 1980s was a cause for introspection and concern in the department, and these concerns were compounded by the very serious budget problems that the university faced at the time. However, by the very end of the period in 1985, the department had largely regained its momentum and had success in recruitment and retention of faculty. As noted, the university's budget problems had been resolved by 1985. The department maintained roughly the same size, about 65 FTE faculty, with a headcount, including joint appointments and 0% appointments, of about 75. During this period the department was heavily tenured, and while the number of assistant professors on the faculty varied from year to year, the average was under three, so that the departmental faculty was on average 95% tenured.

Beginning in the 1960s there had been a growing concern nationally in the American mathematics community, and in particular on the Berkeley campus and in the mathematics department, about the lack of women and minority faculty members. By the early 1970s, this issue crystallized and had a heightened impact on the recruiting process. It was noted in Chapter 16 that the department had made assistant professor offers to Karen Uhlenbeck and Michele Vergne in 1971 and 1972, but that unfortunately both offers had been declined.

There were discussions in department meeting about steps that might be taken to increase the number of women faculty members. There were proposals to devote some fraction of future appointments to women and minority candidates, but there was no consensus on such a plan. One step that was taken was the establishment of a new departmental committee, "Committee W," with a charge to seek out and recommend women and minority candidates for appointment. As the percentage of women among new PhDs nationally was starting to increase, there would be an increase in the number of women candidates. The campus administration, with leadership from the new executive vice chancellor, Mike Heyman, raised the issue on a campus-wide basis, and the federal government initiated investigations and was monitoring faculty composition by ethnicity and gender. The campus began development of affirmative action plans that examined the availability of women and minorities in candidate pools and compared these percentages with the outcomes of recruiting. These issues have remained a concern to this day, and one theme of this chapter is the efforts to increase the number of women on the faculty of the department

One of the most important developments for the mathematics department during this period, and one that was decoupled from state budget issues, was the establishment of the National Science Foundation–funded Mathematical Sciences Research Institute (MSRI) at Berkeley in 1982. The siting of MSRI in Berkeley resulted from a national competition; this competition and the interaction between the department and MSRI after its establishment will be discussed in the following chapter. The mathematics department was reviewed by external review committees twice during this period, once at the beginning of the period in 1975 and again in 1985 at the end of the period. The reports of these committees were important, for they identified problems that the department was able successfully to address.

In 1975, another milestone in the department passed. After nearly forty years of service, Sarah Hallam, the assistant to the chair and also in effect departmental manager, retired. She had seen the department grow and go through major transformations during her tenure, and the department and many chairs were greatly in her debt for her service and assistance in discharging their duties. She died in 1996, and in her will she made a very generous bequest to the department to endow the Sarah Hallam graduate fellowship.

In 1975, the administration initiated a review of the mathematics department, apparently the first ever, save for the ad hoc review done in 1932, which was described in Chapter 5 and which led to the hiring of Evans. Although the campus custom was to use internal review committees consisting of faculty from other departments, it was decided that an external review committee would be more useful in the case of mathematics. A committee consisting of Samuel Eilenberg, Andrew Gleason, Irving Kaplansky, Stephen Kleene, and Jürgen Moser visited the department and wrote a report. The report was quite favorable on all aspects of the department except for the graduate program, which was the object of considerable criticism. First, the committee noted that graduate enrollment had crept up over time and that the ratio of graduate students to ladder faculty was over 6 to 1, whereas "no other important graduate school in mathematics has a ratio much exceeding 3 to 1. It was noted that 85% of applicants were admitted, with entering classes of 70. This practice reflected a longstanding philosophy in the department that as a public university, there was a responsibility to provide broad access to qualified applicants to the highest quality of mathematics education that could be offered.

The committee noted that students seemed to flounder after passing their qualifying exams, usually at the end of the second year, and that there appeared to be a high attrition rate, with only about half of those passing the qualifying examination completing their degrees. It was noted that of the 64 students who passed the qualifying exam in 1970–1971, only 15 has completed their degrees by 1975. Students likewise reported that they had difficulty finding dissertation supervisors and complained about the unpredictability of financial support. The review committee strongly urged that the department take action to fix these serious problems in the graduate program, but did not specify what specifically the department should do. Although the committee did not mention it, their views on the problems with the graduate program were most likely influenced by the 1964 National Research Council ranking, which ranked the effectiveness of Berkeley's graduate program in mathematics lower than the quality of the faculty.

The department took quick action on this report. For several decades, the qualifying examination had consisted of three one-hour oral examinations in algebra, analysis, and a third topic, pitched more or less at the level of first-year graduate courses. Each oral examination was conducted by two faculty members, and these exams were administered twice a year on a Saturday with all the faculty in attendance. This practice was discontinued and replaced by a two-tier examination structure. The first examination was the preliminary examination, which was to consist of two three-hour written examinations covering algebra and analysis at the Berkeley honors undergraduate level. The idea was to make sure that the student had mastered honors-level undergraduate work. The expectation was that many students, including the stronger ones, would take and pass this preliminary examination upon entry to graduate school. The expectation was that all students should pass it within a year of entry. The Berkeley physics department had this kind of exam, and its proven effectiveness in a sister department gave encouragement that it would likewise be effective in mathematics.

The second level examination, called the qualifying exam, was an oral examination on a set of topics given by a committee selected by the student in collaboration with a faculty advisor, with the topics oriented toward a possible dissertation subject. New departmental rules stipulated that the examination committee could not pass the student unless a faculty member (almost always a member of the committee) had agreed to serve as the student's dissertation advisor. This reorientation of the qualifying examination toward a more advanced and focused topic, plus the requirement that the student come out of the exam with a dissertation supervisor, was the response to the problems identified by the review committee. The department did not at the time choose to limit graduate enrollment and become more selective, one possibility identified by the committee. This new examination format has been in effect for thirty years, and was certainly an improvement over the old system. The department maintained the overall size of the graduate program, and indeed, in some years during the 1980s, the entering class was as large as 90 students.

It was only in the 1990s that circumstances combined to make further changes necessary. The declining job market, the limited amount of graduate-student support available, rising fees, and a desire for a smaller program resulted in a decision in the department to shrink the program by about 50%, so that the entering class would be generally between 30 and 40, and at the same time to promise each admitted student five years of support contingent on good progress toward the degree. This latter innovation, which was common in other top programs, was a major morale booster for the graduate students.

Let us now return to the regular hiring during the period. As noted, almost all of the ladder-rank hires, 18 out of 22, were either at the non-tenured level or, if at the tenure level, were mathematicians of recent PhD vintage. In the first two years, 1974 and 1975, there were seven appointments, all at the assistant professor level, including the first woman to be appointed since the retirement of Sophia Levy McDonald in 1954. All of these assistant professors were later promoted to tenure. In 1974, the department hired three assistant professors: Alberto Grünbaum, Ole Hald, and Arthur Ogus.

Alberto Grünbaum was born July 14, 1943, in Córdoba, Argentina, and came to the United States for doctoral work at Rockefeller University. He completed his degree in 1969 under Henry McKean with a dissertation entitled "Some Mathematical Problems Connected with the Boltzmann Equation." He spent the following years until 1973 at the Courant Institute of New York University and then moved to Caltech. His work came to the attention of faculty at Berkeley, and this led to his appointment as a senior-level assistant professor in 1974. He was promoted to associate professor after only two years in 1976, and then to full professor in 1980. Grünbaum's research has ranged broadly over many areas of applied mathematics including signal processing, tomography, and other inverse problems including the phase problem in crystallography, special functions, and orthogonal polynomials. His appointment further strengthened applied mathematics in the department. He was elected by his colleagues as department chair in 1989 and served for three years. Overall, he has supervised the work of 11 doctoral students so far.

Ole Hald was born September 16, 1944, in Give, Denmark, and after college came to the United States for his doctoral work at New York University. There he completed his work in 1972 under Olaf Widlund with a dissertation entitled "On Discrete and Numerical Inverse Sturm–Liouville Problems." After two yeas of post-doctoral work, he came to Berkeley in 1974 as assistant professor and was promoted to associate professor in 1977 and to full professor in 1982. His research focused initially on numerical methods for solution and analysis of ordinary and partial differential equations, but has broadened to include other aspects of differential equations and inverse problems. Hald has won awards for his fine teaching and is one of two faculty members in the department to have won the Campus Distinguished Teaching Award. He was also awarded the Noyce Prize for distinguished undergraduate teaching in the physical sciences, a prize that three other members of the department have also won. The Noyce Prize was established by Robert Noyce, one of the originators of the integrated circuit and cofounder of Intel and a UC regent, in honor of his brother Donald Noyce, a professor in the department of chemistry. Hald has supervised the work of nine doctoral students so far.

Arthur Ogus was born February 12, 1946, in Washington, D.C., and received his doctoral degree in 1972 from Harvard University working under Robin Hartshorne with a dissertation entitled "Local Cohomological Dimension of Algebraic Varieties." After two years of postdoctoral work at Princeton, he was appointed as an assistant professor at Berkeley in 1974, joining his dissertation mentor, who had come to Berkeley two years earlier. Promotion to associate professor followed in 1977 and then promotion to full professor in 1983. Ogus's research has focused on the cohomology of algebraic varieties, and in particular, he has been one of the leaders in developing and applying crystalline cohomology. He has supervised the work of nine doctoral students so far in his career.

In the following recruitment cycle, in 1975, recruitment was more programmatically focused on particular areas and programmatic needs. Altogether, five offers were made, four of which were accepted: Michael Klass, Andreu Mas-Colell, Leo Harrington, and Marina Ratner.

Michel Loève, who had been hired by Evans and Neyman in 1948, had moved over to the statistics department when it separated from mathematics in 1955 but then decided to split his appointment 50/50 between the two departments. When he retired in 1974, the two departments decided that in order to maintain the strong intellectual connections between the two departments, they should seek to find a candidate to replace Loève who would fit comfortably into both departments, as Lester Dubins had some years earlier. The candidate who emerged was Michael Klass, who was at the time serving as a Miller fellow.

Michael Klass was born November 11, 1946, in Sioux City, Iowa, and received his doctoral degree at UCLA in 1972 working under Bruce Rothschild with a dissertation entitled "Enumeration of Partition Classes Induced by Permutation Groups."

He did postdoctoral work for two years at Caltech before coming to Berkeley in 1974 as a Miller fellow. He was appointed as an assistant professor in 1975, but did not take up teaching duties until 1976 at the end of his fellowship. He won promotion to associate professor in 1978, and to full professor in 1984. Klass's research has concerned stochastic processes, including the general properties of sample functions or trajectories and versions of the law of the iterated logarithm. He has supervised the work of six doctoral students so far. Klass suffered a diving accident with spinal cord injury while in his early twenties and has been confined to a wheel chair since then.

In another direction, the growing scientific collaboration with economics indicated the possibility of adding to the faculty someone who would occupy a joint appointment and comfortably span the two departments. Such a person emerged in Andreu Mas-Colell, who was already in the economics department. Mas-Colell was born in Barcelona, Spain, on June 29, 1944, and came to the United States for doctoral work in 1968. Enrolling at the University of Minnesota, he received his doctorate in economics there in 1972. He then came to Berkeley in a postdoctoral position in the Department of Economics, and in 1975 he was appointed as assistant professor of mathematics (50%) and assistant professor of economics (50%). One of his first achievements was to prove the existence of a competitive equilibrium without the traditional hypotheses of completeness and transitivity of consumer preferences. He also explored the conditions for existence of equilibria in infinite-dimensional contexts. His work won him rapid promotion to associate professor in 1977 and then to full professor in 1979. Unfortunately, Harvard lured him away in 1981, and in 1995 he returned to Catalonia. His work was recognized by election to the American Academy of Arts and Sciences.

With the resignation of Ronald Jensen, the logic group in the department was seeking to add a new member to bring their number back up to seven, and Leo Harrington emerged as the leading candidate. Leo Harrington was born May 16, 1946, in San Francisco, California, and received his doctoral degree at MIT in 1973 working under Gerald Sacks with a dissertation entitled "Contributions to Recursion Theory of Higher Types." After two years of postdoctoral work at SUNY Buffalo, he was appointed as an assistant professor at Berkeley in 1975. He won promotion to associate professor in 1977 and to full professor in 1982. Harrington's research started in recursion theory, but over the years he has broadened and deepened his work to include aspects of descriptive set theory and model theory. He is a very popular dissertation advisor, and 30 students have completed their doctoral work under his supervision.

Combinatorics as an area of research had never been adequately represented in the department, although Dick Lehmer and Raphael Robinson had some interests in this subject, as did David Gale and Elwyn Berlekamp. However, both Lehmer and Robinson had recently retired, and the department recognized the need to build strength in this important area. The name of Richard Stanley, an assistant professor at MIT and a 1971 student of Gian-Carlo Rota, came to the fore as a strong candidate at the junior level. The department offered Stanley a beginning associate professorship, but after thinking about it a bit, he declined the invitation and remained at MIT.

The fifth offer in that year's cycle was made to Marina Ratner. Rufus Bowen was aware of Ratner's work and was the first to suggest her name. Her candidacy was brought forth to the department by Committee W, according to its charge, and the department faculty voted in favor of the appointment. Marina Ratner was born October 30, 1938, in Moscow, USSR, and received her doctoral degree in 1969 from Moscow State University working under Yakov Sinai with a dissertation entitled "Geodesic Flows on Unit Tangent Bundles of Compact Surfaces of negative Curvature." She was one of the first scientists to petition to emigrate out of Russia, and in 1971 her petition was granted. She and her young daughter then settled in Israel, where she was employed in temporary positions at the Hebrew University of Jerusalem. Her published work in ergodic theory was viewed with favor at Berkeley, and although Berkeley already had a number of faculty in ergodic theory, the prospect of adding someone trained in the Sinai School in Moscow, with which the West had not been that much in contact because of the Cold War, was scientifically attractive. She was appointed to the faculty at Berkeley in 1975.

Ratner's research focused on horocycle flows on homogeneous spaces, and it won her promotion to associate professor in 1979 and to full professor in 1982. She established important results about the orbit equivalence of those horocycle flows with the standard Kronecker flows on the 2-torus and then established a striking rigidity theorem for them. Then, in a series of papers, she established both the topological and the measure-theoretic versions of the Raghunathan conjecture for unipotent flows, from which generalizations of the Oppenheim conjecture about the values of quadratic forms on integer points followed. She also proved the uniform distribution theorem for unipotent flows as well as conjectures of Margulis extending both versions of the Raghunathan conjecture to groups generated by unipotent elements. Finally, she also proved all these results for products of real and p-adic Lie groups (the S-arithmetic setting), which allowed Borel and Prasad to extend further results about the Oppenheim conjecture to the S-arithmetic setting. Her work won wide recognition, and she was elected to the National Academy of Sciences and the American Academy of Arts and Sciences. She won the Ostrowski Prize in 1993 and the John Carty Prize of the NAS in 1994. Her superb teaching was recognized by the award of the Noyce Prize for excellence in undergraduate teaching in the physical sciences. She has supervised the work of one doctoral student.

In 1975, in light of the increasing interest in mathematical economics, reflected in the appointment of Andreu Mas-Colell, the department added Gerard Debreu, of the economics department, as a 0% professor of mathematics. Debreu was a distinguished scholar of mathematical economics, whose work was subsequently honored by award of a Nobel Prize in economics and his appointment as a university professor. The following year, Steve Smale was given a reciprocal 0% appointment in economics. In 1980, Richard Karp, of the Computer Science Division of EECS, joined the mathematics department as a 0% appointee. Karp had been a member of the letters and sciences computer science department, and is a distinguished scholar of theoretical computer science known particularly for work on the P/NP problem. His work has been recognized by election to the National Academy of Sciences and the American Academy of Arts and Sciences, by award of a National Medal of Science, and by appointment as university professor.

The only recruitment in 1976 came at the very end of the year. On April 27, 1976, it was announced that Julia Bowman Robinson had been elected to the National Academy of Sciences, the first woman to be elected to the mathematics section of the academy. Kelley, as chair, decided that the department should immediately seize this opportunity to take what many regarded as the long overdue step of appointing Julia Robinson as a professor of mathematics.

Julia Bowman was born December 8, 1919, in St. Louis, Missouri, and shortly thereafter her family moved to Arizona and then to San Diego, California. She attended San Diego State College from 1936 to 1939 and then transferred to Berkeley for her senior year, where she received her bachelor's degree in mathematics in 1940. She continued on for graduate work at Berkeley, receiving a master's degree in 1941. Raphael Robinson, with whom she had taken a course in her first year at Berkeley, subsequently courted her successfully, and they were married on December 22, 1941, after which she was known as Julia Robinson. As a child she had suffered from rheumatic fever, an illness that had damaged her heart and kept her out of school for nearly two years. The illness had a lifelong effect on her health, but heart surgery in 1961 followed by two other major operations in the 1960s allowed her to enjoy a more active life [Reid 1996, p. 68].

After receiving her master's degree she continued to study and work in the department, and for some time she worked in Neyman's statistical laboratory. Robinson became interested in mathematical logic first under the influence of her husband, and then under the direction of Alfred Tarski. She completed her doctoral work under Tarski in 1948 with a dissertation entitled "Definability and Decision Problems in Arithmetic," in which she proved that the notion of an integer can be defined arithmetically in terms of the rational numbers. This was a very significant result that had important consequences for other decision problems. After her doctoral work she became interested in Hilbert's tenth problem, which asks whether there is a decision procedure for determining whether a Diophantine equation with integer coefficients has a solution in integers. This was a topic that occupied her attention for the rest of her career. She published a number of significant contributions to the problem, first in 1952, and then in 1961 (jointly with Martin Davis and Hilary Putnam), and in 1969 she published an improvement on the 1961 result. She formulated what was called by others the Robinson hypothesis, and she in fact was closer to a solution of the tenth problem (in the negative) than she imagined. It was in early 1970 that a 22-year-old Russian mathematician, Yuri Matiyasevich, upon reading her 1969 paper, filled in the missing piece in a few weeks of work to finally resolve the tenth problem in the negative. Matiyasevich and the mathematical community accorded Robinson substantial credit for her role in the solution. Martin Davis also deserves a piece of the credit, as does Hilary Putnam. Robinson and Matiyasevich subsequently collaborated on some further refinements to the solution.

Robinson's contributions to the resolution of this Hilbert problem brought her great recognition, with election to the NAS in 1976 being one of the major ones. She had never had a regular faculty position at Berkeley but had taught part-time in the department on a number of occasions. Nepotism rules in place at the time would not have permitted her appointment, but Raphael took early retirement in 1971, so that nepotism was no longer an obstacle. In any case, nepotism rules were eliminated in 1971 as an antiquated relic of the past. As she states in her "autobiography," "In fairness to the University, I should explain that even after the heart operation, I would not have been able to carry a full-time teaching load" [Reid 1996, p. 79]. After a conversation with the dean, who endorsed the proposal, Kelley approached Robinson shortly after her election to the academy in the Spring of 1976 to ask whether she was interested in an appointment as professor in the department, where it was made clear that the appointment could be a part-time one with the percentage time of the appointment completely at her discretion. Her response was positive, and her choice was for a 25% appointment. It took several months to assemble the paperwork for the appointment and to gain approval for it through the various levels of review, and her appointment as professor of mathematics at 25% time was approved over the summer retroactive to July 1, 1976.

Many other honors for Robinson followed, including selection as colloquium lecturer of the American Mathematical Society in 1980 and then election as president of the society in 1983, the first woman to serve in that capacity. She was selected as prize fellow of the MacArthur Foundation in 1983. But in the summer of 1984, Robinson learned that she had leukemia, and on July 30, 1985, she died of this ailment just weeks after her retirement from the university on July 1, 1985. She had very much hoped to return to her research after service as president of the AMS, but this was not to be. Her husband, Raphael, established the Julia Robinson Graduate Fellowships in Mathematics at Berkeley in her honor with an initial contribution, and after his death in 1995, the bulk of their estate came to the department to provide very generous funding for these fellowships.

The following academic year, 1976–1977, the Department made three major tenured appointments effective July 1, 1977: Shing-Tung Yau, Isadore Singer, and Kenneth Ribet.

Shing-Tung Yau was born April 4, 1949, in Shanyou, Guangdong Province, China, and came to Berkeley for doctoral study in 1969. He worked under S. S. Chern and Blaine Lawson, receiving his degree in 1971 at age 22 after two years of graduate study with a dissertation entitled "On the Fundamental Group of Compact Manifolds of Non-Positive Curvature." In 1974 he was appointed as an associate professor at Stanford, and in 1976, he made a great breakthrough by solving the Calabi conjecture, a conjecture dating from 1954, which conjectured the existence on any compact Kähler manifold of a Kähler metric with a prescribed volume form. This result attracted considerable attention, and led the geometers in the department to try to lure him away from Stanford to Berkeley. The department approved the appointment, and the case was sent forward for review and was approved. Given the historical agreements about raiding between the two institutions, and Yau's unwillingness to cut permanently his ties to Stanford, he accepted the offer but only on a trial basis, initially as visiting professor. Everyone in the department hoped that he would decide to remain at Berkeley permanently, in some sense as a successor to Chern, who was scheduled to retire in 1978. But this was not to be, and Yau returned to Stanford after a year. Yau was honored for his work by a Fields Medal in 1982 as well as many other honors and awards. He subsequently left Stanford, and after serving at several institutions, settled at Harvard.

Isadore Singer was born May 4, 1924, in Detroit, Michigan, and received his doctoral degree in 1950 from the University of Chicago under Irving Segal with a dissertation entitled "Lie Algebras of Unbounded Operators." After a combination of junior faculty and postdoctoral positions, he was appointed as full professor at MIT in 1956. He was known for many notable achievements in geometry, analysis, and mathematical physics, but most notably the Atiyah–Singer index theorem. In the spring of 1977, Berkeley arranged, using a variety of campus fund sources, for Singer to be a visiting professor in the department. His visit quickly led to a proposal to appoint him more permanently to the faculty, which was consummated effective July 1, 1977. Singer's appointment was of immense benefit to the department in many ways. His interest in mathematical physics and its interactions with geometry was the topic of a well-attended seminar that generated new intellectual activity; his presence raised the visibility of the department, and he played a key role in the founding of MSRI. Unfortunately, MIT did not take his loss easily, and was very eager to have him return. In 1984, MIT made him a very attractive offer that included appointment as MacArthur Institute professor. Singer decided to return to MIT but deferred cutting his ties to Berkeley. Although he was lost to the department after 1984, with only the slimmest hope that he might return, he remained on leave from Berkeley until 1986, when he retired, taking the title professor emeritus. Singer won many awards for his work, most notably the Abel Prize in 2004, coincident with his 80th birthday. His seven years at Berkeley were significant for the department, and four of the 33 doctoral students that he supervised were Berkeley students.

The department had for some time seen a need to add faculty in number theory, especially the number theory described as arithmetic algebraic geometry. Andew Ogg and to a lesser extent Ichiro Satake worked in this area, and attention turned to the possibility of attracting Ken Ribet to Berkeley. Kenneth Ribet was born June 28, 1948, in New York City, and received his doctoral degree from Harvard University in 1973 with a dissertation entitled "Galois Actions on Division-Points of Abelian Varieties with Many Real Multiplications" written under John Tate. He served on the faculty at Princeton University from 1973 until 1977, at which point Berkeley attracted him with an offer of an associate professorship. He took his initial year on leave in Paris, assuming teaching duties in Berkeley in 1978. He was promoted to full professor in 1982. Ribet's research focused on Galois representations, Abelian varieties defined over number fields, and modular forms, and his work won wide recognition. In the late 1980s, Ribet proved a conjecture of Serre concerning the level of an irreducible modular Galois representation and thereby was able to show that Fermat's last theorem would be a consequence of the famous Taniyama–Shimura conjecture, a result that brought Fermat's last theorem into the mainstream of modern number theory. When Andrew Wiles, with an assist from Richard Taylor, established the Taniyama–Shimura conjecture a few years later (or at least a special case of it sufficient to apply Ribet's result), this 350-year-old problem was finally solved by the combination of the work of Wiles and Ribet. Ribet's work was recognized by election to the American Academy of Arts and Sciences and to the National Academy of Sciences. He has so far supervised the work of 16 doctoral students, including his future Berkeley colleague Bjorn Poonen. Number theory has thrived at Berkeley.

The next recruitment cycle during the year 1977–1978, Kelley's last year as chair, brought three proposed appointments, two of which were successful. The unsuccessful one was an attempt to attract to Berkeley Gregg Zuckermann, a student of Eli Stein, who worked on representation theory of semisimple groups. Zuckerman declined the Berkeley offer of an associate professorship to remain at Yale University. The two successful recruitments were Andrew Majda and Jenny Harrison.

Andrew Majda was born January 30, 1949, in East Chicago, Indiana, and received his doctoral degree from Stanford University in 1973, working under Ralph Phillips with a dissertation entitled "Coercive Inequalities for Non-elliptic Symmetric Systems." He did postdoctoral work at the Courant Institute for three yeas and then was appointed as an assistant professor at UCLA in 1976. He was promoted to associate professor very quickly the following year. His work attracted the attention of faculty at Berkeley with interests in partial differential equations, and efforts were undertaken to lure him away from UCLA to Berkeley. These were successful, and he came to Berkeley as an associate professor in 1978, and was promoted to full professor in 1979. His work has spanned many areas of partial differential equations—scattering theory, acoustics, fluid flow, vorticity, combustion, and shock waves, among others—and has won him wide acclaim. Other universities came calling, and in 1984 Princeton University was able to attract him there. This was a serious loss for Berkeley. Majda's work was recognized by many awards and honors, and he was elected to the National Academy of Sciences and the American Academy of Arts and Sciences. Although he remained at Berkeley for only six years, he attracted many doctoral students, and 7 out of the 19 doctoral students he has supervised received their degrees from Berkeley.

Jenny Harrison was born February 9, 1949, in Georgia, and upon graduation from college in 1971, she received a Marshall scholarship, which she used to attend the University of Warwick for doctoral study in mathematics. She received her doctoral degree at Warwick in 1975 working under Christopher Zeeman and Colin Rourke with a dissertation entitled "Solution to the Denjoy Problem," in which she constructed counterexamples to the smoothing of certain kinds of diffeomorphisms. This work won her an appointment at the Institute for Advanced Study as John Milnor's assistant for a year and then an appointment at Princeton University. In 1977 she was awarded a two-year appointment as a Miller fellow at Berkeley. During her first year at Berkeley she announced that she had found a twice-differentiable counterexample to the Seifert conjecture. Posed in 1950, this conjecture stated that every vector field on the three-sphere has a closed orbit. In 1974 Paul Schweitzer constructed a counterexample that was once differentiable, but his method seemed to foreclose the possibility of counterexamples with higher differentiability. Harrision's result, which used different methods and which seemed to have the possibility of leading to an infinitely differentiable example (everyone's goal), was of considerable interest. This achievement led the department to appoint her as an assistant professor, where she would assume her teaching duties in 1979 after the second year of her Miller fellowship. Her work was a good fit with a large group of faculty in the department interested in dynamical systems and related topics. Harrison continued to refine her Seifert construction to get additional differentiability in which she had some success (three minus epsilon derivatives), but it was several years before her manuscript was in final form and fully checked over by experts.

When Harrison came up for tenure in 1985–1986, the department was deeply split on the merits of her case, with many faculty arguing for her promotion to tenure while many others opposed it. As a plurality of the tenured faculty opposed tenure, the case went forward with a departmental recommendation against tenure. This recommendation was reviewed and approved up through the usual successive levels of campus review. Harrison appealed this decision on grounds of gender discrimination, and after exhausting the university's internal grievance procedure, she then filed suit in California Superior Court against the university, challenging the denial of tenure of grounds of gender discrimination. After negotiations, and with the assistance of a mediator, the university and Harrison agreed to a settlement of the lawsuit very similar to the settlement that had been recently reached between the university and a female faculty member in another department on campus who had likewise challenged her denial of tenure. Under the settlement, a special committee would be appointed consisting of individuals who had not been involved in the first review, and this committee was charged to conduct a de novo review of Harrison's file as it stood then (1992) as supplemented with additional publications since the original review, and to make a recommendation to the chancellor concerning her qualifications for a tenured position at Berkeley. This review process resulted in a favorable outcome for Harrison, and consequently she was reinstated with the rank of full professor in the department effective July 1, 1993.

During the period of appeals and litigation, Harrison had continued her research program, developing new lines of research in geometric measure theory, and her work was supported by federal grants. After her reinstatement as professor, she has continued to develop her research program in geometric measure theory aimed at understanding multivariable calculus on domains with very irregular boundaries. In this she was building on and expanding ideas of Hassler Whitney. More recently, she has published work on a theory of domains based upon an extension of the Cartan exterior calculus to a normed space of domains that includes soap films, manifolds, rough domains, and discrete atomic structures. The analysis involved in her earlier work on the Seifert conjecture counterexamples with three minus epsilon derivatives had pointed the way to these more recent results. She has supervised the doctoral work of two students.

In the following year, 1978–1979, even though more than two appointments were authorized by the dean, only one appointment, that of Richard Schoen, was made. Richard Schoen was born October 23, 1950, in Celina, Ohio, and received his doctoral degree at Stanford University in 1976 working under S.-T. Yau and Leon Simon with a dissertation entitled "Existence and Regularity Theorems for Some Geometric Variational Problems." He came to Berkeley as lecturer for two years immediately after his degree, during the second of which he carried on an active collaboration with S.-T. Yau, who was then in residence. He then went to the Courant Institute as an assistant professor, and the following year, 1979, he was attracted back to Berkeley as assistant professor. Virtually immediate promotion to full professor followed the next year. Schoen's research in partial differential equations and geometry first focused on results about minimal surfaces and hypersurfaces, then moved into mathematical physics with his joint work with S.-T. Yau establishing the positive mass conjecture and the positivity of the Bondi mass in general relativity. He then moved on in joint work with Karen Uhlenbeck on regularity theorems for harmonic maps. His work won him wide recognition including many outside offers, award of a MacArthur Prize in 1983, and subsequent election to the National Academy of Sciences and the American Academy of Arts and Sciences. Beginning in 1983 he was on leave from Berkeley, and in 1985 he ultimately resigned to accept a permanent position at UC San Diego. This was a major loss for Berkeley. Six of the 23 doctoral students he supervised received their degrees at Berkeley. He was subsequently lured back to Stanford.

This was the last successful recruitment by the department for several years, since there was a run of unsuccessful attempts at recruitment. In addition, the department was suffering a series of losses. For instance, of the seven faculty recruited in the period 1976–1979, only two would remain active in the department after 1985, the department having lost Yau, Singer, Majda, and Schoen to outside offers and Julia Robinson to retirement and death. There were other losses, too, and we shall return presently to how the department responded to these problems.

In spite of considerable efforts, the department was not successful in any recruitment efforts during the following three years, 1980, 1981, and 1982. This was a run of very bad luck and a cause for self-evaluation. The first recruitment effort was a very bold one, to attract Dennis Sullivan and William Thurston to Berkeley together. This was reminiscent of the efforts in 1958–1960 to try to recruit leading scholars in pairs. If successful, this pair of recruitments would have made a huge difference for geometry and topology at Berkeley. But it did not work out, since among other factors there were delays in the campus review process and in getting the formal offers out. The following year, the department made an effort to recruit Michael Freedman, who had just resolved the Poincaré conjecture in dimension four. This effort was almost successful, but came undone at the very end because of some technical issues concerning the intercampus transfer: the plan was that he would leave UC San Diego and transfer to Berkeley. This was also a missed opportunity.

The department tried to recruit Robert Zimmer from the University of Chicago. Zimmer was a George Mackey student at Harvard working in ergodic theory and Lie groups and would have added a great deal to the department, complementing the research programs of faculty already at Berkeley. Zimmer thought about the offer but declined it to remain at Chicago. In 2006, Zimmer was appointed as president of the University of Chicago. The department also made an effort to recruit Luis Caffarelli, who also would have added enormous strength in the area of partial differential equations and complement the research of faculty already at Berkeley. He too declined the offer. At this point, the department was zero for five in senior-level recruitments in this time period. These circumstances led to the appointment of a special internal departmental committee on hiring practices that examined these cases and offered recommendations.

It is important to note that at about the same time, the department was losing or about to lose a number of senior faculty to offers from other institutions. Actual resignations many times followed after a period of years of the faculty member being on leave without salary, but in the following, the dates mentioned are those when the faculty member physically departed Berkeley. Blaine Lawson left in 1979 for Stony Brook, Andreu Mas-Colell left in 1981 for Harvard, Oscar Lanford left in 1982 for IHES and then ETH, Richard Schoen left in 1983 for UC San Diego, Andrew Majda left in 1984 for Princeton, I. M. Singer left in 1984 for MIT, and Clifford Taubes went back to Harvard in 1985 after two years at Berkeley. This does not count the sudden death of Rufus Bowen in 1979 and the illness and death of Julia Robinson in 1985. Thus, the failures in recruiting at the senior level were coupled with a loss of senior faculty to other institutions, and this was not a healthy situation.

The dean of physical sciences was clearly aware of these problems, and in response to them and also to follow up on the 1976 review of the department, he appointed in 1984 an external review committee that would visit and evaluate the department. The committee consisted of Raoul Bott, Leo Falicov (UCB physics), Theodore Gamelin, Frederick Gehring, Ronald Graham, and Louis Nirenberg, and the main thrust of their report concerned senior-level recruiting. The report concurred in the findings of the departmental committee on hiring practices, and was quite blunt:

> Recently mathematics has been singularly unsuccessful in its attempts to attract distinguished mathematicians from outside at the tenure level. Three reasons seem to have emerged from our informal interviews with some of the candidates...
>
> [1] inordinate time delay between the initial contact and the final offer.... [2] offers made by the Department were way below counteroffers from the home institutions

... [3] a perception on the part of some candidates that the faculty as a whole was not really enthusiastic to have them join the Department.

The report also noted that a further impediment to recruitment and retention was a feeling that inadequate staff support was available.

The report concludes in this way:

> Our findings ... are in substantial agreement with those described so well in the Report of the Departmental Committee on Hiring Practices. It seems probable that the Department will continue to have problems hiring at the senior level until it can
>
> 1. reduce the time required to make a firm commitment
>
> 2. offer competitive salaries
>
> 3. make candidates feel they are genuinely wanted
>
> 4. improve staff services in the Department.

In other comments of note, the review committee opined that the department "seems to have solved many of the problems with the Graduate program that were cited in the Report of the 1976 Visiting Committee" and praised the Mathematics Opportunity Committee (MOC) program.

As a part of the context it needs to be said that the university as a whole was going through a very difficult budgetary period. The economy was going through a time of recession: tax receipts were down and state spending had been reduced. In addition, the state had had two governors of different parties over a period of 16 years—Reagan, from 1967 to 1975, and Jerry Brown from 1975 to 1983—who for very different reasons were not supportive of the university. Reagan's initial budgetary plans for the university when he took office in 1967 have already been described in a previous chapter, and Jerry Brown was a very different person from his father, Pat Brown, who as governor (1959–1967) had been an enthusiastic supporter of the university. In 1982, average faculty salaries in the university had lagged 23% behind the market, and funding for the entire infrastructure of the university had eroded over these 16 years. The problems noted by the visiting committee concerning noncompetitive salaries and inadequate staff support were no doubt related to these larger budgetary problems that the university was facing.

In 1983, a new governor, George Deukmejian, a Republican, took office, and a few months later there was a new university president, David Gardner. Changes in governorship and in the UC presidency were for a while closely correlated in time. Instead of proposing an increase of approximately 3% in the university state budget for 1984–1985, as had been the plan within the university before he took office, Gardner conceived a bold new strategy to recoup in a single year the entire loss in operating funds from the state over the past 16 years. He consequently proposed a 32% increase in state funding for the university, and to everyone's surprise he was successful [Gardner 2005, p. 200]. George Deukmejian had become and would continue to be an enthusiastic supporter of the university. Faculty salaries went up very steeply in 1984, and funding for infrastructure immediately became more plentiful.

Thus, the foundations for solving two of the problems noted by the visiting committee had been laid even before their report was written. On the other two issues, the visiting committee report confirmed and added weight to the self-diagnosis of the problems that the department had formulated. Thus, even though the recruiting problems noted were well on their way to solution before the report was submitted, the report was immensely helpful in keeping the focus on the solutions to the problems. Recruiting went well in 1983, with four strong junior appointments (Robert Coleman, John Neu, Clifford Taubes, and Robert Anderson). In 1984, Irving Kaplansky, the new director of MSRI, was added to the faculty as a 0% appointment, and in 1985 the department made three strong appointments (Vaughan Jones, Ronald Di Perna, and James Sethian), with only one failed recruitment.

Robert Coleman was born November 22, 1954, in Glen Cove, New York, and received his doctoral degree at Princeton University under Kenkichi Iwasawa with a dissertation entitled "Division Values in Local Fields." Following his degree he did postdoctoral work at Harvard as a Benjamin Peirce instructor and then came to Berkeley as an assistant professor in 1983. His work won him rapid promotion to associate professor in 1985 and to full professor in 1988. Coleman's research lies in number theory, especially the study of curves over number fields and the Diophantine properties of sets of torsion points, where he has developed techniques of p-adic integration, thus producing analogues of abelian integrals; he also introduced p-adic Banach spaces into his study of modular forms. His work is widely recognized and appreciated, and he was awarded a MacArthur fellowship. He has supervised the work of 11 doctoral students so far. A few years after coming to Berkeley, Coleman was diagnosed with multiple sclerosis, a condition that has confined him to a wheelchair.

John Neu was born June 9, 1952, in Oakland, California, and received his doctoral degree at the California Institute of Technology under Donald Cohen with a dissertation entitled "Nonlinear Oscillations in Discrete and Continuous Systems." After his degree, he spent several years on the faculty at Stanford University, and his work came to the attention of Berkeley faculty in part through his participation in a program at MSRI. He was appointed as an assistant professor at Berkeley in 1983, although he was on leave the first year and did not take up faculty duties until 1984. He was promoted to associate professor in 1985 and then to full professor in 1990. His research is in applied mathematics and deals with the analysis of a variety of physical phenomena, including chemical oscillators, wave propagation, reaction–diffusion phenomena, and a number of physiological processes. He works collaboratively with scientists in other fields, especially in the biological sciences. So far he has supervised the work of four doctoral students in mathematics.

Clifford Taubes was born February 21, 1954, in New York City, and received his doctoral degree in 1980 at Harvard University under Arthur Jaffe with a dissertation entitled "The Structure of Static Euclidean Gauge Fields." He then joined the Society of Fellows at Harvard as a junior fellow, and at the end of this three-year fellowship was appointed at Berkeley as an acting associate professor. This was the closest that Berkeley could come to a non-tenured associate professorship. Appointment as assistant professor did not make sense for several reasons, one of which was that the maximum salary range for this rank was too low to be competitive (recall the Berkeley salaries were on average 23% below market at this time). Taubes's work ranged widely over geometry, partial differential equations, and mathematical physics, and was widely acclaimed. The department put Taubes up for promotion to full professor in 1985, but Harvard came calling, and Taubes returned to Harvard as full professor. It is not clear whether there was anything Berkeley could have done to retain Taubes in these circumstances. Taubes has received wide recognition and many awards and honors for his work.

After the resignation of Andreu Mas-Colell, the mathematics and economics departments had been searching for someone with an appropriate mix of mathematics and economics who could serve to bridge the two departments. The work of a young researcher, Robert Anderson, came to the attention of the two departments, and continued discussions led to a proposal for his appointment. Robert Anderson was born February 15, 1951, in Toronto, Ontario, and received his doctoral degree in mathematics from Yale University in 1977 working under Shizuo Kakutani with a dissertation entitled "Star Finite Probability Theory." He joined the faculty at Princeton University as assistant professor and then came to Berkeley in 1983 as a beginning associate professor. After considerable discussion about the distribution of percentage of time, it was decided to distribute it as 75% economics and 25% mathematics. His research began by applying Abraham Robinson's nonstandard models of the real numbers to questions in probability theory, functional analysis, and mathematical economics. Over time, his interests have moved more into standard mathematical economics and economic theory, and his work is well recognized. Promotion to full professor came in 1987. Anderson subsequently served as chair of the economics department, and he has supervised the doctoral work of three students in mathematics, and many more in economics.

When Irving Kaplansky became director of the Mathematical Sciences Research Institute (MSRI) in 1984, succeeding Shiing-Shen Chern, he retired from his professorship at the University of Chicago and was appointed as professor at 0% time in the Berkeley mathematics department. Bringing this distinguished mathematician into the department benefited the department even though he did not assume any formal departmental duties, not least in that the appointment helped ensure close and cordial relations and cooperation between the department and MSRI. Irving Kaplansky was born March 22, 1917, in Toronto, Ontario, and received his doctoral degree at Harvard University in 1941 working under Saunders Mac Lane with a dissertation entitled "Maximal Fields with Valuations." After a stint as a Benjamin Peirce instructor at Harvard and some war-related work, he joined the faculty at the University of Chicago in 1945, and remained there until his retirement. In his work in algebra, he made many contributions to the theory of groups, fields, and rings, including C^* algebras. His work was recognized by many prizes and awards including election to the National Academy of Sciences and the American Academy of Arts and Sciences. He supervised the work of 55 doctoral students, all at the University of Chicago. After retiring as director of MSRI in 1992, he remained an active daily presence there for many years until illness slowed him down only rather recently. He moved to southern California to live with his son and died in Sherman Oaks on June 25, 2006, at the age of 89.

In 1985, the department made an attempt to recruit John Thompson, a Fields Medalist for his work in finite group theory, to Berkeley. But Thompson declined the offer. At the same time, the department proceeded with other possibilities and was successful with two other excellent tenure appointments, those of Vaughan Jones and Ronald Di Perna.

Vaughan F. R. Jones was born December 31, 1952, in Gisborne, New Zealand, and received his doctoral degree at the University of Geneva in 1979 working under Andre Haefliger with a dissertation entitled "Actions of Finite Groups on the Hyperfinite II_1 Factor." Jones would also consider himself a student of Alain Connes, since Connes had a great influence on his work. After postdoctoral work including a year at UCLA, he joined the faculty at the University of Pennsylvania in 1981. He was a major participant in the MSRI programs in 1984–1985, and the department was successful in convincing him to remain at Berkeley with a 1985 appointment as full professor. Jones's results on von Neumann algebras and the structure of subfactors and their index were breakthroughs, and these were followed by his discovery of the relation between these algebras and knot theory and the construction of a new polynomial invariant of knots, now called the Jones polynomial. His subsequent work ranged over a wide variety of topics concerning von Neumann algebras with applications to many areas, including statistical mechanics. He has supervised the doctoral work of 18 students so far. His research accomplishments have won him wide recognition, and he has received many awards, including most notably a Fields Medal in 1990 and election to the National Academy of Sciences and the American Academy of Arts and Sciences. In 2002, Queen Elizabeth II bestowed on him the title Distinguished Companion of the New Zealand Order of Merit, which is the functional equivalent of a knighthood.

Ronald Di Perna was born February 11, 1947, in Somerville, Massachusetts, and received his doctoral degree in 1972 at New York University working under James Glimm with a dissertation entitled "Global Solutions to a Class of Nonlinear Hyperbolic Systems." He served on the faculties of Brown University, the University of Michigan, the University of Wisconsin, and then Duke University, winning promotion to tenure in 1976 and promotion to full professor in 1978. He was offered appointment as full professor at Berkeley in 1985 and was one of only a small number of new appointees during this period who came to Berkeley at a point in their careers beyond six years after the doctorate. His research on basic properties of nonlinear partial differential equations included the first proof of the existence of solutions for nonlinear hyperbolic equations with arbitrary data, local smoothness theorems for conservation laws, and the first existence results for hyper-elasticity with arbitrary data. This was a welcome appointment, one that strengthened the area of applied PDEs, especially when it looked as though Andy Majda was not going to be returning from Princeton. But sadly, a few years after coming to Berkeley, Di Perna fell ill with lymphatic cancer and died of this ailment at age 41 on January 8, 1989. This was a tragic loss to his family, to the department, and to mathematics. Soon afterward, the department instituted the Di Perna lectureship to honor his memory. Every year, a distinguished scholar in applied mathematics is invited to Berkeley as a Di Perna lecturer.

The third appointment in this year was at the assistant professor level, and the candidate selected was James Sethian. Sethian was born May 10, 1954, in Washington, D.C., and received his doctoral degree at UC Berkeley in 1982, working under Alexandre Chorin with a dissertation entitled "An Analysis of Flame Propagation." After three years of postdoctoral work, he was appointed as an assistant professor in 1985 and was one of the few Berkeley doctoral graduates since the 1950s to be appointed to the faculty. He won promotion to associate professor in 1988 and to full professor in 1990. His research has spanned several areas of applied mathematics, including turbulent combustion, the vortex method for fluid mechanics, and most especially the study of how interfaces or boundaries between regions of physical interest move or propagate over time using the level set method, which he helped develop. His work has won wide acclaim, including the award of the Norbert Wiener Prize by the American Mathematical Society in 2004, and it has proven useful in industrial applications. He has supervised the doctoral work of seven students so far, including Jon Wilkening, who joined the faculty in 2005. Sethian is recognized as a talented teacher and has won the Noyce Prize for excellent undergraduate teaching in the physical sciences.

With these appointments, the department seemed to be back on track with its recruiting even before the release of the visiting committee report, which was issued in the late spring of 1985. The restoration of competitive salaries in 1984 as a result of the large increase in state appropriations for the university no doubt helped, as did more careful attention to the courting of candidates. In 1985, the department proposed a formalization after many years of the status of postdoctoral faculty. It took some time to win the concurrence and approval of the campus and the university-wide administration. The end result was that the department instituted the Charles B. Morrey, Jr. (or more simply the Morrey) assistant professorships. These were two-year appointments (later extended to three years) with a teaching load of three semester courses per year. These were meant to compete more effectively with the many other named assistant professorships that were becoming increasingly common at the time. This type of appointment for new doctorates had been proposed twenty-five years earlier, in 1959, to be named after Evans, but the proposal was turned down by the administration, as recounted in Chapter 13. In the intervening years the department had used the catchall title of lecturer without any name attached to it. As already noted, the idea of term assistant professorships or instructorships ran counter to some fundamental principles of academic personnel policy in the University of California, but these reservations had now been overcome.

One final point on curriculum and teaching responsibilities: The entire university had converted to the quarter system in 1967. In the early 1980s, the Berkeley campus received permission to return to the semester system if there was a consensus on campus to do so. There was, and the campus made the reversion in 1983. Under the quarter system, the teaching responsibility for a regular faculty member was five quarter courses per year. When the semester system was reinstated, the department designed a point system, whereby 20 points were expected of each faculty member every year, and a standard upper-division or graduate course was worth 6 points, with additional points awarded for teaching a large lower-division course. The curriculum had to be repackaged into 15-week semester blocks from 10-week quarter blocks. The revamped curriculum did not look that much different from what had appeared in the 1966 course catalogue.

Chapter 19

MSRI: The Initial Years, 1978–1985

The establishment of the Mathematical Sciences Research Institute (or MSRI, which is sometimes pronounced "misery," especially by some Berkeley campus administrative officials who were often frustrated by the problems and challenges that had to be overcome in the process of establishing the institute) was a signal event for the Berkeley mathematics department. The object here is not to write a history of MSRI, as others will do that, but rather to relate the efforts that members of the mathematics department undertook to plan for and create this organization and nurture it through its first few years, from 1978 through 1985. Interactions with the mathematics department on campus and with campus administrative officers and with National Science Foundation officers will be part of the story.

In early 1978, the Division of Mathematical Sciences (DMS) of the National Science Foundation (NSF) issued a request for proposals for the creation of a mathematical sciences research institute. The idea of funding such an institute went back many years within the foundation and DMS, but had come to the fore in 1977. One reason this idea resurfaced was that NSF, through its Physics Division, had recently created the Institute for Theoretical Physics (ITP), which had been sited as result of a national competition at the University of California Santa Barbara. Although controversial, this institute was seen as a success, and it revived earlier ideas for a mathematics institute with a similar national competition for its siting. The idea of a mathematics institute was immediately controversial, and many individual mathematicians opposed the idea on grounds that it would drain money from support of mathematics departments and compete with individual investigator grants. In spite of this initial opposition, the DMS gained approval to issue the request for proposals with a due date of August 1979.

Berkeley was seen by nearly everyone as an obvious candidate to compete for this institute. There were other obvious candidates as well, and there was also concern that there might be some predisposition in Washington to not award this prize to a large established department that already had a large share of federal research dollars. One observation leading to this idea was that UC Santa Barbara had beat out more prestigious and richer institutions for the physics institute. At a mathematics department meeting in the spring of 1978, shortly after the call for proposals had been issued, the Berkeley mathematics faculty decided that Berkeley should compete and then asked a three-person executive committee consisting of Shiing-Shen Chern, Isadore ("Is") Singer, and Calvin Moore to take the lead in developing a proposal.

Two main issues confronted this executive group; the first was how to organize the institute administratively, and second how to organize it scientifically. The first in some ways was the more complex task of the two. The first fundamental decision, which drove everything else administratively, was not to submit the proposal through the campus, as a sponsored project, but rather to form an independent entity, a 501(c)3 not-for-profit California corporation, to submit the proposal. One model that was influential in the planning was the School of Mathematics at the Institute for Advanced Study and its relationship with the Princeton mathematics department. There were many reasons for the choice of a nonprofit corporation. First, with an independent corporation with its own governing board, ownership and control of the institute could be vested in the mathematical community. It was reasoned that this structure would both increase the chances of the proposal's success, and would resonate better with the mathematical community. The institute would have greater visibility in this mode, and would not be seen as just a small part of a larger Berkeley operation. The institute would benefit intellectually to an immense degree from the proximity of the Berkeley mathematics and statistics departments, but it would not be part of them. This structure was different from the structure of the Santa Barbara Physics Institute (ITP), which had been submitted as a proposal through that campus, but because of the nature of the campus and the relationship of the institute to the campus, it was able to achieve a considerable degree of visibility that would not be possible at Berkeley.

An additional advantage of forming an independent corporation was that the institute would be out from under the bureaucratic apparatus of the Berkeley campus, which was known to be both inefficient and oppressive. Also, the institute would have more flexibility on such issues as indirect-cost rates and would not be confined by the University's indirect-cost rate structure. Finally an advantage of being independent that was not fully appreciated at the time, but which was crucial later on, was the ability to negotiate the arrangements with the campus for the construction of the building that was to house the institute permanently. It is highly unlikely that the building ever would have been built if MSRI had been a campus unit.

The first step in building the administrative structure was to assemble the governing board of the institute, known as the Board of Trustees. West-coast research universities including all campuses of the UC system as well as the Lawrence Berkeley National Laboratory (LBNL) were approached and asked to become sponsors of the institute. They would participate in the governance by nominating a member of the Board of Trustees, would receive yet to be determined scientific benefits, and would be affiliated with what was hoped would be a premier research institute. Nine institutions, including of course UC Berkeley and LBNL, signed on, while a number of others held back. In addition to the trustees nominated by the sponsors, there would a roughly equal number of individual trustees, who would be selected from among the mathematical community nationwide. This structure based the governance in a group of institutions from the broad region as well as the national mathematical community. It was clear that a director had to be designated as a part of the proposal, and Chern was the obvious candidate. Is Singer was the one most responsible for successfully twisting Chern's arm to get him to agree to serve as director. Calvin Moore agreed to serve as deputy director. Hyman Bass, one of the individual trustees, was asked to serve as chair of the board, and he generously agreed to do so.

With these elements of the governance structure in place, attention turned to the all-important question of locating physical space for the institute. As an independent entity, it was clearly understood that the institute would pay rent for its space. Permanent space on the central campus was at a premium, and it was exactly at this time that the administration was trying to take space away from the mathematics department in Evans Hall, as recounted in Chapter 13. Campus policy gave priority for central campus space to units that involved students, so the institute would have low priority for such space. The search therefore turned to nearby off-campus locations, and at this point in preparing a proposal to send to NSF, all that could be done was to identify several possibilities. However, at this point, Chancellor Bowker played a key role. He warmly endorsed the efforts to bring this institute to Berkeley, as well as the administrative structure that had been constructed. He wrote a letter to the three-member executive team in which he promised that the campus would locate and provide suitable space for the institute with rent at market rate or below. It was this commitment that really made the administrative structure that had been constructed a credible one. Even though Bowker had retired when the time came for it to be fulfilled, his successor, Chancellor Ira Michael Heyman, honored the pledge.

The scientific structure was of course crucial to the proposal's success. The fundamental decision was that there would be no permanent members, in sharp distinction to IAS at Princeton and to ITP at Santa Barbara. There would be a director and deputy director who would serve for periods of time, but the scientific structure was built around programs. Each year, two areas of the mathematical sciences would be identified that were particularly ripe for advancement, and which could especially benefit from a year-long program in which leaders of the field could come together, freed of their university responsibilities, together with a group of postdoctoral mathematicians who were just beginning work in the area. Other mid-career mathematicians working in the area would be included in the program if they could obtain sabbatical or other leave from their home institutions. The hope was that for the period of the program, the institute would be the worldwide center of activity for this particular field. The vision was therefore that the senior faculty of the institute would be rotating and on loan from their home institutions.

It has been observed that a whole generation of mathematicians had spent a year or more near the beginning of their careers at the IAS Princeton, and that these experiences had been career-shaping for them. But the mathematics community had grown in size, in diversity of fields, and geographically to an extent that the IAS could no longer provide such postdoctoral experiences to enough mathematicians. In principle, of course, one could just expand an existing institute to achieve these goals, but it was generally agreed that such an organization has a size limit, beyond which it loses focus and effectiveness. What exactly defines this limit is a bit fuzzy, but it is somewhere in the neighborhood of 70 to 80 mathematicians in residence, with a larger number for short periods during conferences. These considerations of size entered into the planning for the Berkeley institute and its scientific organization.

This argument about career-shaping postdoctoral experiences and the size limit of an institute was made to the many opponents in the mathematical community of the whole idea of an NSF-funded institute, and the argument was incorporated into the proposal. The goals of the institute were in fact threefold: excellent and focused postdoctoral training experiences, the research results that would come out of bringing together a large number of experts in an intense experience, and finally the renewal benefits that would accrue to mid-career researchers attending the programs. In addition to the main programs, a certain portion of the scientific budget would be set side to provide support to mathematicians who were not part of one of the main programs, but who could benefit from the postdoctoral experience at the institute or a mid-career experience. The program leaders and the programs themselves were to be chosen together as a package, and the program leaders would choose the post doctorates and other participants in the program under the overall supervision of the director and deputy director. One important aspect of the process for selecting participants was that participation of Berkeley faculty members was to be limited to those who would be part of the leadership team of the various programs so as to avoid the perception or the reality of the institute serving as a vehicle to provide academic-year research support to Berkeley faculty. As the institute developed, protocols were put in place so that Berkeley faculty could be granted teaching relief by the department in order to facilitate their participation as program leaders at the institute.

The proposal incorporating these features was submitted to NSF in August 1979, with Chern, Singer, and Moore as principal investigators on behalf of the Mathematical Sciences Research Institute, Inc., a California public-benefit corporation. The institute was to be called rather unimaginatively the Mathematical Sciences Research Institute (MSRI). Funding was requested for a startup year and then funding for five years of full operation with $1.6 million budgeted for the first year of full operation. This level of funding was regarded as essential, and the leadership believed that funding the institute at a lower level would put its success at serious risk, and they had no interest is presiding over an unsuccessful venture. Nothing happened for almost a year and a half, and then late in 1980, NSF announced that they were sending a site-visit team out to Berkeley. In preparation for this visit, the principal investigators had to convince the campus administration to make specific commitments as to how the institute would be housed in accord with former chancellor Bowker's 1979 letter.

When the site-visit team arrived, the campus advanced several possibilities for housing the institute, including the top two floors of the extension building on Oxford Street at the west end of campus, and the possibility of a building on the School for the Deaf and Blind site on Piedmont Avenue. The campus was negotiating with the state to obtain this site and its buildings, which had been vacated when the school moved to new quarters. But the outcome of that negotiation was far from clear, and so this second possibility was quite uncertain. The site-visit team and NSF were not satisfied with these original plans for the MSRI housing, and it required substantial discussion with UCB leadership before a satisfactory plan for housing the institute finally emerged. This plan was to house the institute temporarily in the extension building, and then to construct a permanent facility for MSRI on campus land in the Berkeley hills above the central campus and next to the existing Space Sciences Laboratory. MSRI would pay the costs of remodeling and rehabilitating the space in the extension building in lieu of rent. For the permanent building, the university would borrow the money to construct the building, and the institute would pay rent to the campus sufficient to cover the debt service on the loan over a 25-year period. When the loan had been paid off, the agreement was that the institute would occupy the building rent-free, assuming that the institute was still in existence. This kind of arrangement would have been impossible if the institute were to be funded by a grant to the campus. The campus was taking a risk, since NSF could pull its funding virtually at any point, and if that happened before the debt was paid off, the campus would be left with debt service and no obvious tenant.

The site visit seemed to go well on issues other than housing. The site-visit committee noted that the all-important process for selecting programs and program leaders had not been sufficiently thought through in the original proposal. On reflection, the principal investigators agreed with this very helpful criticism, and remedied it by adding to the governance structure a Scientific Advisory Council (SAC) to consist of a small group of about six eminent mathematical scientists who would review proposals for programs and program leaders and advise the director and deputy director. The principal investigators saw an additional advantage in adding the SAC to the governance structure in that it would insulate the scientific decisions being made from the administrative and political interests of the trustees. Is Singer was the principal architect of this council, and he provided leadership to it over the first few years of the institute's life. The NSF accepted this as an amendment to the original proposal.

After the site visit and negotiations with the campus in regard to space for the institute, DMS put forward their proposal to NSF management. In the spring of 1981, the National Science Board approved the funding for MSRI at the requested level, as well as a second institute, the Institute for Mathematics and Its Applications (IMA), sited at the University of Minnesota. The vote in the science board was far from unanimous, reflecting the lingering concern in the mathematics community about the whole idea of mathematics institutes.

The startup year began in the summer of 1981 with a meeting of the SAC to refine scientific plans for the first two years of full operation. Support staff were hired and financial management systems were put in place as well as a contract with an audit firm. NSF sent a very helpful analyst from their financial management section to advise MSRI on financial transactions with the federal government. It was at this point that MSRI established that its indirect cost rate would be set at 0%. Indirect costing is a method designed by accountants for allocating a variety of campus infrastructure costs to the hundreds or thousands of units or projects that rely on this infrastructure, in a pro rata manner based on direct expenditures. MSRI was an insular project, and all the costs that would normally go into the indirect cost pool were clearly identifiable and could be included as direct costs. Hence 0% overhead! Renovation of the two floors in the extension building was underway and was completed in time for MSRI to move in during August 1982. The Mathematical Sciences Research Institute opened on time for full scientific operation in September 1982.

During the startup year 1981–1982, planning for the permanent facility began. One interesting issue arose with respect to the building site: A facility called the Animal Behavior Research Station was located several hundred yards back into the hills from the proposed building site. The station housed several self-breeding animal colonies whose behavior scientists observed in conditions as close to the wild state as possible. These researchers were concerned that noise from the institute would disrupt their observations. Mathematicians do not make a lot of noise proving theorems, although voices from people talking outdoors might carry. Construction noise, however, was a genuine concern. The animals then in residence were primates, but there were plans to import a colony of hyenas to the station, and then the issue might be who was going to be disturbing whom! These objections were dealt with, and stipulations were inserted into the building contract for noise mitigation during construction. Today, the primates are long gone, but the hyenas are still there.

The NSF was concerned about how large a fraction of the grant funds would be required for rent, that is, to pay the debt service on the loan that would be taken out to cover the construction costs. The normal pattern in a grant was that cost of space was buried in the indirect cost pool, but here the rent was a direct cost and quite visible. Analysis of the amount of space required and an estimate of construction costs yielded a project cost of about $2 million for the building, of which NSF was apprised. At the time, 1981–1982, interest rates were at stratospheric levels because the Federal Reserve Bank was trying to suppress inflation. The rates were so high that financing the building at the expected cost was incompatible with a reasonable limit on how much rent MSRI could pay. It was only when rates came down shortly thereafter that it was possible to proceed with the building project.

The campus decided that this project should be undertaken following a design/build method, whereby a detailed project description is issued and teams consisting of an architect and a building contractor submit designs for the building together with a bid price. One reason for following such a process was that it generally resulted in lower project costs. However, such an approach had not been used on a project of this complexity, and some delays occurred while details were worked out. The initial project description placed a cost limit of $2 million, and when the submissions were reviewed, none was found to be satisfactory. Since it was known that several teams had dropped out of the competition because of the $2 million bid limit, MSRI and the campus decided to raise the bid limit to $2.5 million and reopen the competition. When NSF was informed of this development, there was a bit of grumbling about the implicit 25% increase in rent. However, this decision proved to have been the right one, for in the second round, the team of Shen-Glass Architects (principals Carol Shen-Glass and William Glass) and Amoroso Construction Co. submitted a truly exceptional design with a price at the bid limit, and their entry easily won the competition.

Construction got underway in January 1984 and the building was ready for occupancy in April 1985, nearly four years after the initial award in 1981. During the spring of 1985 there was a minor dustup over bus service from the central campus to the hill area where the MSRI building was located. Even though MSRI is not that far from Evans in terms of direct line-of-sight, the route up the hill, Centennial Drive, is a winding switchback road, and the rise in elevation from Evans Hall is some 800 feet. There was no disagreement that bus service was essential; the only outstanding question was the headway. MSRI argued for 15-minute headway, but the final outcome was 30 minutes between buses, with a promise to reconsider if passenger traffic warranted more frequent service. Bus service and rent were part of a larger disagreement between NSF and the campus. Namely, officials in NSF felt that the campus had not contributed enough of its own resources toward the funding of MSRI in comparison to the benefits that the campus would accrue from having MSRI located on campus. This issue of cost sharing was raised by the site-visit team that NSF had sent out to MSRI in the spring of 1985 that was to advise NSF on the five-year renewal proposal for 1987–1992 that MSRI had submitted. The site-visit team raised the issue of the adequacy of campus cost sharing in a meeting with Chancellor Ira Michael Heyman, and the meeting ended in some acrimony. Nevertheless, the grant was renewed, but the issue was by no means dead, just dormant, for it was to arise again in 1998–1999 in the recompetition.

The scientific programs at MSRI won wide acclaim for their excellence, and this was no doubt due to the fine judgment and leadership provided by the Scientific Advisory Council. The senior scientific leadership of the institute was renewed each year. It is a truism, but one based on observation, that traditional research institutes tend to have an aging problem with research programs led by a single director that end up going stale. Such institutes have great difficulties in renewal. When the author was making a presentation on the founding of MSRI to a group of senior faculty from across the campus, Emilio Segre, a Nobel laureate physicist, who was well aware of these problems in a research institute, exclaimed of MSRI, "Aha! You have discovered the fountain of youth." But the question of biological renewal also entered in the issue of the directorate of the institute. S. S. Chern stepped down as director in 1984 according to his original plan to serve for three years. Is Singer returned to MIT in 1984, and although he continued to be involved in MSRI, it was not the same as having him on site. Calvin Moore resigned in 1985 to accept an administrative position in UC. The renewal process worked very well: Irving Kaplansky assumed the directorship in 1984 and served until 1992. He was followed by Bill Thurston, who served from 1992 to 1997, and David Eisenbud, who has served since 1997. The successive directors have been assisted by many fine deputy and associate directors.

Thus, by 1985, MSRI had established a strong record with its scientific programs, had passed its first review, and had been approved for a five-year renewal of its grant.

It had settled into its permanent quarters and had made a successful transition to the second generation of leadership. MSRI is generally regarded as a great success, and it certainly has been of great benefit to the Berkeley mathematics department. It is an interesting question how one should measure the success of such an enterprise. Is Singer early on formulated a simple criterion for success. MSRI, he said, will be a success when people cannot remember what it was like before MSRI was created. It is fair to say that this criterion had been satisfied even by 1985.

Chapter 20

End Note: Highlights after 1985

This brief concluding chapter will sketch some of the highlights in the evolution of the Berkeley mathematics department after 1985. A more detailed history of this period is perhaps premature and in any case will likely be written by someone else. The most momentous event of these years had its origins outside the university. In 1990, the California economy entered its worst recession since the Great Depression of the 1930s. Revenues dropped dramatically and state appropriations to the university went through unprecedented decreases. Part of the response to these cuts was increased student fees, but of greater import for academic departments was a Voluntary Early Retirement Incentive Program (called VERIP) that was offered to all employees in an effort to reduce salary expenses. In fact, it was offered three times, in 1991, in 1992, and in 1994. This program was particularly attractive to faculty over about age 55 to 57, and many mathematics faculty took advantage of VERIP. In fact, in the earlier narratives about faculty, the reader will have noticed frequent mention of individual faculty members who "VERIPed." In 1990 the department had reached a size of 67 FTE, but in 1994 after the final VERIP program, it had shrunk to under 50 FTE faculty in spite of recruitments during this period.

Campus-wide academic planning resulted in a decision to reduce permanently the size of the ladder-rank faculty on campus by 10%. Since most departments on campus had suffered losses similar to those in mathematics, the campus leadership seized this opportunity to correct the distribution of faculty positions among departments to something closer to optimal. Departments were assigned so-called omega numbers, and were allowed to grow back over time to these numbers. Some departments had omega numbers that represented a reduction of well over 10% from their previous levels, while others had omega numbers that called for growth above and beyond their original complement. Mathematics was in the middle ground, receiving a 10% reduction with an omega number set at 60 FTE. The department began an intense rebuilding program that brought the faculty size close to the omega number in 2002, but it has fallen back since then. As this book goes to press, the faculty FTE of 53.25 is well below the department's omega number owing to a number of retirements in the recent past and to inadequate numbers of replacements authorized by the administration.

Over the 20-year period since 1985, recruiting has gone generally very well. It was always a disappointment when an offer was declined, but there were 47 successful recruitments during this period, representing a mixture of tenured and non-tenured appointments. These are listed below by year of appointment. Some 13 of these 47 faculty resigned to take positions elsewhere, retired, or died, and the year of separation is listed as well as the destination.

1986	Andrew Casson (resigned 2000, Yale)
	Hendrik Lenstra (retired 2003, to University of Leiden)
1987	Dan-Virgil Voiculescu
1988	Sun Yen Alice Chang (resigned 1992, UCLA)
	Andreas Floer (deceased, 1991)
	Mariusz Wodzicki
	Hugh Woodin
1989	James Demmel [50% math, 50% EECS]
	Craig Evans
	Paul Vojta
1990	Alexander Givental
	Curtis McMullen (resigned 1998, Harvard)
1991	Fraydoun Rezakhanlou
	Nicolai Reshetikin
	William Thurston (resigned 1997, UC Davis, now at Cornell)
1992	Vera Serganova
	John Strain
	Thomas Wolff (resigned 1997, Caltech)
1993	Richard Borcherds
	Maxim Kontsevich (resigned 1997, IHES-Paris)
1994	Bernd Sturmfels
1995	
1996	Michael Christ
	Theodore Slaman
	John Steel
1997	David Eisenbud
	Edward Frankel
	Bjorn Poonen
	Grigory Barenblatt

1998	Andrei Okounkov (resigned 2003, Princeton)
	Maciej Zworski
	Alan Edelman (resigned 1999, MIT)
1999	Allen Knutson (resigned 2005, UC San Diego)
	Ai-Ko Liu (resigned 2006)
	Thomas Scanlon
2000	Ming Gu
2001	Mark Haiman
	Michael Hutchings
	Lior Pachter
	Daniel Tataru
2002	Thomas Graber (resigned 2006, Caltech)
2003	
2004	Peter Teichner
	Peter Ozsvath (resigned 2006, Columbia)
	Mina Aganagic [50% Math, 50% Physics]
2005	Jon Wilkening
2006	Ian Agol
	Martin Olsson
	Constantin Teleman

In addition, four probabilists in the statistics department (David Aldous, Stephen Evans, Yuval Peres, and James Pitman) were appointed in 1999 as 0% professors in the department. Subsequently, Yuval Peres's appointment was adjusted to 25% math, 75% statistics. Also, Christina Shannon, a mathematical economist in the economics department, accepted a 0% appointment in mathematics in 2002. There were a number of disappointments when offers were declined. These were Demetrios Christodoulou (1988), Yakov Eliashberg (1989), Ingrid Daubechies (1990), Robert Calderbank [50% math, 50% EECS] (1990), Laurent Clozel (1990), Tai-Ping Liu (1990), Nikolai Makarov (1991), Tomasz Mrowka (1992), Gang Tian (1994), Peter Kronheimer (1994), Carlos Kenig (1996), Peter Constantine (1997), Rahul Pandharipande (2001), Yair Minsky (2003), and Elena Mantovan (2005).

The group in logic was seriously depleted by retirements, dropping from eight to three in a period of three years. Aggressive recruitment has brought the group back up to six, representing again approximately ten percent of the faculty. The department has regained strength in partial differential equations during this period, and applied mathematics is reasonably well represented. Geometry and topology have suffered by retirements and resignations, and there has been a substantial number of unsuccessful recruitments in this area as well. The department now has a presence that it never had before in algebraic combinatorics, with Mark Haiman and Bernd Sturmfels. The appointment of Mina Agananic, who works in string theory, is an effort to develop this area and to create additional contact with the physics department. The department has entered computational biology with the appointment of Lior Pachter. The dotcom boom of the late 1990s brought increased resources to the university from the state, but then the dotcom bust of 2001 resulted in steep cuts in state funding, increased student fees, and a decrease of faculty and other resources available to the department.

The faculty of the mathematics department is a congenial group, quite remarkably free of feuds and animosities that have troubled other departments. Pure and applied mathematics and mathematicians form a unified whole, where there is a free flow of ideas and students across the entire spectrum. This unity is one of the strengths of the department, as is the broad coverage of fields of mathematics. In a sense, this structure and breadth are a legacy that Evans left to the department. In the 1950s and 1960s, there were proposals for creation of a separate department of applied mathematics, but these got nowhere then and have not resurfaced. The department has always had a strong egalitarian tradition and has, for instance, firmly resisted the creation of any endowed chairs in the department.

The academic distinction of the faculty has grown over time. For instance, in 1960, even after the period of very successful recruitment in hiring in 1958–1960, Griffith Evans was the only one out of the fifty-plus active and emeritus faculty who was a member of the National Academy of Sciences. He was also the only one who was a member of the American Academy of Arts and Sciences. In 2005, out of about 100 active and emeritus members of the department, 14 are members of the National Academy of Sciences, 21 are members of the American Academy of Arts and Sciences, and 3 are members of the National Academy of Engineering. Three current members are Fields Medalists (Smale, Jones, and Borcherds), as are five former members (Yau, Thurston, McMullen, Kontsevich, and Okounkov). In fact, three of the four Fields Medalist in 1998 were Berkeley faculty members in 1997, and Yau and Thurston were PhD graduates of Berkeley. In the 1996 NRC rankings of academic departments, the Berkeley mathematics department was tied for first place with Princeton in rankings based on quality of faculty, but was ranked sixth behind Princeton, Chicago, Harvard, MIT, and Stanford in effectiveness of the graduate program. This kind of disparity in rankings of faculty versus effectiveness of the graduate program was also present in the 1964 ACE rankings that were described in Chapter 17.

The Mathematical Sciences Research Institute (MSRI) has continued to thrive. As Irving Kaplansky's term as director was coming to an end, William Thurston was identified as his successor. So, on the third try, the department was successful in recruiting Thurston to Berkeley. Thurston was appointed as professor in 1991, and then assumed the directorship of MSRI the following year. He served a five-year term ending in 1997, at which time David Eisenbud was appointed as the next director. At the same time, Eisenbud was appointed as professor in the department. When the NSF announced in 1997 that the foundation would be recompeting the mathematics institutes, Calvin Moore, then the department chair, realized that the long dormant issue of campus cost sharing or matching funds for MSRI would be a major issue in the recompetition. He persuaded the campus leadership to respond favorably on this point, with the result that the campus agreed to "cancel" the out-standing debt on the loan taken out to finance the construction of the building, so that MSRI would henceforth occupy the building rent-free. The outstanding debt at the time was approximately $1.2 million, so this represented a substantial campus contribution. Under the original agreement, MSRI was to occupy the building rent-free after the debt was retired, so the net effect was to advance the time of rent-free occupancy by many years.

The campus also agreed to an increase in the amount of released time provided at campus expense to Berkeley faculty so that they can take on scientific leadership at MSRI in the scientific programs. The new director, David Eisenbud, assisted in these negotiations. MSRI was successful in the recompetition, and under Eisenbud's leadership has set out on a course to raise substantial private funds to help support MSRI. MSRI succeeded in raising $11 million to finance the construction of a much-needed addition to the building, which has recently been completed as this manuscript goes to press. Approval was sought and obtained to name the facility Chern Hall. The goal now is to seek to create a substantial endowment to provide a permanent income stream to assist in the support of the institute.

In 1995, one of the PhD graduates of the department, Robert Uomeni, won the California state lottery, with a prize in excess of $20 million. One of his first actions was to contact the department to say that he would like to create an endowment in honor of Professor Chern, to whom Uomeni felt a deep debt of gratitude for his great contribution to Uomeni's success as a graduate student and for his kindness and support. Because of the department's reluctance to embrace endowed chairs, Uomeni and his wife, Louise Bidwell, endowed the Chern visiting professorship, whereby the income from the endowment would be used to bring distinguished visitors to the department either to give a series of lectures or in some cases to give an entire course. This endowment has proven to be invaluable to the department in attracting visitors, especially since other funding for this kind of invitation has always been scarce.

In 1995, the department was reviewed for a third time (the cycle seems to be about every ten years). The external committee consisted of Peter Lax, Barry Mazur, and Dusa McDuff, and the main conclusion from their visit was that the size of the faculty had to be rebuilt. "If something is not done soon there is, in our opinion, a real danger that the Mathematics Department will collapse." This was a time when the size of the faculty had shrunk from 67 FTE to under 50, and the rebuilding process had not yet started. The committee's recommendation was of course directed much more at the administration than at the department, and to the extent that it helped establish the omega number at 60, it was helpful. The committee also opined that "The Administration and the Department are woefully out of touch." Both the rebuilding of the faculty after VERIP and the establishment of effective communications between the department and administration have in good measure been achieved.

If the ten-year cycle of external reviews is the pattern, then a 2005 review is already a bit overdue. In fact as this book goes to press, the administration has scheduled a review of the mathematics department for the academic year 2006–2007. What would such a review committee find? One item that it would notice is that by improving its advising and outreach activities, the department has nearly quadrupled the number of undergraduate majors, from a graduating class of about 70–80 per year to one of about 265. This figure represents about five percent of all bachelor's degrees granted at Berkeley, as contrasted to the national average of one to one-and-a-half percent of bachelor's degrees awarded in mathematics. The committee would also notice that the department does not have enough faculty to carry out its teaching mission, especially in light of the large number of undergraduate majors, and would urge the administration to provide relief. And it would be noted that the department's postdoctoral program has been sharply curtailed by budget cuts and the committee would urge restoration of the program. The committee would certainly urge the department to continue and expand its efforts to include more women in the faculty, and finally it would decry the inadequacy of the level of graduate-student support, which presently is not competitive. The answer to the final issue is fundamentally a vigorous private fundraising campaign, which is already underway.

The bottom line is that as a result of recurring state budgetary problems and the current statewide structural deficit and political gridlock, state resources flowing to the university have steadily declined over time, both in constant dollars and as a fraction of the share of the campus's core operating budget. The university and the Berkeley campus are consequently more and more dependent on student fees, fundraising, and other revenue streams. The campus is thus becoming in some respects more like a private university, with state appropriations now just over $380 million out of core operating budget of over $900 million, and out of a total budget including research grants and auxiliary enterprises of over $1.3 billion. While state funds are an indispensable core, the financial model both for the campus and for departments and schools within the campus is shifting, and will likely continue to shift in years to come, with all the consequences to the mathematics department thus implied.

Appendices

Appendix 1 (Letter from Welcker to Hallidie, March 28, 1881)

March 28, 1881

My dear Hallidie,

I called some weeks ago to talk with you somewhat on the general aspect of University affairs but did not find you in. I came today to say something concerning them but with more personal relation to myself.

I was informed a short time ago by a gentleman that in an interview he had with Mr. George Davidson, who is a Regent, that Davidson said a good deal derogatory to the University and in an especial manner showed hostility to the Dept. of Mathematics; saying that its tone was very low, far inferior to Yale College or Harvard, and that the system of examinations was improper and that he— Davidson—as a member of the Visiting Committee was going to look after the Dept. of Mathematics specially—i.e. that mathematics would be his special charge. My informant said that his hostility to me was so open that he deemed it his duty to inform of it.

Now I can not descend to defend my department against such attacks; twelve years of successful labor in the organization and conduct of that department have immensely elevated the tone and degree of mathematical instruction over the entire State, and scattered over it several hundred young graduates who are my living vouchers. But I deem it my duty to state that this man who, as a Regent, is thus discussing a Department of the University has never been present at any exercise or examination of that Department and has consequently never had any opportunity to form a judgement either favorable or otherwise.

I desire that you and other Regents should know this fact so as to be able to set a proper valuation on anything that Mr. Davidson may say against the Department.

Very truly your friend
W. T. Welcker

To Hon. A. S. Hallidie
San Francisco

Appendix 2 (Report of the Committee on Instruction and Visitation, May 31, 1881)

The Committee on Instruction and Visitation respectfully offer the following report: The Committee have discharged the duties assigned to them during the past year, and have found much to engage their attention, awaken interest, inspire confidence, and suggest careful and serious criticism. They have made repeated visits to the University, equivalent to about thirty visits of a single individual; had conferences with members of the Faculty, and held frequent meetings among themselves.

There are some important general facts that should be presented at the beginning. The personnel of the University, as represented in the teaching force and discipline of the institution consists, of thirty-five men—president, Professors, Instructors and Assistants. This body includes, theoretically, a wide range of experience and learning, from young men recently graduated to those who have been engaged as teachers for twenty-five years, and studies extending from classics to mathematics, from natural science to philology, and from agriculture to engineering and mechanic art. To give intellectual symmetry to this body of instruction, a Chair of Mining, a Chair of Philology, and a Chair of Intellectual and Moral Philosophy are required; and in the opinion of the committee, the efforts of the Regents to increase the intellectual and teaching force of the University should be directed mainly to these three points.

The pupilage of the University includes upon the rolls two hundred and forty-five students. Of these fifty-four are absent leaving one hundred and ninety-one in attendance. Of this number one hundred and forty are men and fifty-one are young women. There are eight special students, leaving one hundred and thirty-two in the full four year course, including those classified as students at large and special course students. In the Military, or Battalion Drill, there are one hundred and sixteen on duty, nine being excused permanently and seven temporarily. Of the one hundred and ninety-one in actual attendance, twenty-eight are in the Classical Course, seventy-four are in the Literary Course, fourteen are in the Department of Chemistry, nine are studying Mining, eight mechanics, ten Civil Engineering, twelve Agriculture, eight are studying German, one studying Spanish, and twenty-seven are pursuing scientific studies in connection with a general course.

The University was organized and received its first class in 1869. At the close of the last academic year, June 1880, the total number of graduates, including three classes received from the College of California, was two hundred and forty-eight, being an average number of thirty-six per annum, out of a total number of entered of six hundred and eighty; that is, less than thirty percent of those who enter the University continue through the course, or the course regularly prescribed.

There are several causes to be assigned for this somewhat striking fact: First, there are probably some who, on entering the University find that they have no vocation for study; who were influenced to undertake college more by the wishes of friends than of any love of learning or fitness of studious habits. Such find themselves out of place, and even agreeable companionships cannot make them at home in a community of learning. They naturally withdraw after awhile with the conviction that some other field is more appropriate to them. Another cause, doubtless, is change of circumstance, making it impossible to sustain the expense of residence at the University. There is yet another cause, more efficient perhaps than either of these. It is low terms of admission and inadequate preparation. As a general fact, inadequate preparation is a great evil in college life and experience. The youth who is not well "fitted" enters the University handicapped and is fagged through his whole course. The great falling off in the classes of the University, more than sixty percent, cannot be accounted for except on this ground. The student who is poorly fitted finds himself in a position that he cannot hold, loses courage, falls back into a lower rank, or gives up. His previous training has given him no muscle. The device of lowering standards of admission is sometimes resorted to as a remedy, and as a way to increase pupilage. But nothing can be done in that way. The better the standard, the better the man will go through. The University can render no better service to good learning than to maintain a clear and honorable standard of admissions, thus not only doing better work within its walls, but giving room for the preparatory schools of the State to do better work and indirectly infuse new life into the common school system.

The committee are of the opinion that a more intimate and vital relation should be established between the University and the schools of the State, but are not prepared at present to project a scheme. A very important consideration pertaining to the intellectual discipline of the University is the question of optional studies. To what extent shall optional studies be allowed? As a general theory, the youth seeks the advantages of the University for intellectual guidance, and to ask him what studies he will take, is like a physician asking his patient what medicine he will have. The physician is called to decide that very thing. The University ought to be wiser than the youth who resorts to it.

In the opinion of the committee, courses of study should be firmly maintained, and option allowed only where the student displays unusual force and liking, or for some cause, such as ill health, is unable to take the prescribed course. The optional plan thus strictly pursued, preserves the intellectual tone of the University by putting the direction of studies in the hands of masters, and at the same time allowing sufficient room for every aptitude or necessity, where that aptitude or necessity exists. It is also especially important in an institution that admits young women, as it practically avoids the controversy regarding equal or parallel studies, and at the same time holds an honorable intellectual standard. The attention of the committee has been unavoidably attracted to the manifest want of efficient administration and unity of idea and purpose in the University. For the President of the University the committee entertain every sentiment of personal respect, but they would neglect a plain duty if they failed to bring this manifest want of energy and organizing force to the notice of the Board.

As a consequence of this want of administration the University is drifting, and a want of order, activity, and enthusiasm pervades the whole. The Faculty, not well united, need direction; studies require to be relatively adjusted; and the whole intellectual and moral force of the University brought to bear on one point: the increase of intellectual activity and scholarly manners.

The committee do not find that the University has any vital sympathy or connection with the progress of education, and in some of the departments outgrown methods are still adhered to. The want of a guiding sympathetic mind is manifest in the University as a whole, and in all of its parts. Impressed as the committee are by these facts, and feeling the imperative necessity of change for the better, they have carefully considered what changes ought to be made. They have bestowed upon the whole subject much thought and attention, and the confidence they feel in the recommendations they present to the Board, is only equaled by the sense of duty that urges them.

The committee therefore makes the following recommendations:

1. That the Presidential office be separated from any special Chair in the University and that the duties of the President be mainly administrative and executive.

2. That the Presidential office be declared vacant.

3. That the Chair of Physics be declared vacant.

4. That the office of Instructor of Physics and Mechanics be declared vacant and that Mr. F. Slate, Jr., now Superintendent of the Physical Laboratory be appointed to the vacancy, and that he also perform the duties of the Superintendent of the Physical Laboratory.

5. That the Chair of Mathematics be declared vacant.

6. That the office of Instructor in Spanish be declared vacant and that it remain vacant.

7. That the office of Instructor in Hebrew be declared vacant and that it remain vacant.

8. That the office of Instructor in Chemistry, now held by S. August Harding, be declared vacant, upon his proposed departure for Europe.

9. That the vacancies herein proposed shall, unless otherwise specified, take effect on the first day of August, 1881.

10. That the following schedule of salaries be adopted from the first day of August, 1881. [This table proposed a reduction in salary for all 11 professors from $3600 per year to $3000 per year. It proposed an increase in salary for the instructor in mathematics from $1500 per year to $1800 per year and an increase in the salary for the assistant instructor in mathematics from $1200 per year to $1600 per year (instructor and assistant instructor were the only other instructional titles in the university at the time, and the total number of faculty was 27, and there were 33 university employees all told). The librarian of the university was given a raise from $1200 to $1600 per year. The president's yearly salary was to be set at $4500—a reduction from $5360—and the salary of the secretary of the Board of Regents was to be reduced from $3600 to $3000.]

All of the foregoing recommendations we deem so important and material to the future of the University, that we respectfully ask that their consideration be postponed for final action to the annual meeting to be held at the University on the first Tuesday in June.

Horatio Stebbins
George Davidson
Wm Ashburner
B. B. Redding
Fred M. Campbell

Appendix 3 (*Extract of the Report of the Committee on Instruction and Visitation, May 21, 1882*)

To the Board of Regents of the University of California

1. [nonmathematics matters]

2. [nonmathematics matters]

3. In relation to the matter of filling the chair of Mathematics, your Committee has for the months past been diligently in correspondence and enquiry. We have examined and carefully considered the testimonials of applicants.

In our correspondence with the heads of various educational institutions in the East, Mr. W. I. Stringham was so highly spoken of and strongly recommended for the position if his services could be secured that we entered into correspondence with him and learn that he will accept the position if tendered.

With regard to the opinions entertained of this gentleman's qualifications and fitness for the position, we will quote as follows:

President Gilman on being applied to suggest names for this chair answered as follows: "Of those having exceptional ability, I should name W. I. Stringham now in Europe holding a Parker Fellowship from Harvard. He was in Baltimore for more than two years and received a degree of Ph.D. and wrote some noteworthy articles on Quaternions for the American Academy proceedings, Boston and on Dimensions of space in American Journal of Mathematics. You can learn about him from Prof. Jas. Peirce or Prof. Chas. Norton of Cambridge. He has had exceptional opportunities and I think he has borne himself well."

President Eliot and Prof. J. M. Peirce on being applied to expressed their opinion of him in the strongest and most forceful manner and both answered that they "knew of no other man whom we could get who was his equal."

After carefully considering all the testimonials and recommendations of all the applicants, we unhesitatingly recommend for election to this chair Mr. W. I. Stringham.

Horatio Stebbins
B.B. Redding
Wm. Ashburner
Fred M. Campbell

Appendix 4 (Kelley's White Paper, Spring 1957)

1. The current crisis in mathematics

There is an incredible shortage of mathematicians. This shortage is even more severe than in the war years, and there is compelling reason to believe that this situation will get worse. Mathematicians are now quite as difficult to hire as physicists and engineers; they will shortly be scarcer than the latter. The reasons for this situation are the following:

a. There are only about 250 Ph.D.'s in mathematics awarded each year in this country (253 in 1954 and 252 in 1955). Even this production is an enormous expansion over the pre-world war II output; in 1940 there were 88 Ph.D.'s granted in mathematics. This must be compared, for example, with over 90 requests for mathematics Ph.D., as teachers that come to the Department of the University of California alone.

b. The expanding enrollment of college students has been discussed elsewhere, and we only remark that this appalling flood of students is not requiring less mathematics in proportion to their other studies. As an example of the effect of this expansion, we may note that (as of February) San Jose State had four vacancies in mathematics and not one application. Other schools are in a similar position.

c. There is an expanding demand for mathematics instruction per student. It is only necessary to note that in 1952–53, when the enrollment of the University was still decreasing, the enrollment in mathematics courses rose from 8973 student credit hours to 9140 student credit hours. In the next few years, while the enrollment in the University increased by small percentages, Mathematics increased by 5.5% in 1953, 7.8% in 1954, and 18.5% in 1955. In 1954 and 1955, upper division Mathematics increased by 20.1% and 34.3%.

d. Industry is requiring an increasing number of mathematicians. General Electric and other large industrial concerns have now adopted the policy of visiting Mathematics Departments in the fall to recruit the spring crop of students. Such arrangements are commonplace in Chemistry and Physics, but these are new in Mathematics—a discipline where the output of Ph.D.s was traditionally balanced with the teaching requirement of the colleges. As an example of the present state of affairs, we mention that one of our finishing graduate students (M. Lees) was offered $10,000 a year at Boeing Aircraft, with a stipend of $8,000 for his wife. Representatives of industry and government laboratories are interviewing students two and three years prior to their Ph.D. degrees, in effect to "get a line" on them for future employment.

e. In addition to the previous factors, the advent of high speed computing machines has created a remarkable demand for mathematical talent. The number of machines now on order at IBM would require the entire output of mathematicians for three years; this ignores the future demand for machines, the output of other companies, and the fact that the demand for sub-Ph.D. personnel has a debilitating effect on graduate training. The salary scale for beginning computing personnel is now at the junior executive level. We wished to appoint a young man (J. Franklin) to an Assistant Professorship in the Department. He is now three years past his Ph.D., but he found it difficult to see why he should trade his $13,000 a year position for our Assistant Professorship. And this is typical.

2. Current Position of the Department

a. Scientific position.

It seems necessary to give some estimate of the current scientific standing of the Department, although we realize that self analysis is apt to be optimistic, and that the practice of self adoration is not confined to superior departments. To begin with, we attempt to give reasonable coverage to all fields of mathematics. This seems essential in a university of this sort, although certain mathematics departments deliberately limit the scope of their coverage to attain concentration. Thus Stanford has a fine department, primarily devoted to classical analysis, and NYU has an excellent department, strongly favoring analysis and applied mathematics.

Our position in the several fields may be summarized: in foundations and meta-mathematics, very good; in analysis, very good; in geometry and algebraic topology, weak; in algebra, weak; in applied mathematics, weak; in statistics (now a separate department), very good. This is not intended to indicate that the individuals in the Department in the fields labelled "weak" are not good; it is simply that in program, in numbers, and in group effort these fields are weak.

The scientific position of our Mathematics Department relative to other mathematics departments in the country may be summarized: we are not in the first three (Chicago, Harvard, and Princeton), and we are in the first ten.

Our program of graduate training is in a much better position than the above would indicate. For the last two years we have trained as many Ph.D.'s as any American University, and more than any save Princeton. (See appended statistics.) Steadily increasing numbers of National Science Foundation fellows and visiting scholars come to Berkeley. We believe we have a reputation of a vigorous Department, on the way up. There is no member of our Department who dos not publish systematically.

3. Recruiting Problems.

a. Junior staff.

In this time of shortage the recruiting of junior staff is difficult indeed, since most universities rely primarily on junior appointments to fill vacancies, just as we do. Young mathematicians seek departmental strength (particularly in their own field), a vigorous graduate school, a low teaching load, a high salary, and pleasant living conditions. The first of these depends on the general position of the department, described above, and on the appointment of senior staff, which is discussed below. Our teaching load is nine hours, compared to six or less at Chicago, Harvard, and Columbia. Moreover the actual load is heavier. Because lower division courses are taught primarily by teaching assistants, there is a great deal of supervision necessary. Five of our fifteen junior appointments are supervising three to sixteen assistants.

Moreover, we give no teaching credit for supervision of Ph.D. students, and we have not been able to give teaching credit for all seminars. Two seminars were given by the junior staff during the fall semester without teaching credit. The situation on salary and promotion is similar to that in other physical sciences. A brief summary of our recruiting troubles this spring will indicate the general situation.

We were unsuccessful in trying to appoint; S. Abhyankar (one year past the Ph.D. went to Cornell at $7000), A. Nijenhuis (better salary and lighter teaching load at Washington), Froelicher took a chair at Fribourg, S. Sternberg (went to Chicago, at lighter teaching load and presumably higher salary), Shepardson (remained at Bristol, England with much lower teaching load), J Franklin (3 years past Ph.D. remained in industry at $13,000). There were other candidates who were lost at the pre-negotiation stage to the combination of higher salary and lower load.

We are having difficulties repelling raiders on our junior staff; within the last few weeks there have been substantial offers: To Helson from Rochester, to Kostant from Yale, to Eells from Washington and others, and to Bishop from Carnegie Tech and Dartmouth. Our continued progress depends on holding our junior staff, and maintaining a reputation of vigorous development; indeed, in the present crisis we must go up or down; we cannot mark time.

b. Recruiting of graduate students.

We are fortunate in that we always have a rather large number of applicants for graduate assistantships; many schools are finding that the current crisis is drying up the source of supply.

The vast majority of our graduate students are supported by teaching assistantships. The remainder are supported primarily on contracts and grants with various government agencies. There is about $152,000 a year in NSF, ONR, OSR, and OOR contracts in our mathematics department, and, since there is no need of equipment purchase, this is a considerable budget. It is possibly unfortunate that we depend so much on these contracts; however, this is hardly the time to be fastidious.

The stipends for assistantships are dangerously low. Many schools are now making substantially better offers.

c. Senior staff.

We recognize that most new appointments must be made at the junior level; nevertheless, unless one credits a department with superhuman prescience, an occasional appointment at tenure grade must be made to maintain balance and power.

Our mathematics department has had less than the necessary minimum of tenure appointments. Of the present positions in the department, only one and a half were appointed at the tenure level (Kelley, beginning associate professor in 1947, and Huskey, half time in engineering, third level associate professor, 1954). Since the appointment of Professor G. C. Evans in 1934, there has been only one tenure appointment in addition to these—that of Professor J. Neyman, now of the statistics department. We anticipate the appointment of B. Friedman at the rank of full professor, this fall.

Since 1954 Stanford has appointed five men at tenure rank in mathematics, and is now negotiating a sixth; our own statistics department has appointed three full professors in addition to Neyman; UCLA has appointed at least three, Chicago four, Harvard two or three; three of four of Princeton's mathematics department were appointed at the professorial level.

Appointments at high level are very difficult to make. The position vacated by Professor G. C. Evans was offered to S. Chern (Chicago), N. Steenrod (Princeton), B. Eckmann (Zurich), and J. H. C. Whitehead (Oxford). Men of this calibre are neither starving nor maltreated, and substantial inducements are necessary. It is almost essential that not one but several such appointments be made simultaneously. Chern expressed an unwillingness to move from Chicago unless 3 or 4 similar appointments were made at the same time.

4. Teaching problems.

Our fundamental problem is very simple: too many students and too few staff. The lower division courses are taught primarily by teaching assistants: of 315 class hours per week of instruction, 236 are taught by teaching assistants. Of those 79 hours which are met by staff, 43 are in courses which cannot be met by assistants. These assistants are, with an occasional catastrophic exception, eager, responsible, and intelligent. The staff makes a conscientious effort to supervise their teaching. Nevertheless our assistants are and will be subject to constant criticism from outside the department. Some will be merited; but in teaching calculus, which serves as a qualifying course for the physical sciences, many of the students who fail (17 percent in 3A, 9.7 percent in 3B, 20.9 percent in 4A, 8.8 percent in 4B) will blame their failure on the most obvious scapegoat: the teaching assistant.

Our position is actually much improved in the last two and a half years. Additional positions assigned the department have made it possible to reduce the size of assistant-taught classes from 40 to 28, and some of the large upper division courses have been split into sections. It is obvious that there is still a great deal to be done.

The department has made strenuous efforts to improve course organization; we have dropped college algebra from the curriculum, replacing it by remedial sections of calculus (five hours class for three hours credit); mathematics of finance has, with the consent of the school of business administration, been dropped; the single remaining pre-calculus algebra course offers two units of credit for three hours work. We are now near the limit of improvement in the lower division that can be made by course organization. We intend to experiment cautiously with large lecture sections—not as an economy measure, but more to give the students contact with mature mathematicians.

The following gives an overall view of the use of our teaching resources,

Number of teaching hours used in

	LD	UD	Grad
Staff	79	102	66 + Ph.D. supervision
Assistant	236	24	

Student class hours per week taught
LD: 8223 UD: 2823 Grad: 860

In conclusion it should be said that our staff is very small for a university of this size. Michigan has 58 staff members (including statistics) and approximately the same number of teaching assistants as we, Illinois has a staff of 49, with 66 assistants and Minnesota has a staff of 57. In comparison, our present staff size is 29 [Kelley is using staff to include faculty in professorial ranks—the number we have been counting—plus faculty with the rank of Instructor].

5. Plan of action

a. Maintenance of scientific position.

It is absolutely essential that we keep our current junior staff and that we continue to attract good young men. Our department has been built by hiring promising young men and the current crisis threatens both our supply line and our current supply.

A very decisive factor in maintaining our hiring-and-holding program is the teaching load. We now have a nine hour load for instructors and assistant professors, a seven and a half hour load for associate professors, and a six hour load for full professors. There is no lightening of these duties for supervision of the assistants, university committee work, administrative work, supervision of Ph.D. students, or (in many cases) for giving a seminar. This austerity program has been necessitated by the shortage of staff. We recommend that this load be reduced to six hours across the board. This is, with the load of graduate students, and the supervision of teaching assistants, an entirely reasonable recommendation. It will leave us in a slightly worse position than the mathematics departments with which we wish to compete, and (to be blunt about it) in a slightly better position than certain of the departments we wish to raid. We note that the teaching load will still be at least as great as that of competing departments, and still greater than that of our own Physics department. This reduction in teaching load would unquestionably be the greatest single item in improving our recruitment of young men. Even if this recommendation is approved, it would be patently impossible to accomplish a six hour load in 1957–58; a commitment in this direction, with a reasonable chance of obtaining a six hour load in 1958–59 would improve our position immeasurably.

In order to retain the young men now hired, it is necessary to meet outside offers. This will mean that one, two, and even three year accelerations over the "normal" rate will be necessary.

b. Improvement of scientific position.

The greatest single factor needed for improvement of our position is the appointment of distinguished senior staff. In addition to their scientific impact on the department, a few such appointments would simplify our recruiting of junior staff. There are three categories in which appointments are needed.

First, the central core of pure mathematics needs strengthening. We need a distinguished appointment in algebraic topology or geometry, and another in pure algebra. Another, in an area determined by the precise interests of the first two appointments, is also needed.

Second, our applied mathematics program needs at least one senior appointment. The forthcoming appointment of Friedman has met with approbation in the physics department and the college of engineering. But this appointment will still leave us very short handed for the task of scientific liaison with the departments using mathematics. The increasing use of mathematics will make the establishment of an Applied Mathematics Laboratory a necessity. The next step in this direction is the appointment of a senior professor in this area.

Third, our program in numerical analysis and computing requires a senior appointment. Student enrollment in this area is very high, and our current resources are scarcely adequate to meet the classes involved. The necessary expansion of curriculum in this direction will make our position desperate. With the Computing Center now in operation, this is an opportunity which must not be neglected.

e. Improvement of teaching position.

Improvement of our teaching position depends primarily on the assignment of new staff. Our representations in this direction have recently been well received by the Budget Committee and by the administration; we hope for continued understanding of the problem. In spite of recent improvements, there still seems to be very little chance of teaching any substantial part of the lower division offerings with staff. Consequently, we hope to obtain funds from one of the foundations to inaugurate a program of training for new teaching assistants. A serious one semester training program might well solve a number of problems which trouble us here. We shall need administrative approval of this proposal at the appropriate time.

Summary

There is now a completely unprecedented shortage of mathematicians. Our department, which has been built on the basis of shrewd junior appointments, needs authority to go to a six hour teaching load and sympathetic cooperation on the matter of accelerated promotions in order to maintain its current position. The advance in scientific position which should be expected of us depends in large part on new senior appointments. Three such appointments are needed in pure mathematics, to bolster the fundamental disciplines. One is needed in applied mathematics—the next step in a program for the eventual establishment of an applied mathematics laboratory. One senior appointment is needed in numerical analysis and computing, to serve our large group of students and to take advantage of our current opportunities. At least three of these five appointments are needed immediately, and the other two within two years.

In conclusion: We have, with meager resources, built a respectable mathematics department. Without some extraordinary assistance, we cannot, in this time of crisis, even maintain our position. We petition for an understanding and help in our effort to make a department of the calibre which this University deserves.

[The original document had an appendix consisting of a table with the number of mathematical sciences Ph.D.s granted by institution for 1954 and 1955, taken from the Bulletin of the AMS. We do not reproduce this table. Suffice it to say that Berkeley and Princeton granted 16 doctorates in each of these two years. Michigan, Harvard, Chicago, Illinois, Stanford, North Carolina, and Wisconsin all averaged between 9 and 13 doctorates for these two years. This white paper was included as an attachment to many personnel cases that the Department sent forward to the administration in 1958 and 1959 and so copies of it reside in many departmental files. A copy may be found in the University Archives [UA, CU 5. 21, Box 24, Henkin file].]

References

[AMSC] Peter Duren et al., editors. *A Century of Mathematics in America* (3 volumes). Providence: American Mathematical Society, 1989.

[Arc] R. C. Archibald. "Benjamin Peirce." *American Math. Monthly* 32 (1925), pp. 1–30.

[Ba] Steven Batterson. *Stephen Smale: The Mathematician who Broke the Dimension Barrier*. Providence: American Mathematical Society, 2000.

[Ber] Benjamin Bernstein. Papers. Bancroft Library. UC Berkeley.

[Caj] Florian Cajori. *The Teaching and History of Mathematics in the United States*. Washington, D.C.: US Government Printing Office, 1890.

[Con] Lincoln Constance. *Strengthening the College of Letters and Science, Berkeley at Mid-Century*. Berkeley: Public Policy Press, Institute of Governmental Studies, 2002.

[Coo] J. L. Coolidge. "The Story of Mathematics at Harvard." *Harvard Alumni Bulletin* 26 (January 1924), p. 376.

[Cur] J. H. Curtiss. "A Federal Program in Applied Mathematics." *Science* 107 (March 12, 1948), pp. 257–262.

[Dav] George Davidson. Papers. Bancroft Library, UC Berkeley.

[DeG] Morris DeGroot. "A Conversation with David Blackwell." *Statistical Science* 1 (1986), pp. 40–53.

[Der] John Derbeyshire. *Prime Obsession*. New York: Plume, 2004.

[Doug] John Douglass. *The California Idea and American Higher Education*. Stanford: Stanford University Press, 2000.

[Evans] Griffith C. Evans. Papers. Bancroft Library, UC Berkeley.

[Fef] Anita and Solomon Feferman. *Alfred Tarski*. New York: Cambridge University Press, 2004.

[Fer 1930] William Warren Ferrier. *The Origin and Development of the University of California*. Berkeley: Sather Gate Book Store, 1930.

[Fer 1937] William Warren Ferrier. *Ninety Years of Education in California, 1846–1936*. Berkeley: Sather Gate Book Store, 1937.

[Fon] Joseph Fontenrose. *Classics at Berkeley: The First Century*. Berkeley: Department of Classics, 1982.

[For] Sidney Forman. *West Point.* New York: Columbia University Press, 1950.

[Gardner 1967] David P. Gardner. *The California Oath Controversy.* Berkeley: UC Press, 1967.

[Gardner 2005] David P. Gardner. *Earning My Degree.* Berkeley: UC Press, 2005.

[Hus] Harry D. Huskey. "SWAC: The Pioneer Day Session at NCC July 1978." *IEEE Annals of the History of Computing* 19 (1997), pp. 51–61.

[IM] In Memoriam (biographical essays). UC History Digital Archives. Online at sunsite. berkeley.edu/uchistory/archives_exhibits/in_memoriam/index4.html.

[Jolly] William L. Jolly. *From Retorts to Lasers: The Story of Chemistry at Berkeley.* Berkeley: UCB College of Chemistry, 1987.

[Kerr 2001] Clark Kerr. *The Gold and the Blue, Vol. 1: Academic Triumphs.* Berkeley: UC Press, 2001.

[Kerr 2003] Clark Kerr. *The Gold and the Blue, Vol. 2: Political Troubles.* Berkeley: UC Press, 2003.

[Kyte] George C. Kyte '15. "The Bourdon that Never Burned." *California Monthly* (November 1947), p. 22.

[Lewy] Hans Lewy. Papers. Bancroft Library, UC Berkeley.

[Lehmer] Lehmer Conference Program, August 24–26, 2000. UC Berkeley Department of Mathematics.

[Pa-Ro] Karen Hunger Parshall and David E. Rowe. *The Emergence of the American Mathematical Community.* Providence: American Mathematical Society, 1994.

[RMCS IV] S. Mac Lane, editor. *Reports of the Mid-West Category Seminar, IV.* Lecture Notes in Mathematics 137. Berlin: Springer-Verlag, 1970.

[Reid 1976] Constance Reid. *Courant.* New York: Springer-Verlag, 1976.

[Reid 1982] Constance Reid. *Neyman.* New York: Springer-Verlag, 1982.

[Reid 1996] Constance Reid. *Julia: A Life in Mathematics.* Washington, D.C.: Mathematical Association of America, 1996.

[Ryder] Robin Ryder. "An Opportune Time: Griffith C. Evans and Mathematics at Berkeley." In Peter Duren et al., editors. *A Century of Mathematics in America*, volume 2. Providence: American Mathematical Society, 1989, pp. 283–302.

[RM] Minutes of meetings of the Board of Regents of the University of California, Office of the Secretary of the Regents of the University of California, Oakland California.

[Sh] Millicent Shinn. "The University of California, I." *Overland Monthly* XX:118 (October 1892), pp. 337–362.

[Sc] Michael Scanlon. "Who Were the American Postulate Theorists?" *Journal of Symbolic Logic* 56 (1991), pp. 981–1002.

[SMCP] Files of the Office of Space Management and Capital Programs, Berkeley Campus, UC Berkeley.

[Sp] Pauline Sperry. "Formula for Happiness at Eighty." *Smith Alumnae Quarterly* 56, p. 154.

[St] Verne Stadtman. *The University of California, 1868–1968.* New York: McGraw-Hill, 1970.

[Ste] George Stewart. *The Year of the Oath.* Garden City, NY: Doubleday, 1950.

[Str] *In Memoriam: Dean Stringham.* Berkeley: University of California Press, 1910 (also reprinted from University of California *Chronicle* XII:1).

[UA] University Archives, Bancroft Library, UC Berkeley.

[UCCR] Verne Stadtman et al., editors. *The Centennial Record of the University of California.* Berkeley: UC Press, 1968.

[Wil] Samuel H. Willey. *The History of the College of California.* San Francisco: California Historical Society, 1887.

[Wood] Sally B. Woodbridge. *John Galen Howard and the University of California.* Berkeley: University of California Press, 2002.

Photograph Credits

Following page 110

College of California: Call number Pic 19:01, reproduced courtesy of University Archives, The Bancroft Library, University of California, Berkeley.

Early campus scene: Call number Neg 3:20b, reproduced courtesy of University Archives, The Bancroft Library, University of California, Berkeley.

William Welcker: Reproduced courtesy of the UCB Mathematics Department.

Irving Stringham: Reproduced courtesy of the UCB Mathematics Department.

Mellen Haskell: Reproduced courtesy of the UCB Mathematics Department.

Derrick N. Lehmer: Reproduced courtesy of the UCB Mathematics Department.

George Edwards: *California Monthly*, November 1947; reproduced courtesy of *California Magazine*.

George Edwards: Call number UARC 13:3488, reproduced courtesy of University Archives, The Bancroft Library, University of California, Berkeley.

Pauline Sperry: Call number UARC 13:3075, reproduced courtesy of University Archives, The Bancroft Library, University of California, Berkeley.

Emma Trotskaya Lehmer: Reproduced courtesy of the Lehmer family.

D. N. and D. H. Lehmer: Reproduced courtesy of the Carnegie Institution of Washington.

Congruence machine: Reproduced courtesy of the Carnegie Institution of Washington.

Sophia Levy McDonald: Call number UARC 13:2535, reproduced courtesy of University Archives, The Bancroft Library, University of California, Berkeley.

John McDonald: Call number UARC 13:2598, reproduced courtesy of University Archives, The Bancroft Library, University of California, Berkeley; photograph by Kee Coleman.

Benjamin Bernstein: Call number UARC 13:853b, reproduced courtesy of University Archives, The Bancroft Library, University of California, Berkeley; photograph by Kee Coleman.

Bing Wong: Call number UARC 13:3303a, reproduced courtesy of University Archives, The Bancroft Library, University of California, Berkeley; photograph by Kee Coleman.

Griffith Evans: Reproduced courtesy of the UCB Mathematics Department.

Julia Robinson: Reproduced courtesy of the UCB Statistics Department.

David, Scott, Blackwell, and Fix: Reproduced courtesy of the UCB Statistics Department.

Jerzy Neyman: Reproduced courtesy of the UCB Statistics Department.

Charles Morrey: Reproduced courtesy of the UCB Mathematics Department; photograph by G. Bergman.

Derrick H. Lehmer: Reproduced courtesy of the UCB Mathematics Department; photograph by G. Bergman.

Hans Lewy: Reproduced courtesy of the UCB Mathematics Department; photograph by G. Bergman.

Alfred Tarski: Reproduced courtesy of the UCB Mathematics Department; photograph by G. Bergman.

John Kelley: Reproduced courtesy of the UCB Mathematics Department; photograph by G. Bergman.

Following page 238

Tosio Kato: Reproduced courtesy of the UCB Mathematics Department; photograph by G. Bergman.

Abraham Taub: Reproduced courtesy of the UCB Mathematics Department; photograph by G. Bergman.

Julia Robinson: Reproduced courtesy of the UCB Mathematics Department; photograph by G. Bergman.

Raphael Robinson: Reproduced courtesy of the UCB Mathematics Department; photograph by G. Bergman.

S. S. Chern: Reproduced courtesy of the UCB Mathematics Department; photograph by G. Bergman.

Chern and Mrs. Chern with President Ford: Reproduced courtesy of the UCB Mathematics Department.

Chern and President Ford: Reproduced courtesy of the UCB Mathematics Department.

Rufus Bowen: Reproduced courtesy of the UCB Mathematics Department; photograph by G. Bergman.

Gerard Debreu: Reproduced courtesy of the UCB Mathematics Department; photograph by G. Bergman.

Evans Hall: Reproduced courtesy of the UCB Mathematics Department; photograph by the author.

"Galois": Reproduced courtesy of the UCB Mathematics Department; photograph by the author.

Reeb foliation: Reproduced courtesy of the UCB Mathematics Department and the artist; photograph by the author.

MSRI groundbreaking ceremony: Reproduced courtesy of Irving Kaplansky.

MSRI from Grizzly Peak: Photograph by the author.

MSRI looking southwest: Photograph by the author.

Index

Printed and bound by CPI Group (UK) Ltd, Croydon, CR0 4YY

23/10/2024

01777671-0005